T0378073

Shallow Foundations and Soil Constitutive Laws

Shallow Foundations and Soil Constitutive Laws

Swami Saran

Professor (Retd.), Department of Civil Engineering
Indian Institute of Technology
Roorkee, India

CRC Press
Taylor & Francis Group
Boca Raton London New York

CRC Press is an imprint of the
Taylor & Francis Group, an **informa** business
A SCIENCE PUBLISHERS BOOK

CRC Press
Taylor & Francis Group
6000 Broken Sound Parkway NW, Suite 300
Boca Raton, FL 33487-2742

Printed on acid-free paper
Version Date: 20171013

International Standard Book Number-13: 978-0-8153-7487-9 (Hardback)

Visit the Taylor & Francis Web site at
http://www.taylorandfrancis.com

and the CRC Press Web site at
http://www.crcpress.com

Dedication

I dedicate this book to my respected mother
(Late) Smt. Kanti Devi

The heart of my mother was a deep abyss at the bottom of which I will always find forgiveness. The education I have is thanks to her. Her heart was my classroom.

Preface

This book is intended primarily for use senior undergraduate students, post graduate students, Ph.D students, research scholars, teachers and academicians. It will serve as a textbook for courses on footings design or constitutive laws usually taught at post graduate level. Attempt has been made to present the analysis of footings located in different situations in the field in lucid manner. Footings located in seismic regions have also been covered.

The author has been actively engaged in teaching graduate and postgraduate level courses since 1964 at IIT Roorkee. More than half a dozen Ph.D theses have been guided by the author on the similar idea. The approach has been to deal with a topic in totality, starting from the initial stage of the determination of the constitutive law of soil, and selection of appropriate stress equations, and ending with the analysis which includes determination of the bearing capacity and settlement.

The methodology has been demonstrated through illustrative examples included in each chapter. All formulas, charts and examples are given in SI units.

Some of the materials included in the book has been taken from the works of the other authors specifically of Ph.D students who worked under author's supervision.In spite of sincere efforts, some contributions may not have been acknowledged. The author apologies all the authors and publishers for their kind permission to reproduce tables figures and equations.

The author wishes to express his appreciation to Mr. Tinku Biswas for his continuous help in preparing and typing the manuscript and Shri Rajesh Shukla for his suggestions. Thanks are also due to many colleagues, friends and students for their encouragement.

The author will be failing in his duty if he does not acknowledge the support from his family members, especially his wife and grandson.

Contents

About the Book

The book offers a systematic analysis of footings (i.e. shallow foundations) in a realistic way, using constitutive relationships of the soil. The aim of the book is to deal with the theme holistically, involving the determination of the constitutive law of the soil, and then proportioning the footing occurring in different situations in actual practice.

The book has eleven chapters. After giving an introduction and scope of the book in the first chapter, second and third chapters are respectively devoted to constitutive laws of soil and basic stress equations. In the third chapter analysis of strip footings subjected to central vertical load has been dealt. This analysis has been extended for eccentric –inclined load in the fifth chapter. Since problems of shallow foundations resting adjacent to a slope are of prime importance, this aspect has been dealt in sixth chapter. In the seventh chapter, analysis pertaining to square and rectangular footings have been presented. Effect of interference between adjacent footing is covered in chapter eight. Since ring footings are usually provided for tanks, silos, towers etc., ninth chapter is devoted to this. Added attraction of the book is its chapter ten in which footings located in seismic regions have been covered. Effect of embedment below the ground surface on the behavior of footings located both in non-seismic and seismic regions has been dealt in the chapter eleven.

The book is intended for senior undergraduate, postgraduate and Ph.D. students of civil engineering, research scholars, practicing engineers, teachers and academicians. The analyses are based on the latest information available. A number of illustrated examples have been included in the text. SI units have been used in the book.

About the Author

Dr Swami Saran, Emeritus Fellow (retd.) from the Department of Earthquake Engineering, Indian Institute of Technology, Roorkee, obtained a Ph.D. in 1969 from the University of Roorkee. An established teacher, researcher and active consultant, he is the recipient of Khosla Research Awards (four times), IGS Awards (six times) including the prestigious Kuckulmann Award and also awards from I.S.E.T., I.S.T.E. and I.S.C.M.S. He has guided 29 Ph.D. theses, 84 masters theses, published 207 research papers and seven books. Dr. Saran has initiated research work on reinforced soil, analysis of foundations using constitutive laws and displacement-dependent static/seismic analysis of retaining walls. He has provided consultancy to more than 300 projects of national importance, including multistoreyed buildings, cement and tyre factories, thermal plants, machine foundations, towers and chimneys, bridges, oil storage tanks, historical monuments, ground improvement problems, etc. He visited UK in 1974 under an exchange programme and AIT Bangkok as a Visiting Professor in 1987. He has also visited USA, Australia and Nepal to attend conferences. He is a member of several national and international professional bodies. A national conference (NCFRS – 2007) was organized at IIT, Roorkee to honour him. Considering his contribution, in 2011, Indian Geotechnical Society conferred the honorary fellowship on him.

CHAPTER 1

Introduction

1.1 General

Shallow foundations are the common type of foundations designed for many important structures. In general, spread footings, combined footings and rafts or mat foundations are included under shallow foundations. Wherever possible, spread footings are used because these are more economical. The foundations may be subjected to a vertical load, moment and shear at the base which induces stress and deformation in the supporting soil. For proportioning such foundations, bearing capacity, settlement and tilt are the main criteria.

Stresses and deformations are two inseparable features which are observed when a material is subjected to external forces. The evaluation of stresses and deformations in elastic, isotropic and homogeneous mass has been done on the basis of the theory of elasticity. Soil mass is formed in Nature in such a way that the above three criteria are never fulfilled. Therefore, determination of stresses and deformations for a real soil is very difficult.

In the design of shallow foundations, the estimation of bearing capacity, settlement and tilt is currently done in separate, independent steps. However, for determining these, the rational approach is to use continuous pressure-settlement and pressure-tilt characteristics of the actual foundation. Considering all the situations that occur in practice while using footings as foundation, attempts are being made to evaluate the pressure-settlement and pressure-tilt characteristics of footings using non-linear stress-strain relation of soil from these, bearing capacity, settlement and tilt of a footing can be obtained in a single step more rationally.

1.2 Current Methods of Proportioning Shallow Foundation

The bearing capacity of a centrally loaded footing is currently estimated by Terzaghi's theory (1943) or Meyerhof theory (1951). These theories are applicable to homogeneous and isotropic soils. Terzaghi (1943) developed his theory by considering the soil above the base of foundation as a uniform surcharge, while Meyerhof (1951) considered the shear strength of the soil. Their solutions are based upon assumptions of approximate rupture surfaces which are neither statically nor kinematically possible (Saran, 1969). No solutions are available to indicate a surface of rupture which can be proved to be both statically and kinematically possible. Hence, the exact value of bearing capacity is still unknown.

The bearing capacity of an eccentrically loaded footing is currently estimated by using Meyerhof's theory (1953), according to which an eccentrically-loaded footing is treated arbitrarily as a centrally loaded footing of width $B-2e$ where B is the width of the footing and e is the eccentricity of load. Saran (1969) proposed an analytical solution using limit equilibrium approach considering the failure of surface to develop in one direction only. He presented non-dimensional bearing-capacity factors in the form of charts for computation of ultimate bearing capacity. Agarwal (1986) extended the work with Saran (1969) for footings subjected to eccentric inclined load.

The settlement of a foundation subjected to central vertical load is generally estimated by Terzaghi's (1943) theory of one-dimensional consolidation for clay, and by performing plate load tests, standard penetration tests and static penetration tests for sand (Saran, 2006). The results of plate load test are extrapolated to actual size of footing. Terzaghi's one-dimensional consolidation theory is valid only when the lateral yield is negligible; for example, in the case of a clay layer sandwiched between two sand layers or a large raft resting on a thin clay layer supported by a rigid base. This theory leads to an underestimate of settlement to a large extent for thick layers of clays where lateral yield is significant. In this theory, the vertical strains are computed from stresses worked out by the elastic theory which may be true for smaller surface loads; but at higher loads, soil becomes plastic and exhibits non-linear stress-strain characteristics (Sharan, 1977). Empirical charts and correlations are available to obtain the settlement from standard penetration tests and static cone penetration data (Peck et al., 1974; Saran, 2006).

Analytical solutions based upon the elastic theory are available for determination of tilt of an eccentrically-loaded footing. Non-dimensional correlations for footings on sand, based upon experimental values, are available (Saran, 1969; Agarwal, 1986) for computation of tilt for the cases footings subjected to eccentric vertical load and eccentric-inclined load respectively. No analytical procedure is available which uses non-linear stress-strain relations of soil for estimation of the tilt of footings.

The above discussions lead to the conclusion that the method currently in use for estimating bearing capacity, settlement and tilt of centrally and eccentrically obliquely loaded footings are approximate. Therefore, the analytical results are checked by field observations. The method currently in use for determining the above quantities in field for centrally-loaded footings is by plate load test, but no field method is prescribed for eccentrically obliquely loaded footings. A plate load test is performed with a standard size of plate and then the results are extrapolated to the actual size of footing by using the relationship proposed by Taylor (1948). These relationships are at best semi-empirical.

The pressure bulb under a footing is affected by the size and shape of the foundation and the nature of supporting soil strata. The extrapolation of results of plate load test with smaller size of plate can give reliable results for footings resting on uniform homogeneous soils but where change in the nature of supporting soil is expected, the extrapolation may lead to erroneous results. Besides, the plate load test is a very costly and slow process and, therefore, attempts have been made by Duncan and Chang (1970), Desai (1971) and Basavanna (1975) to obtain pressure settlement curve analytically by finite element techniques using non-linear stress-strain relation for soil.

Pressure settlement curves for uniformly loaded circular and square footings in clay have been obtained successfully by finite element technique by Desai (1971) and Basavanna (1975) respectively, using soil parameters from undrained triaxial compression test. The predicted and experimental curves compare very well at all stress levels, but the pressure settlement curve for uniformly loaded strip and square footings in sand obtained by Duncan and Chang (1970) and Basavanna (1975) respectively by finite element method taking soil parameters from undrained triaxial compression test had large variations with the experimental results. Recently, some researchers analysed this problem using finite element and finite difference techniques (DeBorst and Vermeer, 1984; Shiau, 2003; Loukidis and Salgado, 2009; Dai et al., 2013). Method of characteristics has been utilized by few investigators (Kumar, 2009; Han et al., 2016; Cascone et al., 2016; Sun et al., 2016).

1.3 Scope of the Book

The book offers a systematic treatment of the analysis of footings (i.e. shallow foundations) in a more realistic way by using constitutive relationships of the soil. The aim of the book is to deal with the theme in its entirety, involving the determination of the constitutive law of the soil, and then proportioning of the footing occurring in different situations in actual practice.

The book is intended to be used by civil engineering senior under-graduate and post-graduate students, Ph.D. research scholars, practising engineers, teachers and academicians.

The book has eleven chapters. After giving an introduction and scope of the book in the first chapter, second and third chapters are devoted to constitutive laws of soil and basic stress equations. In the third chapter, analysis of strip footings subjected to central vertical load has been covered. This analysis has been extended for eccentric-inclined load in the fifth chapter. Since problems of shallow foundations resting adjacent to a slope are of prime importance, this aspect has been dealt with in the sixth chapter. In the seventh chapter, analysis pertaining to square and rectangular footings have been presented. Effect of interference between adjacent footing is covered in chapter eight. Since ring footings are usually provided for tanks, silos, towers, etc., ninth chapter is devoted to this. Added attraction of the book is its chapter ten in which footings located in seismic regions have been covered. Effect of embedment below the ground surface on the behaviour of footings located both in non-seismic and seismic regions has been dealt with in chapter eleven.

The analyses are based on the latest state-of-the-art available on the subject. A number of illustrative examples have been included in the text and SI units have been followed throughout.

REFERENCES

Agarwal, R.K. (1986). Behaviour of Shallow Foundations Subjected to Eccentric-inclined Loads. Ph.D. Thesis, University of Roorkee, Roorkee.

Basavanna, B.M. (1975). Bearing Capacity of Soil under Static and Transient Load. Ph.D. Thesis, School of Research and Training in Earthquake Engineering, University of Roorkee, Roorkee.

Cascone, Ernesto and Orazio Casablanca (2016). Static and Seismic Bearing Capacity of Shallow Strip Footings. *Soil Dynamics and Earthquake Engineering,* **84:** 204-223.

Dai, Zi Hang and Xu, Xiang (2013). Comparison of Analytical Solutions with Finite Element Solutions for Ultimate Bearing Capacity of Strip Footings. *Applied Mechanics and Materials,* **353:** 3294-3303.

De Borst, R. and Vermeer, P.A. (1984). Possibilites and Limitations of Finite Elements for Limit Analysis. *Geotechnique,* **34(2):** 199-210.

Desai, C.K. (1971). Non-linear Analysis Using Spline Functions. *J. Soil Mech. and Found. Div., ASCE,* **97(10):** 1461-1480.

Duncan, J.M. and Chang, C.Y. (1970). Non-linear Analysis of Stress Strain in Soils. *J. Soil Mech. and Found. Div., ASCE,* **96(5):** 1629-1653.

Han, Dongdong, Xinyu Xie, Lingwei Zheng and Huang, Li (2016). The Bearing Capacity Factor $N\gamma$ of Strip Footings on c–ϕ–γ Soil Using the Method of Characteristics. Springer Plus **5(1):** 1482.

Kumar, Jyant (2009). The Variation of $N\gamma$ with Footing Roughness Using the Method of Characteristics. *International Journal for Numerical and Analytical Methods in Geomechanics*, **33(2):** 275-284.

Loukidis, D. and Salgado, R. (2009). Bearing Capacity of Strip and Circular Footings in Sand Using Finite Elements. *Computers and Geotechnics*, **36(5):** 871-879.

Meyerhof, G.G. (1951). Ultimate Bearing Capacity of Foundation. *Geotechnique*, **2(4):** 301-332.

Meyerhof, G.G. (1953). The Bearing Capacity of Footings under Eccentric and Inclined Loads. *Proceedings of 3rd International Conference on Soil Mechanics and Foundation Engineering*, Vol. I: 440-445.

Peck, R.B., Hanson, W.E. and Thornburn, W.T. (1974). Foundation Engineering. John Wiley and Sons, New York.

Saran, S. (1969). Bearing Capacity of Footings Subjected to Moments. Ph.D. Thesis, Geotechnical Engineering Section, Civil Engineering Department, University of Roorkee, Roorkee.

Saran, S. (2006). Analysis and Design of Substructures Limit State Design. Oxford and IBH Publishing Co., New Delhi.

Sharan, U.N. (1977). Pressure Settlement Characteristics of Surface Footings Using Constitutive Laws. Ph.D. Thesis, University of Roorkee, Roorkee.

Shiau, J.S., Lyamin, A.V. and Sloan, S.W. (2003). Bearing Capacity of a Sand Layer on Clay by Finite Element Limit Analysis. *Canadian Geotechnical Journal*, **40(5):** 900-915.

Sun, Jian-Ping, Zhao, Zhi-Ye and Cheng, Yi-Pik (2013). Bearing Capacity Analysis Using the Method of Characteristics. *Acta Mechanica Sinica*, **29(2):** 179-188.

Taylor, D.W. (1948). Fundamentals of Soil Mechanics. John Wiley and Sons, New York.

Terzaghi, K. (1943), Theoretical Soil Mechanics. John Wiley and Sons, Inc. New York.

CHAPTER 2

Constitutive Laws of Soils

2.1 General

Constitutive laws of soil define the mechanical behaviour of soil. They basically represent the non-linear stress-strain characteristics of soil and its strength. They are of prime importance for analyzing almost all applied non-linear problems of soil mechanics, e.g. stability analysis of slopes, of dams and embankments, of rafts, piles and wells. Constitutive laws are known to be function of the

- Physical properties of soil
- Moisture content and relative density
- Confining pressure
- Intermediate principal stress
- Strain rate
- Stress history of soil

A review of all pertinent literature available regarding the influence of the above factors on the mechanical behaviour of soil are summarized below (Sharan, 1977).

2.2 Factors Affecting Constitutive Laws

2.2.1 Influence of Particle Physical Properties

The influence of physical properties of granular soil particles characterized by shape, size and grading have been studied to a limited extent. It has been found that shear strength decreases with increase of particle size and increases with increase in angularity of particle shape. But grading has little effect on shear strength as long as the coefficient of uniformity remains unaltered (Vallerga, 1957; Leslic, 1963; Kolbuszewski, 1963; Kirkpatric, 1965; Marshall, 1965; Leussink, 1965; Lee and Seed, 1967; Fumagalli, 1969; Koerner, 1970). The behaviour of soil composed of flaky (platey) mineral particles, such as mica, talcum, montmorillonite, kalonite, allumina and chlorite, have been found quite different from those composed of bulky particles such as feldspar, iron ore, halocytes and calcite or rod-like particles such as hornblende. The basic change in stress-strain curve, the tremendous volumetric decrease and terminal orientation of particles were noticed in soil composed of flaky particles (Taylor, 1948; Horn and Deere, 1962; Jumikis, 1965; Koerner, 1970).

2.2.2 Effect of Moisture Content and Relative Density

The strength characteristics of clay are dependent upon moisture content, structural formation, i.e. whether it is flocculated, dispersed or honeycombed and also upon the properties of the pore fluid. The moisture content has, however, no significant effect on the strength of granular soil (Kirkpatric, 1965). Higher initial tangent modulus and strength are obtained for denser soils. The increase of density to the extent that there is no structural breakage causes an increase in the strength.

2.2.3 Effect of Confining Pressure

The significant effect of confining pressure on the behaviour of soil has been well recognized. It is generally found that the strength and compressibility of soil varies with confining pressures. At higher confining pressures, structural breakdown and increased compressibility are observed. Mohr's rupture envelope has been found to be curved with more pronounced curvature for dense sand and gravel samples in higher pressure range. In case of partially saturated clay, both the tangent modulus and the compressive strength have been found to be affected by confining pressure, but in case of saturated clay, they remain unaffected (Holtz and Gibbs, 1956; Hall and Gordon, 1963; Leslie, 1963; Vesic and Barkdale, 1963; Bishop et al., 1965; Lee and Seed, 1967; Pounce and Bells, 1971).

2.2.4 Effect of Intermediate Principal Stress

Limited literature is available on the effect of intermediate principal stress on the strength characteristics of soil. The strength and initial tangent modulus of soil under plane strain condition has been found to be higher but strain at peak stress is lower than in the triaxial compression and extension tests (Cornforth, 1964; Ko and Scot, 1967). The failure stress has been found to be maximum when $\sigma_2 = \sigma_3$ and increased with increase in σ_2 keeping σ_3 constant (Bishop, 1953; Southerland and Mesdary, 1969; Duncan, 1970). In general, this behaviour may be shown graphically as seen in Fig. 2.1. The intermediate principal stress has the effect of stiffening

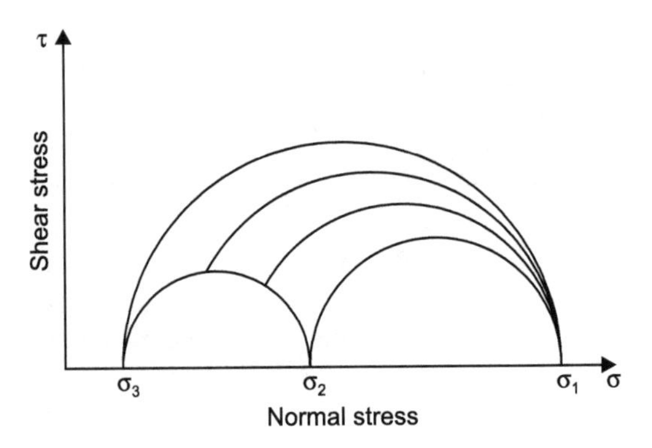

Fig. 2.1. Principal stresses on Mohr's diagram.

and increasing the strength of soil when it moves from the condition of triaxial compression to plane strain condition. The maximum possible shear stress is given by one-half of the difference in these stresses

$$\tau_{max} = \frac{1}{2}(\sigma_1 - \sigma_3) \tag{2.1}$$

The octahedral shear stress is given by

$$\tau_{oct} = \frac{1}{3}\sqrt{(\sigma_1 - \sigma_3)^2 + (\sigma_2 - \sigma_3)^2 + (\sigma_3 - \sigma_1)^2} \tag{2.2}$$

It is shown from the above expression that τ_{oct} varies only between limits of 0.816 (when $\sigma_2 = \sigma_1$ or σ_3) to 0.943 $\left(\text{when } \sigma_2 = \frac{(\sigma_1 + \sigma_3)}{2}\right)$ times the maximum shear stress. There is consequently a change of about 15% only in the relationship between τ_{oct} and τ_{max} and over the entire range of possible values. In many cases, therefore, one may be substituted for the other (Newmark, 1960).

2.2.5 Effect of Rate of Strain

The effect of strain rate on small samples of sand was investigated by Casagrande and Shannon (1947), Skempton and Bishop (1951), Prakash and Venkatesan (1960), Whitman and Healey (1962) and Whitman (1967). It was found that there is no significant effect of the rate of strain on the strength characteristics of granular material. The strength characteristics of clay are affected to a significant extent by the rate of strain. Higher tangent modulus and strength were observed at faster rate of loading. An increase in strain rate from one per cent per minute to 10^{-3} per minute gave 20% increase in strength (Taylor, 1948).

2.2.6 The Effect of Stress History

The shear strength parameters have been found to be independent of the stress history and the use of anisotropic consolidation in case of sand (Bishop and Eldin, 1953). But the shear strength characteristic of clay is greatly influenced by the consolidation history and appear to depend upon the over consolidation ratio (Simon, 1960; Hankel, 1966).

The factors influencing the stress-strain characteristics of soil are quite large in number and it is not possible to establish any general law which can take into account all the above factors. Therefore, laboratory testing under simulated field conditions is done to establish constitutive law for a given foundation soil. The constitutive relationship for simplified practical cases only may be studied with the help of rheological and mathematical models.

It may be mentioned here that in the field of constitutive modelling, more works are available which consider advance aspects related to different practical situations (Gudehus, 1979; Kogho et al., 1993; Darve et al., 1995; Jommi, 2000; Houlsby, 2002; Sheng et al., 2004). Discussion on these is beyond the scope of the book.

2.3 Models for Constitutive Relationships

2.3.1 Rheological Models

Soil is generally composed of solids, liquid and gaseous material. The solid part is composed of numerous grains of various mineralogical composition and is porous. The pore is generally filled with water and air. Soil resists the effect of external forces in a manner different from simple solid continua. In non-cohesive soil, the external force is resisted by the inter-granular friction at the contact surfaces. In cohesive soil, composed of clay minerals, the strength of the films of adsorbed water surrounding the grains accounts for the resistance of the soil to deformation (Sukelje, 1969).

The soil exhibits elasticity as well as creep under constant stress. Creep occurs at a rate which remains either constant or varies with time. Stress relaxation under constant applied strain is also observed in soil. This behaviour of soil can be described by visco-elastic models consisting of rheological elements, namely Hookean elastic body, Newtonian viscous liquid, Saint venant plastic body and Pascal's liquid. The relationship between stress and strain of these bodies is given by Scott (1963). The rheological models are constructed in an intuitive way and the corresponding relationships between stress and strain are deduced and compared with the experimental observations. This comparison controls the applicability of the assumed rheological models. Various rheological models which may be suitable for soil have been proposed by Rao (1967).

The model discussed here and given in Fig. 2.2 is constructed to represent elasto-plastic and viscous behaviour of soil and found to represent soil behaviour (Chakravarty, 1970; Prakash, Saran and Sharan, 1975; Sharan, 1977). The model consists of spring (H) and viscous dash-pot (N) on the left branch of the model and on the right branch of the model are gas or air (G) and incompressible liquid (IP) represented by a cylinder and a tight-fitting piston filled with air and a cylinder with water and tight-fitting piston respectively.

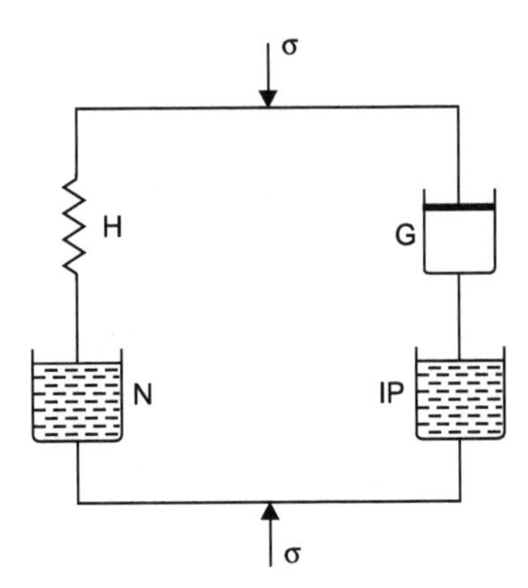

Fig. 2.2. Rheological model.

The relation between stress and strain is given by the following expression (Prakash, Saran and Sharan, 1975; Sharan, 1976):

$$\sigma = \dot{\varepsilon}\lambda(1-e^{-E/\lambda T}) + \frac{u_o\varepsilon}{1-\varepsilon} \tag{2.3}$$

where σ = applied stress

$\dot{\varepsilon}$ = rate of strain

λ = Trouton's coefficient of viscosity

T = time

ε = strain

u_0 = initial pore pressure

At large strains, time T in the above equation becomes large and it reduces to

$$\sigma = \dot{\varepsilon}\lambda + \frac{u_0\varepsilon}{1-\varepsilon} \tag{2.4}$$

Therefore, for a given stress, strain at failure and rate of strain, the value of λ can be obtained.

The values of E at various strains for a given curve can also be calculated. The theoretical stress-strain curves for various values of E and λ for this model are given in Fig. 2.3.

The rheological model discussed herein has been found to predict well the stress-strain curves for cohesive as well as non-cohesive soil by taking E at 1% strain (Chakravarty, 1970; Prakash, Saran and Sharan, 1975). The great drawback of this method is the correct evaluation of the elastic constant E which is done arbitrary at 1% strain. Soil exhibits non-linearity in stress-strain relationship even at low strain level and these curves are sensitive to magnitude of E as shown in Fig. 2.3.

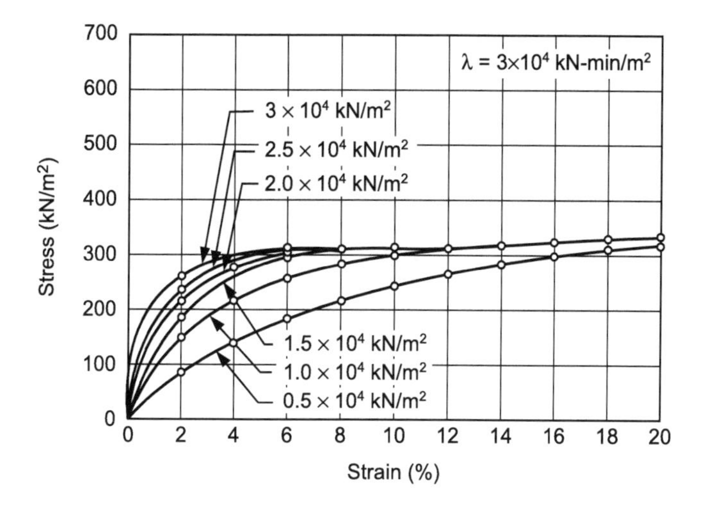

Fig. 2.3. Stress-strain curves (rheological model).

2.3.2 Mathematical Models

The behaviour of soil over a wide range of stresses is non-linear, inelastic and dependent upon the magnitude of the confining pressure used in the test. A review of literature on simplified practical mathematical models which takes into account this nonlinearity, stress dependency and inelasticity of soil behaviour is given below (Sharan, 1977).

(i) Hyperbolic Function

Kondner (1963) and Kondner and Zelasko (1963) have shown that the non-linear stress-strain curves of both clay and sand (Fig. 2.4a) may be approximated by hyperbolae (Fig. 2.4b) of the following form with high degree of accuracy:

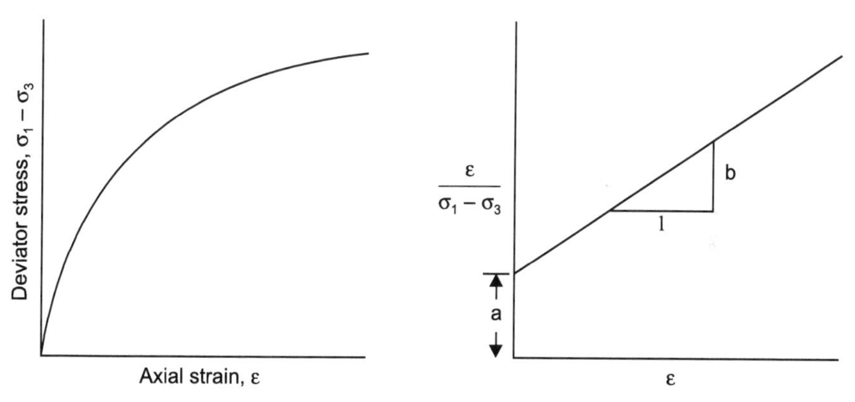

Fig. 2.4. (a) Typical triaxial test data. (b) Transformed hyperbola plot.

$$\frac{\varepsilon}{(\sigma_1 - \sigma_3)} = a + b\varepsilon \tag{2.5a}$$

Or,
$$\varepsilon = \frac{a(\sigma_1 - \sigma_3)}{1 - b(\sigma_1 - \sigma_3)} \tag{2.5b}$$

where ε = axial strain

a, b = constants of hyperbola

The plot $\frac{\varepsilon}{(\sigma_1 - \sigma_3)}$ versus ε gives a straight line where 'a' is the intercept on Y-axis and 'b' is the slope of line (2.4b). Working in limit $\varepsilon \rightarrow \infty$ with Eq. (2.5), the reciprocal of 'b' represents the ultimate compressive strength of the soil which is larger than compressive strength at failure. This is expected because the hyperbola remains below the asymptote at all finite values of strain. The ratio R_f of compressive strength $(\sigma_1 - \sigma_3)_f$ to ultimate compressive value σ_u varies from 0.75 to 1.0 for different soils independent of confining pressure (Kondner, 1963).

A differentiation of Eq. (2.5) with respect to strain ε gives the inverse of 'a' as the initial tangent modulus. Duncan and Chang (1970) have expressed the Kondner's expression in terms of shear strength and initial tangent modulus as given below:

$$(\sigma_1 - \sigma_3) = \varepsilon \left[1 - \frac{R_f (1 - \sin\phi)(\sigma_1 - \sigma_3)}{2c\cos\phi + 2\sigma_3 \sin\phi} \right] E_i \tag{2.6}$$

where c = cohesion

ϕ = angle of internal friction

E_i = initial tangent modulus

$R_f = (\sigma_1 - \sigma_3)_f / \sigma_u$

Janbu's (1963) experimental studies have shown that the relationship between initial tangent modulus and confining pressure may be expressed as follows:

$$E_i = Kp_a \left(\frac{\sigma_3}{p_a} \right)^n \tag{2.7}$$

where p_a = atmospheric pressure, K and n = pure numbers.

The values of K and n can be readily determined from the results by plotting E_i against σ_3 on log-log scales and fitting a straight line to the data.

(ii) Parabolic Functions

Hansen (1963) proposed two additional functional representation stress-strain relationships:

$$(\sigma_1 - \sigma_3) = \left[\frac{\varepsilon}{a+b}\right]^{1/2} \tag{2.8}$$

$$(\sigma_1 - \sigma_3) = \left[\frac{\varepsilon^{1/2}}{a+b}\right] \tag{2.9}$$

The first equation accounts for the possibility of parabolic variations in stress-strain curves at small strains. The second equation is an alternative form to account for the parabolic variation and possesses the property of giving a maximum value of $(\sigma_1 - \sigma_3)$ for finite strain, i.e. it is suitable when the curve shows a decrease after the peak stress. Hansen (1963) used one of the data from Kondner (1963) and compared the stress obtained from the above equations. He observed that if stress-strain curve is initially parabolic, Eq. (2.8) will probably be better and if it is work softening, then Eq. (2.9) should be attempted.

(iii) Spline Function

Desai (1971) compared the above mathematical models with that of the spline functions which approximate the given non-linear stress-strain relations by a number of polynomials of a given degree spanning a number of data points. Details of splines and their mathematical properties are described by Schoenburg (1964), Ahlberg et al. (1967) and Greville (1967). Kondner's (1963) work compared very well with that Desai's (1971) spline functions.

2.3.3 Discussion

The mathematical models discussed above are purely empirical in nature. Therefore, it is likely that one relation may represent the actual stress-strain to a better degree than the other. A comparative study for predicting bearing capacity and settlement was carried out by Desai (1971) for circular footing in clay using the spline function and hyperbolic equation. His study showed that the ultimate bearing capacity computed by using spline function differed by about 0.52% and that using hyperbola differed by 1.58% from the experimental results. The settlements, predicted at 60% of the ultimate load, differed 3 to 6% and 20 to 25% on computation by spline function and hyperbolic equation respectively.

Basavanna (1975) has discussed the cause of this large variation in case of hyperbolic equation of Kondner (1963). Desai had computed the values '*a*' and '*b*' constants of hyperbola by using the criteria suggested by Duncan and Chang (1970) which takes into account the straight line passing through the points of 70 to 95% of mobilized strength. Basavanna (1975) suggests that if due weightage is given to the experimental points in the initial portion of the curve, different values of '*a*' and '*b*' parameters are obtained and the predicted pressure settlement curve is close to the experimental one.

The above discussion leads to the conclusion that the stress-strain relationship represented by the mathematical model suggested by Kondner is simple and acceptable and, therefore, may be adopted for investigation of constitutive laws of soil. Keeping this in view, behaviour of footing under different types of loading discussed in subsequent chapters have been presented using Kondner hyperbolic concept of constitutive laws of soils (Sharan, 1977).

2.4 Constitutive Laws of Typical Soils

2.4.1 For Buckshot Clay

The Buckshot clay used by Caroll (1963) was brought from the field in disturbed condition at natural water content of 30 to 40%, allowed to air dry to water contents of about 10% and then processed to the desired water content for test purposes. Salient properties of Buckshot clay are given in Table 2.1.

Table 2.1. Properties of Buckshot clay

Montmorllonite	25%
Illite	25%
Quartz	20%
Feldspar	20%
Fe_2O_3	2%
Organic matter	1%
Percent finer than No.200 sieve (74 microns)	90 to 95
Percent finer than 2 microns	45 to 50
Liquid limit	55 to 70
Plastic limit	20
Water content	32.2%
Bulk density	18.1 kN/m^3
Poisson's ratio (μ)	0.375
Coefficient of permeability	10^{-11} m/s
Coefficient of consolidation	2×10^{-7} m/sec
Compression index	0.52 ± 0.04
Undrained shear strength	72 kN/m^2
Saturation	90%

The stress-strain curves obtained by undrained triaxial shear tests for Buckshot clay at moisture content of 32.2% are given in Fig. 2.5(a) for confining pressures of 105 kN/m^2 and 211 kN/m^2 as reported by Carrol (1963). Carrol had given stress-strain curves for other water contents also but the degree of saturation obtain for this was in the order of 90%. This causes a very little variation in stress-strain relationship with confining pressure as shown in Fig. 2.5(a). The transformed hyperbolic stress-strain plot for the above curves are given in Fig. 2.5(b).

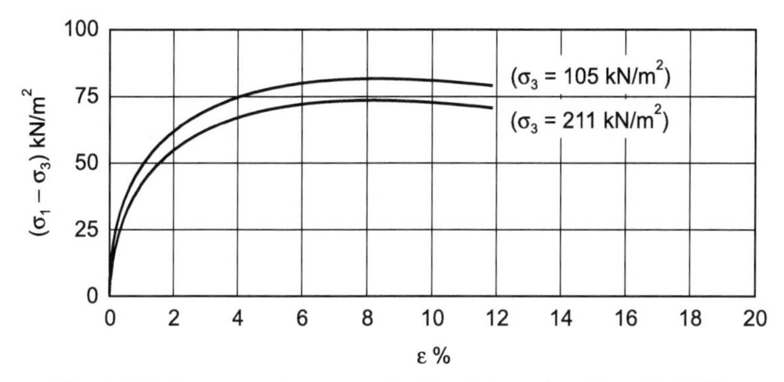

Fig. 2.5(a). Stress-strain curves for Buckshot clay (Carrol, 1963).

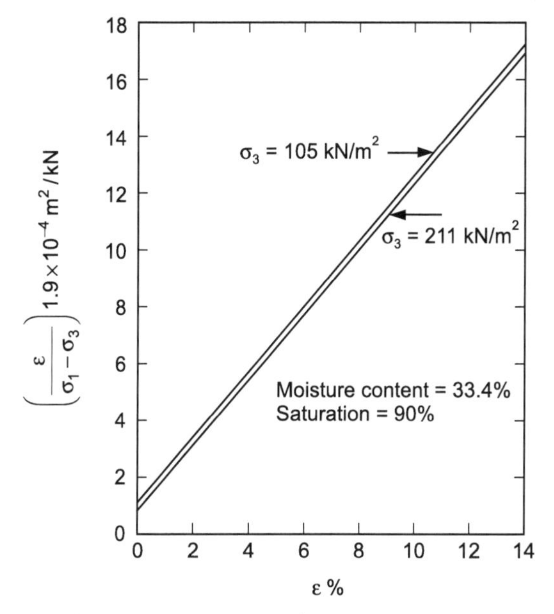

Fig. 2.5(b). Hyperbolic stress-strain curve for Buckshot clay (After Carrol, 1963).

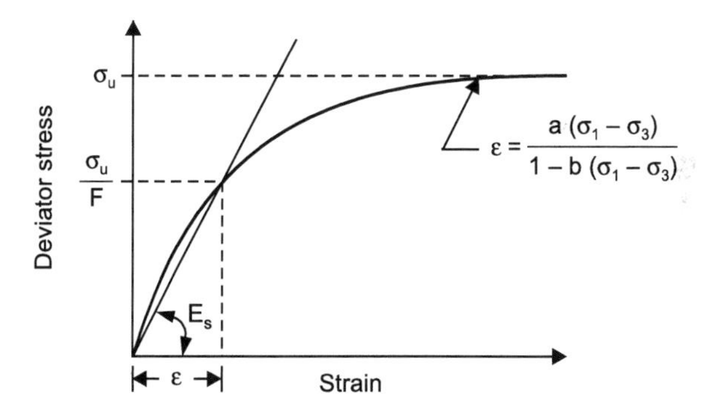

Fig. 2.5(c). Hyperbolic stress-strain relationship.

The ratio of the ultimate strength in hyperbolic representation to the actual failure strength was obtained as 1.10 (Fig. 2.5c). His undrained triaxial test results show negligible influence of confining pressure on tangent modulus E_1 or $\left(\frac{1}{a}\right)$ and shear strength σ_u $\left(\text{or } \frac{1}{b}\right)$. The average values of b/a was found to be 142. The values of a and b were obtained as 1.9×10^{-4} m²/kN and 214×10^{-4} m²/kN respectively.

2.4.2 For Ranipur Sand

The physical properties of Ranipur sand are given in Table 2.2.

Table 2.2. Properties of Ranipur sand

Type of soil	SP
Effective size (D_{10})	0.17
Uniformity coefficient	1.76
Mean specific gravity	2.65
Minimum void ratio	0.58
Maximum void ratio	0.89
Average density	1.63 g/cc
Relative density	84%

To obtain constitutive relationship for Ranipur sand, triaxial tests were conducted with different confining pressures, varying from 150 kN/m² to 500 kN/m² at relative density 84%. The stress-strain curves are shown in Fig. 2.6(a) (Prakash, 1975).

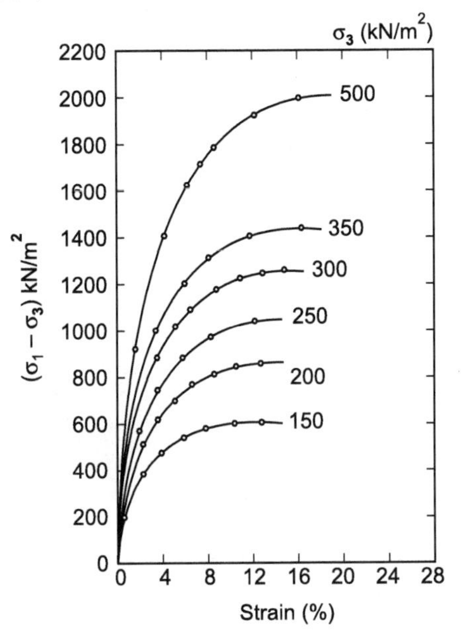

Fig. 2.6(a). Stress-strain curves for Ranipur sand (D_R = 84%).

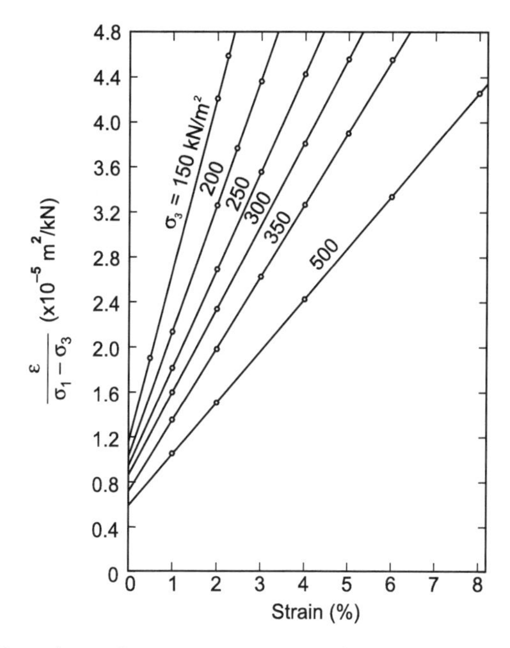

Fig. 2.6(b). Transformed stress-strain curves for Ranipur sand (D_R = 84%).

The transformed hyperbolic stress-strain curves are given in Fig. 2.6 (b). It is found that hyperbolic equation (Kondner, 1963) can be used to represent the stress-strain relationship.

The parameter '*a*' and '*b*' of the hyperbola were correlated with the confining pressures and are plotted in Figs 2.7 and 2.8. The following relationship holds good for Ranipur sand.

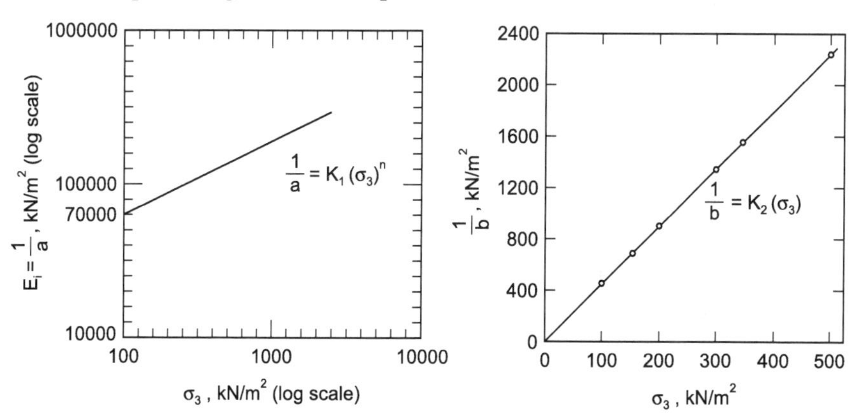

Fig. 2.7. Variation of $E_i \left(= \frac{1}{a}\right)$ w.r.t. confining pressure for Ranipur sand (D_R = 84%).

Fig. 2.8. Variation of Kondner's ($1/b$) w.r.t. confining pressure for Ranipur sand (D_R = 84%).

$$1/a = K_1 (\sigma_3)^n$$
$$1/b = K_2 \sigma_3$$

The values of K_1, n and K_2 are given in Table 2.3.

Table 2.3. Parameters of Constitutive Law for Ranipur sand

Relative density	K_1	n	K_2
84%	700	0.5	4.40

The angle of shearing resistance was determined with the help of triaxial tests. These tests gave the value of the angle of internal friction of sand at 41° at relative density 84%.

Illustrative Examples

Example 2.1

A drained triaxial test was performed on Dhanori clay (LL = 53%, P.I = 25%, S = 91%) giving the deviator stress versus axial strain plot as shown in Fig. 2.9. Plot the data in the transformed pattern as suggested by Kondner (1963), and obtain the values of '*a*' and '*b*' parameters.

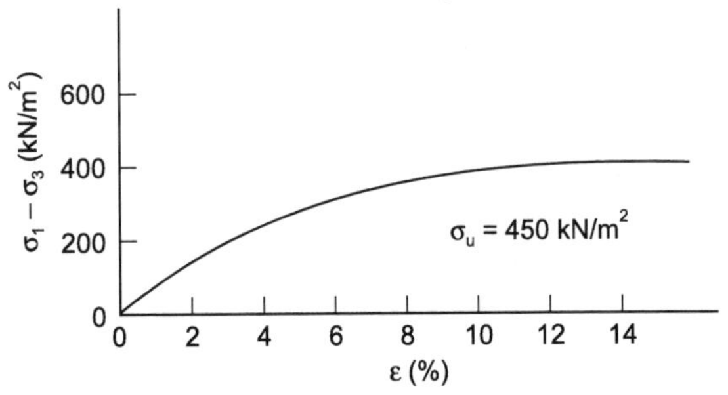

Fig. 2.9. Stress-strain curve for drained triaxial shear test for Dhanori clay (Prakash, 1975).

Solution

1. Transformed plot i.e. $\dfrac{\varepsilon}{(\sigma_1 - \sigma_3)}$ versus ε has been plotted as shown in Fig. 2.10.

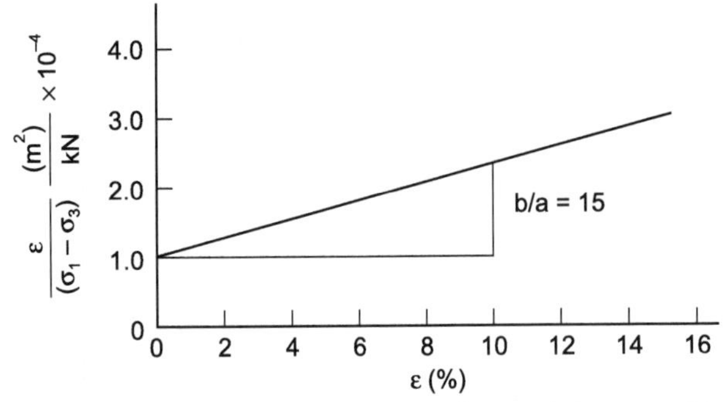

Fig. 2.10. Hyperbolic stress-strain plots for drained triaxial shear test (Prakash, 1975).

2. This plot gives

$$a = 0.826 \times 10^{-4}\ \mathrm{m^2/kN}$$

$$b = \frac{1.5 \times 10^{-4}}{0.1}\ \mathrm{m^2/kN} = 15 \times 10^{-4}\ \mathrm{m^2/kN}$$

$$b/a = 18.16$$

$$\sigma_u = 450\ \mathrm{kN/m^2}$$

$$\frac{1}{b} = 666\ \mathrm{kN/m^2} = \sigma_{um}$$

$$\frac{1}{a} = 12106\ \mathrm{kN/m^2}$$

$$\frac{\sigma_{um}}{\sigma_u} = \frac{666}{450} = 1.48$$

Example 2.2

Drained triaxial test were performed on Amanatgarh sand (SP, D_{10} = 0.15 mm, C_u = 2.0, D_r = 70%) at different confining pressure. The results are shown in Fig. 2.11. Determine the values of parameters *a* and *b*, and their trend with confining pressure.

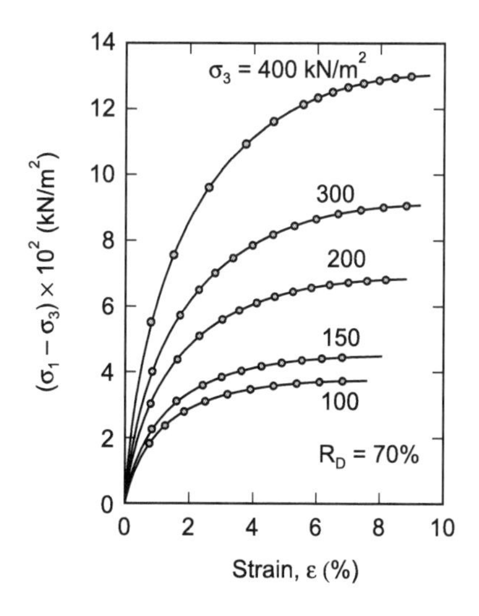

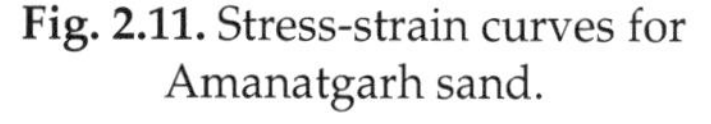

Fig. 2.11. Stress-strain curves for Amanatgarh sand.

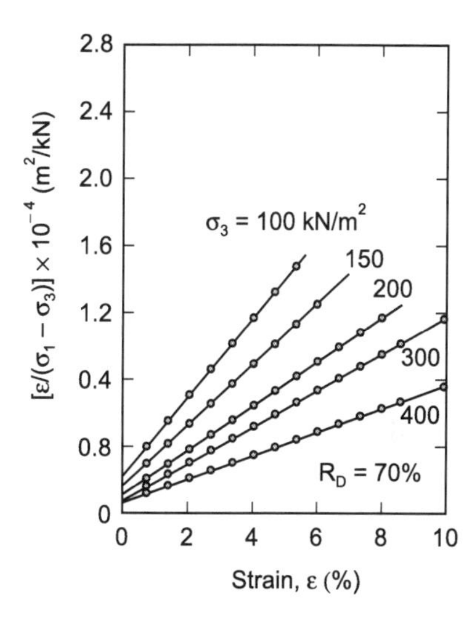

Fig. 2.12. Transformed stress-strain curves for Amanatgarh sand.

Solution

The transformed hyperbolic stress-strain curves were obtained from the triaxial results in Fig. 2.11 and are shown in Fig. 2.12. It has been found that the hyperbolic equation presented by Kondner (1963) can be used to represent the stress-strain relationship.

The parameters a (or $1/E_i$) and b (or $1/\sigma_u$) of the hyperbola were correlated with confining pressure for the relative density of 70% as shown in Figs 2.13 and 2.14 respectively. The following relationships were found to hold good for Amanatgarh sand

$$\frac{1}{a} = K_1 (\sigma_3)^n$$

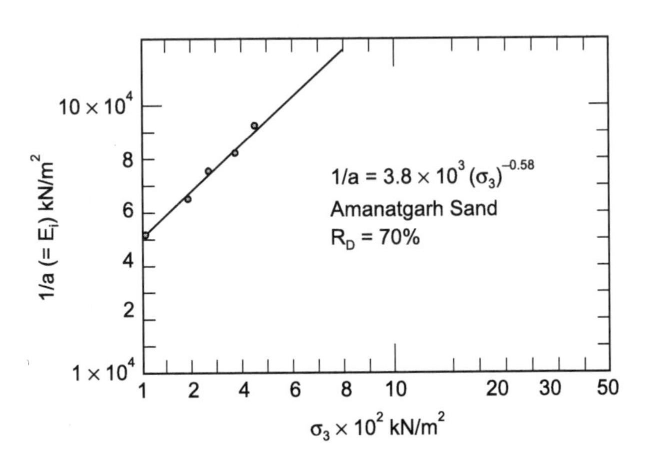

Fig. 2.13. Hyperbolic constant $(1/a)$ versus σ_3.

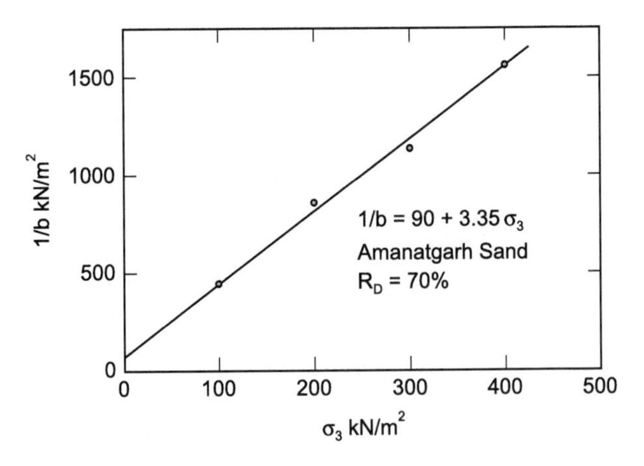

Fig. 2.14. Hyperbolic constant $(1/b)$ versus σ_3.

$$\frac{1}{b} = K_2 + K_2\,(\sigma_3)$$

where K_1, K_2, K_3 and n are as given in Table 2.4 and σ_3 is the confining pressure (kN/m^2).

Table 2.4. Parameters of Constitutive Laws for Amanatgarh sand

Relative density, R_D	K_1	n	K_2	K_3
70%	3.8×10^3	0.58	90	3.35

Practice Problems

1. What is meant by constitutive laws of soils? Describe briefly the factors affecting them.
2. Discuss the salient features of following constitutive models:
   (a) Kondner's mathematical model (1963)
   (b) Duncan and Chang (1970) model.
3. For a typical soil values of parameters *'a'* and *'b'* were found as 0.9×10^{-4} m^2/kN and 20×10^{-4} m^2/kN respectively. Plot (i) transformed stress-strain curves, and (ii) actual stress-strain curves.

REFERENCES

Ahlberg, J.H., Nilson, F.W. and Walsh, J.L. (1967). The Theory of Splines and their Applications. Academic Press, New York.

Basavanna, B.M. (1975). Bearing Capacity of Soil under Static and Transient Loads. Ph.D. Thesis, School of Research and Training in Earthquake Engineering, University of Roorkee, Roorkee.

Bishop, A.W. et al. (1965). Triaxial Tests on Soil at Elevated Cell Pressure. *Proc. of 6th International Conference on Soil Mechanics and Foundation Engineering,* **Vol. I:** 170-174.

Bishop, A.W. and Eldin, A.K.G. (1953). The Effect of Stress History on the Relation between Angle of Shearing Resistance and Porosity in Sand. *3rd International Conference on Soil Mechanics and Foundation Engineering,* **Vol. I:** 100-105.

Carrol, W.F. (1963). Dynamic Bearing Capacity of Soil. Technical Report No. 3-599, U.S. Army Engineers, Waterways Experiment Stations, Corps of Engineers, Vicksburg, Mississippi.

Casagrande, A. and Shanon, W.L. (1947). Research on Stress Deformation and Strength Characteristics of Soil and Soft Rocks under Transient Loading. Publication, Harward University, Grad. School of Engineering. **No. 447:** 132.

Chakaravaty, S.K. (1970). Bearing Capacity in Relation to Stress-Strain Characteristics of Soils. Ph.D. Thesis, Ranchi University, Ranchi.

Cornforth, D.H. (1964). Some experiments on the influence of strain conditions on the strength of sands. *Geotechnique*, **14(2):** 143-167.

Darve, F., Flavigny, E. and Meghachou, M. (1995). Constitutive Modelling and Instabilities of Soil Behavior. *Computers und Geotechnics*, **17:** 203-224.

Desai, C.K. (1971). Non-Linear Analysis Using Spline Functions. *Journal American Society of Civil Engineers, Found. Div.*, **97(10):** 1461-1480.

Duncan, J.M. and Chang, C.Y. (1970). Non-linear Analysis of Stress Strain in Soils. *Journal American Society of Civil Engineers, Found Div.*, Vol. 96, **SM5:** 1629-1653.

Fumagalli, E. (1969). Tests on Cohesionless Materials for Rockfill Dams. *American Society of Civil Engineers*, **SM5:** 313-332.

Greville, T.T.N. (1967). Spline Functions, Interpolation and Numerical Quandrature. *Mathematical Methods of Digital Computers*, Vol. II, A. Ralston and H.S. Wilfeds, John Wiley and Sons.

Gudehus, G. (1979). A Comparison of Some Constitutive Laws for Soils Under Radially Symmetric Loading and Unloading. In: W. Wittke (Ed.). A.A. Balkema Publishers, Rotterdam. 1309 p.

Hall, E.B. and Gordon, B.B. (1963). Triaxial Testing with Large-scale High Pressure Equipment. *Laboratory Shear Testing of Soils*, ASTM. STP **62:** 315-328.

Hankel, D.J. and Wade, N.H. (1966). Plane Strain Test on a Saturated Remoulded Clay. *Proc. American Society of Civil Engineers*, SJ 116, **92:** 67-80.

Hansen, J.B. (1963). Discussion of Hyperbolic Stress-Strain Response of Cohesive Soils. *Proceeding American Society of Civil Engineers*, **SM 4:** 242-242.

Holtz, W.G. and Gibb, H.J. (1956). Triaxial Shear Tests on Pervious Gravelly Soils. American Society of Civil Engineers. *Journal of Soil Mechanics and Foundation Engineering*, **1:** 1-22.

Horn, M.R. and Deere, D.U. (1962). Frictional Characteristics of Minerals. *Geotechnique*, **12:** 319-335.

Houlsby, G.T. (2003). Some Mathematics for the Constitutive Modelling of Soils. *Proc. Conference on Mathematical Methods in Geomechanics*, Horton, Greece. In: Advanced Mathematical and Computational Geomechanics. Springer, Berlin, Heidelberg.

Janbu, N. (1963). Soil Compressibility as Determined by Oedometer and Triaxial Tests. *European Conference on Soil Mechanics and Foundation Engineering Wiesbaden*, Germany, **Vol. I:** 19-25.

Jommi, C. (2000). Remarks on the Constitutive Modelling of Unsaturated Soils. In: Experimental Evidence and Theoretical Approaches in Unsaturated Soils. Tarantino and Mancuso (eds). Balkema, Rotterdam, ISBN 90-5809-186-4.

Jumikis, R.A. (1965). Soil Mechanics. D. Van Nostrand Company, INC. N.Y.

Kirkpatric, W.M. (1965). Effect of Grain Size and Grading on the Shearing Behavior of Granular Materials. *Proceeding of 6th International Conference on Soil Mechanic and Foundation Engineering*, **Vol. I:** 273-277.

Ko, H.Y. and Scott, R.F. (1967). A New Soil Testing Apparatus. *Geotechnique*, **17:** 40-57.

Koerner, R.M. (1970). Behaviour of Single Mineral Soils in Triaxial Shear. *American Society of Civil Engineers*, **SM 4:** 1373-1390.

Kogho, Y., Nakano, M. and Miyazaki, T. (1993), Theoretical Aspects of Constitutive Modelling for Unsaturated Soils. *Soils and Foundations*, **33(4):** 49-63.

Kolbuszewski, J. and Fredrick, M.R. (1963). The Significance of Particle Shape and Size on the Mechanical Behavior of Granular Materials. *European Conference on Soil Mechanics and Foundation Engineering,* Wiesbaden, **Vol. I:** 253-263.

Kondner, R.L. and Zelasko, S. (1963). A Hyperbolic Stress-Strain Formulation for Sands. *Proceedings of 2nd Pan American Conference on Soil Mechanics and Foundation Engineering,* Brazil, **Vol. I:** 289-324.

Kondner, R.L. (1963). Hyperbolic Stress-Strain Response of Cohesive Soils. Proceedings of American Society of Civil Engineers. *Journal of Soil Mechanics and Foundation Engineering,* SM 3, **89:** 115-143.

Lee, K.L. and Seed, H.B. (1967). Drained Strength Characteristics of Cohesionless Soils. *Proceedings of American Society of Civil Engineers,* SM 6, **117:** 5561.

Lee, K.L. and Seed, H.B. (1967). Undrained Strength Characteristic of Cohesionless Soils. Proceedings of American Society of Civil Engineers. *Journal of Soil Mechanics and Foundation Engineering,* **6:** 333-360.

Leslie, D.E. (1963). Large scale Triaxial Tests on Gravelley Soils. *Proceeding of 2nd Pan American Conference,* **Vol. I:** 181-202.

Leussink, H. (1965). Discussion, Effect of Specimen Size on the Shear Strength of Granular Materials. *Proceedings of 6th International Conference on Soil Mechanics and Foundation Engineering,* **Vo1. III:** 316.

Marshall, R.J. (1965). Discussion. *Proceedings of 6th International Conference on Soil Mechanics and Foundation Engineering,* **Vol. III:** 310-316.

Pounce, V.M. and Bell, J.M. (1971). Shear Strength of Sand at Extremely Low Pressure. Proceedings of American Society of Civil Engineers. *Journal of Soil Mechanics and Foundation Engineering,* **No. 4:** 625.

Prakash, C. (1975). Behaviour of Eccentrically Loaded Footings in Clay. M.E. Thesis, Department of Civil Engineering, University of Roorkee (India).

Prakash, S. and Saran, S. (1977). Settlement and Tilt of Eccentrically Loaded Footings. *Journal of Structural Engineering,* **4(4):** 164-176.

Prakash, S. and Venkatesan, S. (1960). Shearing Characteristics of Dense Sands at Different Rates of Strain. *Journal of Scientific and Industrial Research,* **19A(5):** 219-223.

Rao, N.V.R.L.N. (1967). Rheological Models and their Use in Soil Mechanics. *Journal of Indian National Society of Soil Mechanics and Foundation Engineering,* 225 p.

Schoenburg, I.J. (1964). On Interpolation by Spline Functions and its Minimal Properties. *International Series of Numerical Analysis,* Vol. 5, Academic Press.

Scott, R.F. (1963). Principles of Soil Mechanics. Addison Wesley Pub. Co. Inc. Readin Mao. 280 p.

Sharan, U.N. (1977). Pressure Settlement Characteristics of Surface Footings Using Constitutive Laws. Ph.D. Thesis. University of Roorkee, Roorkee.

Sheng, D., Sloan, S.W. and Gens, A. (2004). A constitutive model for unsaturated soils: Thermomechanical and computational aspects. *Comput Mech.* **33(6):** 453-465.

Simon, N.E. (1960). Effect of Over Consolidation on Shear Strength Characteristics of Undisturbed Oslo Clay. Reseema Conference on Cohesive Soils. University of Colorado, Boulder, Colorado, 747-763.

Skempton and Bjerrum (1957). A Contribution to the Settlement Analysis of Foundations in Clay. *Geotechnique*, London, **Vol. 17:** 168-178.

Southerland, H.S. and Mesdary, M.S. (1969). The Influence of Intermediate Principal Stress on Strength of Sand. *Proceedings of the 7th International Conference on Soil Mechanics and Foundation Engineering*, Mexico, **Vol. I:** 391-394.

Sukelje, L. (1969). Rheological Aspects of Soil Mechanics. Wiley Inter Science, N.Y.

Taylor, D.W. (1948). Fundamentals of Soil Mechanics. John Wiley and Sons, New York.

Vallerga, B.A. et al. (1957). Effect of Shape, Size and Surface Roughness of Aggregates Particles on the Strength of Granular Materials. ASTM, STP 212.

Vesic, A.J. and Barkdale, R.D. (1963). On Shear Strength of Sand at High Pressure. *Laboratory Shear Testing of Soils.* ASTM, STP, **361:** 301-305.

Whitman, R.V. and Healey, K.A. (1962). Shear Strength of Sand During Rapid Loading. *Journal of American Society of Civil Engineering, Found. Div.*, SM 2, **88:** 99-132.

Whitman, R.V. (1967). The Behaviour of Soils under Transient Loading. *Proceedings of 4th International Conference on Soil Mechanics and Foundation Engineering,* **Vol. II:** 717.

CHAPTER 3

Stress Distribution in Soil Mass

3.1 General

For many problems of practical interest, it is necessary to estimate settlements under conditions in which the induced stress varies spatially. The first step in the analyses of such problems usually involves estimation of the initial states of stress in the soil and of the changes in these stresses during loading and when the soil approaches equilibrium. Two methods of analyses are commonly followed.

The theoretical solutions give total stresses in a weightless elastic medium because the soil is assumed to be a continuum. Their resolution into effective stresses and pore water pressures is a separate problem not considered in terms of the theory of elasticity. The assumption that the medium is weightless simplifies the theoretical analyses and the form of the resulting equations. The actual state of stress is found by superimposing the stresses calculated for the weightless medium on the initial state of stress in the soil prior to application of the load (Olson, 1989).

The soil is generally assumed to be isotropic and linearly elastic so that its properties can be described by using only two parameters. In most of the solutions it will be convenient to use Young's modulus, E, as one of the parameters and Poisson's ratio, μ, as the other.

The stress distribution in an elastic, isotropic, homogeneous and semi-infinite soil mass due to point load at the surface was derived by Boussinesq (1985) and was extended to other loading conditions by Melan (1919), Carothers (1920), Kolosov (1935) and Mindlin (1936). The equations have been computed on the assumption that the modulus of elasticity for the soil mass is constant. In saturated or nearly saturated clays, the modulus of elasticity is independent of depth or confining pressure (Carrol, 1963) and Poisson's ratio is taken as 0.5, with no volume change. Therefore, Boussinesq's approach holds good in saturated clays. Stress distribution in elastic, anisotropic and semi-infinite mass was found by Westergaard (1938). The stress distribution in elastic, isotropic and non-homogeneous soil mass for plane strain and axi-symmetric cases, where shear modulus G (or Young modulus E) increases linearly with depth, were derived by Gibson (1967) for uniform pressure intensity. The equations are independent of E and are sensitive to Poisson's ratio (μ) and reduce to original equations of Boussinesq where $\mu = 0.5$. Stress equations for triangular or trapezoidal type of loading are not available.

In this chapter, stress distributions of different surface loads that are considered of most practical values, are presented.

3.2 Different Types of Loading

3.2.1 Uniform Vertical Loading (Fig. 3.1), (Carothers, 1920 and Poulos and Davis, 1974)

$$\sigma_z = \frac{q}{\pi}[\alpha + \sin\alpha.\cos(\alpha + 2\delta)] \tag{3.1a}$$

$$\sigma_x = \frac{q}{\pi}[\alpha - \sin\alpha.\cos(\alpha + 2\delta)] \tag{3.1b}$$

$$\tau_{xz} = \frac{q}{\pi}[\sin\alpha.\sin(\alpha + 2\delta)] \tag{3.1c}$$

3.2.2 Uniform Horizontal Loading (Fig. 3.2), (Kolosov, 1935 and Poulos and Davis, 1974)

$$\sigma_z = \frac{q}{\pi}\left[\sin\alpha\sin(\alpha + 2\delta)\right] \tag{3.2a}$$

$$\sigma_x = \frac{q}{\pi}\left[\log_e\left(\frac{R_1}{R_2}\right)^2 - \sin\alpha\sin(\alpha + 2\delta)\right] \tag{3.2b}$$

$$\tau_{xz} = \frac{q}{\pi}\left[\alpha - \sin\alpha\cos(\alpha + 2\delta)\right] \tag{3.2c}$$

3.2.3 Symmetrical Vertical Triangular Loading (Fig. 3.3), (Gray, 1936 and Poulos and Davis, 1974)

$$\sigma_z = \frac{q}{\pi}\left[(\alpha_1 + \alpha_2) + \frac{x}{b}(\alpha_1 - \alpha_2)\right] \tag{3.3a}$$

$$\sigma_x = \frac{q}{\pi}\left[(\alpha_1+\alpha_2)+\frac{x}{b}(\alpha_1-\alpha_2)-\frac{2z}{b}\log_e\left(\frac{R_1R_2}{R_0^2}\right)\right] \tag{3.3b}$$

$$\tau_{xz} = -\frac{qz}{\pi b}(\alpha_1-\sigma_2) \tag{3.3c}$$

3.2.4 Vertical Loading Increasing Linearly (Fig. 3.4), (Carothers, 1920 and Jumikis, 1969; Harr, 1966)

$$\sigma_z = \frac{q}{2\pi}\left[\frac{x}{b}\cdot\alpha-\sin 2\delta\right] \tag{3.4a}$$

$$\sigma_x = \frac{q}{2\pi}\left[\frac{x}{b}\alpha-\frac{z}{b}\log_e\frac{R_1{}^2}{R_2{}^2}+\sin 2\delta\right] \tag{3.4b}$$

$$\tau_{xz} = \frac{q}{2\pi}\left[1+\cos 2\delta-\frac{z\alpha}{b}\right] \tag{3.4c}$$

3.2.5 Vertical Loading Decreasing Linearly (Fig. 3.5), (Carothers, 1920; Jumikis, 1969 and Harr, 1966)

$$\sigma_z = \frac{q}{2\pi b}\left[(x-2b)\cdot\delta-(x-2b)\cdot(\alpha+\delta)+\frac{2bxz}{R_1^2}\right] \tag{3.5a}$$

$$\sigma_x = \frac{q}{2\pi b}\Big[(x-2b)\cdot\delta-z.\log_e(R_2^2)-$$

$$(x-2b)\cdot(\alpha+\delta)+z.\log_e(R_1^2)-\frac{2bxz}{R_1^2}\Big] \tag{3.5b}$$

$$\tau_{xz} = \frac{q}{2\pi b}\left[z\cdot(\alpha+\delta)-z\cdot\delta+\frac{2bz^2}{R_1^2}\right] \tag{3.5c}$$

3.2.6 Horizontal Loading Increasing Linearly (Fig. 3.6) (Gray, 1936)

$$\sigma_z = \frac{q}{2\pi}\left[1+\cos 2\delta-\frac{z}{b}\alpha\right] \tag{3.6a}$$

$$\sigma_x = \frac{q}{2\pi}\left[\frac{3za}{b}+\frac{x}{b}\log_e\left(\frac{R_1^2}{R_2^2}\right)-\cos 2\delta-\delta\right] \tag{3.6b}$$

$$\tau_{xz} = \frac{q}{2\pi}\left[\frac{x}{b}\alpha-\frac{z}{b}\log_e\left(\frac{R_1^2}{R_2^2}\right)+\sin 2\delta\right] \tag{3.5c}$$

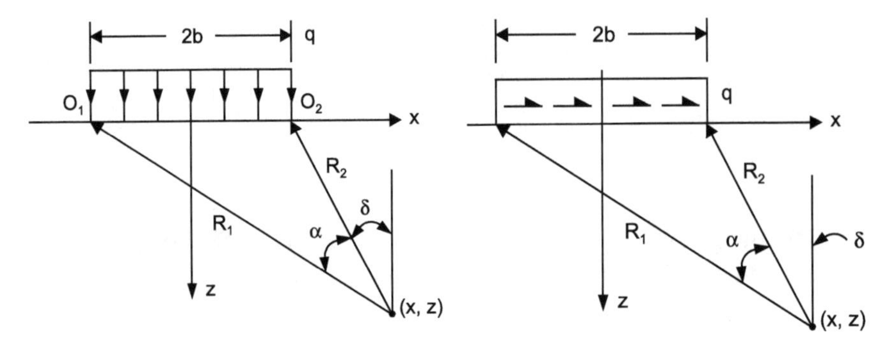

Fig. 3.1. Uniformly distributed loading.

Fig. 3.2. Uniformly distributed horizontal load.

On referring to Figs 3.1 and 3.2, the origin passing through the centre of footing,

$$\tan\delta = \frac{x-b}{z} \tag{3.7a}$$

$$\tan(\alpha+\delta) = \frac{x+b}{z} \tag{3.7b}$$

$$R_1^2 = (x+b)^2 + z^2 \tag{3.8a}$$

$$R_2^2 = (x-b)^2 + z^2 \tag{3.8b}$$

Therefore,

$$\delta = \tan^{-1}\frac{(x-b)}{z} \tag{3.9a}$$

$$\alpha = \tan^{-1}\frac{(x+b)}{z} - \tan^{-1}\frac{(x-b)}{z} \tag{3.9b}$$

On referring to Fig. 3.3, origin passing through the centre of footing, values of R_1^2 and R_2^2 will be the same as given in Eqs (3.8a) and (3.8b)

$$R_0^2 = x^2 + z^2 \tag{3.10}$$

$$\alpha_1 = \tan^{-1}\left(\frac{x+b}{z}\right) - \tan^{-1}\left(\frac{x}{z}\right) \tag{3.11a}$$

$$\alpha_2 = \tan^{-1}\left(\frac{x}{z}\right) - \tan^{-1}\left(\frac{x-b}{z}\right) \tag{3.11b}$$

On referring to Figs 3.4, 3.5 and 3.6 the origin passing through the left edge of footing,

$$R_1^2 = x^2 + z^2 \tag{3.12a}$$

$$R_2^2 = (x-2b)^2 + z^2 \tag{3.12b}$$

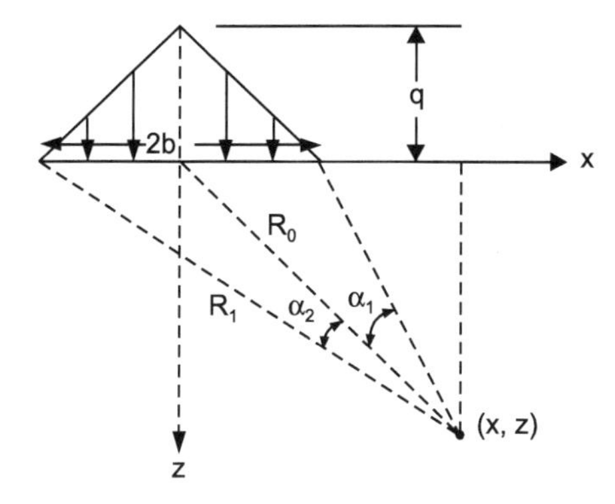

Fig. 3.3. Symmetrical triangular vertical loading.

$$\delta = \tan^{-1}\left(\frac{x-2b}{z}\right) \tag{3.13a}$$

$$\alpha = \tan^{-1}\left(\frac{x}{z}\right) - \tan^{-1}\left(\frac{x-2b}{z}\right) \tag{3.13b}$$

σ_z, σ_x and σ_{xz} are vertical stress, horizontal stress and shear stress respectively in the soil mass at a distance x from origin and at depth z, for the applied stress as shown in Figs 3.1 to 3.5 of strip footing width $2b$.

Using the above Eqs (3.1a) to (3.13b), the values of σ_z/q, σ_x/q and σ_{xz}/q were obtained for different values of (x/b) and (z/b).

Corresponding principal stresses, σ_1 and σ_3 were obtained by using the following equations:

$$\sigma_1 = \frac{\sigma_z + \sigma_x}{2} + \sqrt{\left(\frac{\sigma_z - \sigma_x}{2}\right)^2 + \tau_{xz}^{\ 2}} \tag{3.14}$$

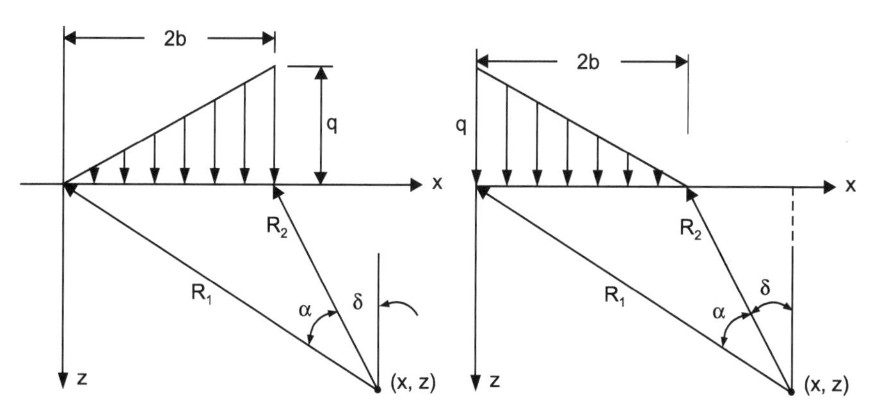

Fig. 3.4. Vertical load increasing linearly.

Fig. 3.5. Vertical load decreasing linearly.

$$\sigma_3 = \frac{\sigma_z + \sigma_x}{2} - \sqrt{\left(\frac{\sigma_z - \sigma_x}{2}\right)^2 + \tau_{xz}^{\ 2}} \tag{3.15}$$

$$\theta = \frac{1}{2}\tan^{-1}\left(\frac{2\tau_{xz}}{\sigma_1 - \sigma_x}\right) \tag{3.16}$$

where θ is the angle which the major principal stress σ_1 makes with the vertical.

In Tables 3.1-3.6 the values of stresses σ_z/q, σ_x/q, τ_{xz}/q, σ_1/q, σ_3/q and θ are given for uniform distributed loading, tangential loading, symmetric vertical triangular load, vertical loading increasing linearly, vertical loading decreasing linearly and horizontal loading increasing linearly respectively for typical values of x/b and z/b.

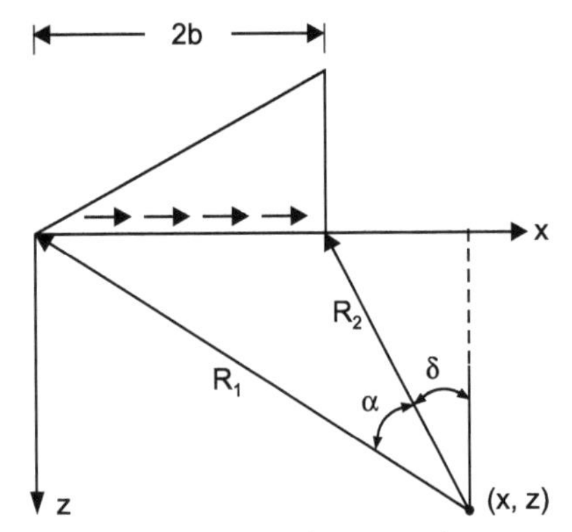

Fig. 3.6. Linearly increasing horizontal loading.

Table 3.1. Values of stresses for uniformly distributed loading (x is measured from the centre of footing, Fig. 3.1)

x/b	z/b	σ_z/q	σ_x/q	τ_{xz}/q	σ_1/q	σ_3/q	θ
0.0000	0.2500	0.9938	0.6942	0.0000	0.9938	0.6942	0.0000
0.0000	0.5000	0.9595	0.4502	0.0000	0.9595	0.4502	0.0000
0.0000	1.0000	0.8183	0.1817	0.0000	0.8183	0.1817	0.0000
0.0000	1.5000	0.6682	0.0805	0.0000	0.6682	0.0805	0.0000
0.0000	2.0000	0.5498	0.0405	0.0000	0.5498	0.0405	0.0000
0.0000	2.5000	0.4618	0.0227	0.0000	0.4618	0.0227	0.0000
0.0000	3.0000	0.3958	0.0138	0.0000	0.3958	0.0138	0.0000
0.0000	3.5000	0.3453	0.0090	0.0000	0.3453	0.0090	0.0000
0.0000	4.0000	0.3058	0.0062	0.0000	0.3058	0.0062	0.0000
0.2500	0.2500	0.9915	0.6780	0.0196	0.9927	0.6768	3.5625
0.2500	0.5000	0.9484	0.4350	0.0540	0.9540	0.4294	5.9443
0.2500	1.0000	0.7981	0.1820	0.0795	0.8082	0.1719	7.2351
0.2500	1.5000	0.6526	0.0849	0.0668	0.6603	0.0771	6.6203
0.2500	2.0000	0.5397	0.0443	0.0502	0.5447	0.0393	5.7247
0.2500	2.5000	0.4553	0.0254	0.0374	0.4585	0.0222	4.9329
0.2500	3.0000	0.3916	0.0157	0.0284	0.3937	0.0136	4.2918
0.2500	3.5000	0.3424	0.0103	0.0220	0.3439	0.0089	3.7795
0.2500	4.0000	0.3037	0.0071	0.0175	0.3047	0.0061	3.3672

(Contd.)

0.5000	0.2500	0.9788	0.6209	0.0551	0.9871	0.6126	8.5514
0.5000	0.5000	0.9022	0.3929	0.1273	0.9323	0.3629	13.2825
0.5000	1.0000	0.7347	0.1862	0.1567	0.7763	0.1446	14.8724
0.5000	1.5000	0.6071	0.0978	0.1273	0.6371	0.0677	13.2825
0.5000	2.0000	0.5105	0.0551	0.0959	0.5299	0.0358	11.4168
0.5000	2.5000	0.4365	0.0332	0.0720	0.4490	0.0207	9.8269
0.5000	3.0000	0.3791	0.0212	0.0551	0.3874	0.0129	8.5514
0.5000	3.5000	0.3339	0.0142	0.0430	0.3396	0.0085	7.5342
0.5000	4.0000	0.2976	0.0100	0.0343	0.3017	0.0059	6.7155
0.7500	0.2500	0.9086	0.5011	0.1528	0.9595	0.4502	18.4349
0.7500	0.5000	0.7704	0.3476	0.2306	0.8719	0.2461	23.7448
0.7500	1.0000	0.6247	0.2007	0.2212	0.7192	0.1063	23.1094
0.7500	1.5000	0.5359	0.1181	0.1749	0.5995	0.0546	19.9682
0.7500	2.0000	0.4653	0.0715	0.1331	0.5061	0.0307	17.0305
0.7500	2.5000	0.4072	0.0451	0.1015	0.4337	0.0185	14.6407
0.7500	3.0000	0.3594	0.0297	0.0786	0.3772	0.0119	12.7464
0.7500	3.5000	0.3202	0.0203	0.0620	0.3326	0.0080	11.2397
0.7500	4.0000	0.2878	0.0144	0.0499	0.2967	0.0056	10.0265
1.0000	0.2500	0.4996	0.4212	0.3134	0.7763	0.1446	41.4375
1.0000	0.5000	0.4969	0.3471	0.2996	0.7308	0.1132	37.9819
1.0000	1.0000	0.4797	0.2251	0.2546	0.6371	0.0677	31.7175
1.0000	1.5000	0.4480	0.1424	0.2037	0.5498	0.0405	26.5651
1.0000	2.0000	0.4092	0.0908	0.1592	0.4751	0.0249	22.5000
1.0000	2.5000	0.3700	0.0595	0.1242	0.4136	0.0159	19.3299
1.0000	3.0000	0.3341	0.0403	0.0979	0.3637	0.0106	16.8450
1.0000	3.5000	0.3024	0.0281	0.0784	0.3232	0.0073	14.8724
1.0000	4.0000	0.2749	0.0203	0.0637	0.2899	0.0052	13.2825
-0.2500	0.2500	0.9915	0.6780	-0.0196	0.9927	0.6768	-3.5625
-0.2500	0.5000	0.9484	0.4350	-0.0540	0.9540	0.4294	-5.9443
-0.2500	1.0000	0.7981	0.1820	-0.0795	0.8082	0.1719	-7.2351
-0.2500	1.5000	0.6526	0.0849	-0.0668	0.6603	0.0771	-6.6203
-0.2500	2.0000	0.5397	0.0443	-0.0502	0.5447	0.0393	-5.7247
-0.2500	2.5000	0.4553	0.0254	-0.0374	0.4585	0.0222	-4.9329
-0.2500	3.0000	0.3916	0.0157	-0.0284	0.3937	0.0136	-4.2918
-0.2500	3.5000	0.3424	0.0103	-0.0220	0.3439	0.0089	-3.7795
-0.2500	4.0000	0.3037	0.0071	-0.0175	0.3047	0.0061	-3.3672
-0.5000	0.2500	0.9788	0.6209	-0.0551	0.9871	0.6126	-8.5514
-0.5000	0.5000	0.9022	0.3929	-0.1273	0.9323	0.3629	-13.2825
-0.5000	1.0000	0.7347	0.1862	-0.1567	0.7763	0.1446	-14.8724
-0.5000	1.5000	0.6071	0.0978	-0.1273	0.6371	0.0677	-13.2825
-0.5000	2.0000	0.5105	0.0551	-0.0959	0.5299	0.0358	-11.4168
-0.5000	2.5000	0.4365	0.0332	-0.0720	0.4490	0.0207	-9.8269
-0.5000	3.0000	0.3791	0.0212	-0.0551	0.3874	0.0129	-8.5514

(Contd.)

Table 3.1. (*Contd.*)

x/b	z/b	σ_z/q	σ_x/q	τ_{xz}/q	σ_1/q	σ_3/q	θ
-0.5000	3.5000	0.3339	0.0142	-0.0430	0.3396	0.0085	-7.5342
-0.5000	4.0000	0.2976	0.0100	-0.0343	0.3017	0.0059	-6.7155
-0.7500	0.2500	0.9086	0.5011	-0.1528	0.9595	0.4502	-18.4349
-0.7500	0.5000	0.7704	0.3476	-0.2306	0.8719	0.2461	-23.7448
-0.7500	1.0000	0.6247	0.2007	-0.2212	0.7192	0.1063	-23.1094
-0.7500	1.5000	0.5359	0.1181	-0.1749	0.5995	0.0546	-19.9682
-0.7500	2.0000	0.4653	0.0715	-0.1331	0.5061	0.0307	-17.0305
-0.7500	2.5000	0.4072	0.0451	-0.1015	0.4337	0.0185	-14.6407
-0.7500	3.0000	0.3594	0.0297	-0.0786	0.3772	0.0119	-12.7464
-0.7500	3.5000	0.3202	0.0203	-0.0620	0.3326	0.0080	-11.2397
-0.7500	4.0000	0.2878	0.0144	-0.0499	0.2967	0.0056	-10.0265
-1.0000	0.2500	0.4996	0.4212	-0.3134	0.7763	0.1446	-41.4375
-1.0000	0.5000	0.4969	0.3471	-0.2996	0.7308	0.1132	-37.9819
-1.0000	1.0000	0.4797	0.2251	-0.2546	0.6371	0.0677	-31.7175
-1.0000	1.5000	0.4480	0.1424	-0.2037	0.5498	0.0405	-26.5651
-1.0000	2.0000	0.4092	0.0908	-0.1592	0.4751	0.0249	-22.5000
-1.0000	2.5000	0.3700	0.0595	-0.1242	0.4136	0.0159	-19.3299
-1.0000	3.0000	0.3341	0.0403	-0.0979	0.3637	0.0106	-16.8450
-1.0000	3.5000	0.3024	0.0281	-0.0784	0.3232	0.0073	-14.8724
-1.0000	4.0000	0.2749	0.0203	-0.0637	0.2899	0.0052	-13.2825

Table 3.2. Values of stresses for tangential loading (x is measured from the centre of footing, Fig. 3.2)

x/b	z/b	σ_z/q	σ_x/q	τ_{xz}/q	σ_1/q	σ_3/q	θ
0.0000	0.2500	0.0000	0.0000	0.6942	0.6942	-0.6942	#DIV/0!
0.0000	0.5000	0.0000	0.0000	0.4502	0.4502	-0.4502	#DIV/0!
0.0000	1.0000	0.0000	0.0000	0.1817	0.1817	-0.1817	#DIV/0!
0.0000	1.5000	0.0000	0.0000	0.0805	0.0805	-0.0805	#DIV/0!
0.0000	2.0000	0.0000	0.0000	0.0405	0.0405	-0.0405	#DIV/0!
0.0000	2.5000	0.0000	0.0000	0.0227	0.0227	-0.0227	#DIV/0!
0.0000	3.0000	0.0000	0.0000	0.0138	0.0138	-0.0138	#DIV/0!
0.0000	3.5000	0.0000	0.0000	0.0090	0.0090	-0.0090	#DIV/0!
0.0000	4.0000	0.0000	0.0000	0.0062	0.0062	-0.0062	#DIV/0!
0.2500	0.2500	0.0196	0.2846	0.6780	0.8429	-0.5388	50.5280
0.2500	0.5000	0.0540	0.2014	0.4350	0.5689	-0.3135	49.8051
0.2500	1.0000	0.0795	0.0780	0.1820	0.2607	-0.1033	44.8793
0.2500	1.5000	0.0668	0.0300	0.0849	0.1352	-0.0384	38.8904
0.2500	2.0000	0.0502	0.0129	0.0443	0.0796	-0.0165	33.5959
0.2500	2.5000	0.0374	0.0062	0.0254	0.0516	-0.0080	29.2498

(*Contd.*)

0.2500	3.0000	0.0284	0.0033	0.0157	0.0359	-0.0043	25.7344
0.2500	3.5000	0.0220	0.0019	0.0103	0.0264	-0.0025	22.8813
0.2500	4.0000	0.0175	0.0012	0.0071	0.0202	-0.0015	20.5434
0.5000	0.2500	0.0551	0.5820	0.6209	0.9930	-0.3560	56.4972
0.5000	0.5000	0.1273	0.3850	0.3929	0.6697	-0.1574	54.0760
0.5000	1.0000	0.1567	0.1474	0.1862	0.3383	-0.0342	44.2874
0.5000	1.5000	0.1273	0.0598	0.0978	0.1970	-0.0099	35.4711
0.5000	2.0000	0.0959	0.0269	0.0551	0.1264	-0.0036	28.9849
0.5000	2.5000	0.0720	0.0134	0.0332	0.0870	-0.0016	24.2796
0.5000	3.0000	0.0551	0.0072	0.0212	0.0631	-0.0008	20.7907
0.5000	3.5000	0.0430	0.0042	0.0142	0.0477	-0.0004	18.1312
0.5000	4.0000	0.0343	0.0026	0.0100	0.0372	-0.0003	16.0501
0.7500	0.2500	0.1528	0.8718	0.5011	1.1290	-0.1044	62.8283
0.7500	0.5000	0.2306	0.5209	0.3476	0.7524	-0.0009	56.3301
0.7500	1.0000	0.2212	0.2057	0.2007	0.4143	0.0126	43.8904
0.7500	1.5000	0.1749	0.0899	0.1181	0.2579	0.0068	35.1004
0.7500	2.0000	0.1331	0.0429	0.0715	0.1725	0.0035	28.8683
0.7500	2.5000	0.1015	0.0222	0.0451	0.1219	0.0019	24.3319
0.7500	3.0000	0.0786	0.0124	0.0297	0.0900	0.0011	20.9347
0.7500	3.5000	0.0620	0.0074	0.0203	0.0688	0.0006	18.3208
0.7500	4.0000	0.0499	0.0046	0.0144	0.0541	0.0004	16.2597
1.0000	0.2500	0.3134	1.0153	0.4212	1.2127	0.1161	64.8999
1.0000	0.5000	0.2996	0.6023	0.3471	0.8296	0.0722	56.7778
1.0000	1.0000	0.2546	0.2577	0.2251	0.4812	0.0311	45.1912
1.0000	1.5000	0.2037	0.1215	0.1424	0.3108	0.0144	36.9459
1.0000	2.0000	0.1592	0.0615	0.0908	0.2135	0.0072	30.8690
1.0000	2.5000	0.1242	0.0332	0.0595	0.1536	0.0038	26.3027
1.0000	3.0000	0.0979	0.0191	0.0403	0.1149	0.0022	22.8015
1.0000	3.5000	0.0784	0.0116	0.0281	0.0886	0.0013	20.0608
1.0000	4.0000	0.0637	0.0074	0.0203	0.0702	0.0008	17.8725
-0.2500	0.2500	-0.0196	-0.2846	0.6780	0.5388	-0.8429	-50.5280
-0.2500	0.5000	-0.0540	-0.2014	0.4350	0.3135	-0.5689	-49.8051
-0.2500	1.0000	-0.0795	-0.0780	0.1820	0.1033	-0.2607	-44.8793
-0.2500	1.5000	-0.0668	-0.0300	0.0849	0.0384	-0.1352	-38.8904
-0.2500	2.0000	-0.0502	-0.0129	0.0443	0.0165	-0.0796	-33.5959
-0.2500	2.5000	-0.0374	-0.0062	0.0254	0.0080	-0.0516	-29.2498
-0.2500	3.0000	-0.0284	-0.0033	0.0157	0.0043	-0.0359	-25.7344
-0.2500	3.5000	-0.0220	-0.0019	0.0103	0.0025	-0.0264	-22.8813
-0.2500	4.0000	-0.0175	-0.0012	0.0071	0.0015	-0.0202	-20.5434
-0.5000	0.2500	-0.0551	-0.5820	0.6209	0.3560	-0.9930	-56.4972
-0.5000	0.5000	-0.1273	-0.3850	0.3929	0.1574	-0.6697	-54.0760
-0.5000	1.0000	-0.1567	-0.1474	0.1862	0.0342	-0.3383	-44.2874
-0.5000	1.5000	-0.1273	-0.0598	0.0978	0.0099	-0.1970	-35.4711

(Contd.)

Table 3.2. (*Contd.*)

x/b	z/b	σ_z/q	σ_x/q	τ_{xz}/q	σ_1/q	σ_3/q	θ
-0.5000	2.0000	-0.0959	-0.0269	0.0551	0.0036	-0.1264	-28.9849
-0.5000	2.5000	-0.0720	-0.0134	0.0332	0.0016	-0.0870	-24.2796
-0.5000	3.0000	-0.0551	-0.0072	0.0212	0.0008	-0.0631	-20.7907
-0.5000	3.5000	-0.0430	-0.0042	0.0142	0.0004	-0.0477	-18.1312
-0.5000	4.0000	-0.0343	-0.0026	0.0100	0.0003	-0.0372	-16.0501
-0.7500	0.2500	-0.1528	-0.8718	0.5011	0.1044	-1.1290	-62.8283
-0.7500	0.5000	-0.2306	-0.5209	0.3476	0.0009	-0.7524	-56.3301
-0.7500	1.0000	-0.2212	-0.2057	0.2007	-0.0126	-0.4143	-43.8904
-0.7500	1.5000	-0.1749	-0.0899	0.1181	-0.0068	-0.2579	-35.1004
-0.7500	2.0000	-0.1331	-0.0429	0.0715	-0.0035	-0.1725	-28.8683
-0.7500	2.5000	-0.1015	-0.0222	0.0451	-0.0019	-0.1219	-24.3319
-0.7500	3.0000	-0.0786	-0.0124	0.0297	-0.0011	-0.0900	-20.9347
-0.7500	3.5000	-0.0620	-0.0074	0.0203	-0.0006	-0.0688	-18.3208
-0.7500	4.0000	-0.0499	-0.0046	0.0144	-0.0004	-0.0541	-16.2597
-1.0000	0.2500	-0.3134	-1.0153	0.4212	-0.1161	-1.2127	-64.8999
-1.0000	0.5000	-0.2996	-0.6023	0.3471	-0.0722	-0.8296	-56.7778
-1.0000	1.0000	-0.2546	-0.2577	0.2251	-0.0311	-0.4812	-45.1912
-1.0000	1.5000	-0.2037	-0.1215	0.1424	-0.0144	-0.3108	-36.9459
-1.0000	2.0000	-0.1592	-0.0615	0.0908	-0.0072	-0.2135	-30.8690
-1.0000	2.5000	-0.1242	-0.0332	0.0595	-0.0038	-0.1536	-26.3027
-1.0000	3.0000	-0.0979	-0.0191	0.0403	-0.0022	-0.1149	-22.8015
-1.0000	3.5000	-0.0784	-0.0116	0.0281	-0.0013	-0.0886	-20.0608
-1.0000	4.0000	-0.0637	-0.0074	0.0203	-0.0008	-0.0702	-17.8725

Table 3.3. Values of stresses for symmetrical vertical triangular loading (x is measured from the centre of footing, Fig. 3.3)

x/b	z/b	σ_z/q	σ_x/q	τ_{xz}/q	σ_1/q	σ_3/q	θ
0.0000	0.2500	0.8440	0.3931	0.0000	0.8440	0.3931	0.0000
0.0000	0.5000	0.7048	0.1925	0.0000	0.7048	0.1925	0.0000
0.0000	1.0000	0.5000	0.0587	0.0000	0.5000	0.0587	0.0000
0.0000	1.5000	0.3743	0.0232	0.0000	0.3743	0.0232	0.0000
0.0000	2.0000	0.2952	0.0111	0.0000	0.2952	0.0111	0.0000
0.0000	2.5000	0.2422	0.0060	0.0000	0.2422	0.0060	0.0000
0.0000	3.0000	0.2048	0.0036	0.0000	0.2048	0.0036	0.0000
0.0000	3.5000	0.1772	0.0023	0.0000	0.1772	0.0023	0.0000
0.0000	4.0000	0.1560	0.0016	0.0000	0.1560	0.0016	0.0000
0.2500	0.2500	0.7196	0.3875	0.1151	0.7556	0.3515	17.3611
0.2500	0.5000	0.6344	0.2026	0.1146	0.6629	0.1741	13.9743
0.2500	1.0000	0.4712	0.0682	0.0756	0.4849	0.0545	10.2790
0.2500	1.5000	0.3608	0.0287	0.0474	0.3675	0.0220	7.9591

(*Contd.*)

0.2500	2.0000	0.2881	0.0142	0.0311	0.2916	0.0107	6.3998
0.2500	2.5000	0.2382	0.0079	0.0216	0.2402	0.0059	5.3124
0.2500	3.0000	0.2023	0.0048	0.0157	0.2036	0.0035	4.5237
0.2500	3.5000	0.1755	0.0031	0.0119	0.1763	0.0023	3.9306
0.2500	4.0000	0.1548	0.0021	0.0093	0.1554	0.0016	3.4707
0.5000	0.2500	0.4949	0.3357	0.1525	0.5873	0.2433	31.2096
0.5000	0.5000	0.4714	0.2152	0.1762	0.5611	0.1255	26.9944
0.5000	1.0000	0.3955	0.0913	0.1299	0.4434	0.0434	20.2537
0.5000	1.5000	0.3238	0.0431	0.0859	0.3480	0.0190	15.7327
0.5000	2.0000	0.2683	0.0227	0.0582	0.2814	0.0096	12.6845
0.5000	2.5000	0.2266	0.0131	0.0412	0.2343	0.0055	10.5518
0.5000	3.0000	0.1951	0.0082	0.0304	0.1999	0.0034	8.9994
0.5000	3.5000	0.1707	0.0054	0.0232	0.1739	0.0022	7.8285
0.5000	4.0000	0.1515	0.0037	0.0182	0.1537	0.0015	6.9183
0.7500	0.2500	0.2621	0.2621	0.1476	0.4097	0.1145	45.0000
0.7500	0.5000	0.2876	0.2160	0.1809	0.4363	0.0674	39.4040
0.7500	1.0000	0.2981	0.1167	0.1529	0.3851	0.0296	29.6621
0.7500	1.5000	0.2720	0.0618	0.1099	0.3190	0.0148	23.1456
0.7500	2.0000	0.2390	0.0348	0.0783	0.2656	0.0082	18.7461
0.7500	2.5000	0.2090	0.0209	0.0572	0.2250	0.0049	15.6497
0.7500	3.0000	0.1838	0.0133	0.0430	0.1940	0.0031	13.3817
0.7500	3.5000	0.1632	0.0089	0.0333	0.1700	0.0021	11.6623
0.7500	4.0000	0.1462	0.0062	0.0264	0.1510	0.0014	10.3204
1.0000	0.2500	0.0768	0.1955	0.0959	0.2490	0.0234	60.8788
1.0000	0.5000	0.1392	0.2006	0.1414	0.3146	0.0252	51.1226
1.0000	1.0000	0.2048	0.1338	0.1476	0.3211	0.0175	38.2348
1.0000	1.5000	0.2160	0.0793	0.1188	0.2847	0.0107	30.0425
1.0000	2.0000	0.2048	0.0477	0.0903	0.2460	0.0065	24.4907
1.0000	2.5000	0.1873	0.0299	0.0687	0.2130	0.0041	20.5453
1.0000	3.0000	0.1695	0.0196	0.0530	0.1863	0.0027	17.6296
1.0000	3.5000	0.1533	0.0134	0.0417	0.1648	0.0019	15.4033
1.0000	4.0000	0.1392	0.0095	0.0335	0.1473	0.0013	13.6562
-0.2500	0.2500	0.7196	0.3875	-0.1151	0.7556	0.3515	-17.3611
-0.2500	0.5000	0.6344	0.2026	-0.1146	0.6629	0.1741	-13.9743
-0.2500	1.0000	0.4712	0.0682	-0.0756	0.4849	0.0545	-10.2790
-0.2500	1.5000	0.3608	0.0287	-0.0474	0.3675	0.0220	-7.9591
-0.2500	2.0000	0.2881	0.0142	-0.0311	0.2916	0.0107	-6.3998
-0.2500	2.5000	0.2382	0.0079	-0.0216	0.2402	0.0059	-5.3124
-0.2500	3.0000	0.2023	0.0048	-0.0157	0.2036	0.0035	-4.5237
-0.2500	3.5000	0.1755	0.0031	-0.0119	0.1763	0.0023	-3.9306
-0.2500	4.0000	0.1548	0.0021	-0.0093	0.1554	0.0016	-3.4707
-0.5000	0.2500	0.4949	0.3357	-0.1525	0.5873	0.2433	-31.2096
-0.5000	0.5000	0.4714	0.2152	-0.1762	0.5611	0.1255	-26.9944

(Contd.)

Table 3.3. (*Contd.*)

x/b	z/b	σ_z/q	σ_x/q	τ_{xz}/q	σ_1/q	σ_3/q	θ
-0.5000	1.0000	0.3955	0.0913	-0.1299	0.4434	0.0434	-20.2537
-0.5000	1.5000	0.3238	0.0431	-0.0859	0.3480	0.0190	-15.7327
-0.5000	2.0000	0.2683	0.0227	-0.0582	0.2814	0.0096	-12.6845
-0.5000	2.5000	0.2266	0.0131	-0.0412	0.2343	0.0055	-10.5518
-0.5000	3.0000	0.1951	0.0082	-0.0304	0.1999	0.0034	-8.9994
-0.5000	3.5000	0.1707	0.0054	-0.0232	0.1739	0.0022	-7.8285
-0.5000	4.0000	0.1515	0.0037	-0.0182	0.1537	0.0015	-6.9183
-0.7500	0.2500	0.2621	0.2621	-0.1476	0.4097	0.1145	-45.0000
-0.7500	0.5000	0.2876	0.2160	-0.1809	0.4363	0.0674	-39.4040
-0.7500	1.0000	0.2981	0.1167	-0.1529	0.3851	0.0296	-29.6621
-0.7500	1.5000	0.2720	0.0618	-0.1099	0.3190	0.0148	-23.1456
-0.7500	2.0000	0.2390	0.0348	-0.0783	0.2656	0.0082	-18.7461
-0.7500	2.5000	0.2090	0.0209	-0.0572	0.2250	0.0049	-15.6497
-0.7500	3.0000	0.1838	0.0133	-0.0430	0.1940	0.0031	-13.3817
-0.7500	3.5000	0.1632	0.0089	-0.0333	0.1700	0.0021	-11.6623
-0.7500	4.0000	0.1462	0.0062	-0.0264	0.1510	0.0014	-10.3204
-1.0000	0.2500	0.0768	0.1955	-0.0959	0.2490	0.0234	-60.8788
-1.0000	0.5000	0.1392	0.2006	-0.1414	0.3146	0.0252	-51.1226
-1.0000	1.0000	0.2048	0.1338	-0.1476	0.3211	0.0175	-38.2348
-1.0000	1.5000	0.2160	0.0793	-0.1188	0.2847	0.0107	-30.0425
-1.0000	2.0000	0.2048	0.0477	-0.0903	0.2460	0.0065	-24.4907
-1.0000	2.5000	0.1873	0.0299	-0.0687	0.2130	0.0041	-20.5453
-1.0000	3.0000	0.1695	0.0196	-0.0530	0.1863	0.0027	-17.6296
-1.0000	3.5000	0.1533	0.0134	-0.0417	0.1648	0.0019	-15.4033
-1.0000	4.0000	0.1392	0.0095	-0.0335	0.1473	0.0013	-13.6562

Table 3.4. Values of stresses for linearly increasing triangular loading (x is measured from left hand side edge of the footing, Fig. 3.4)

x/b	z/b	σ_z/q	σ_x/q	σ_{xz}/q	σ_1/q	σ_3/q	θ
0.0000	0.2500	0.0392	0.1269	-0.0527	0.1516	0.0145	25.1001
0.0000	0.5000	0.0749	0.1506	-0.0868	0.2074	0.0181	33.2222
0.0000	1.0000	0.1273	0.1288	-0.1125	0.2406	0.0155	44.8088
0.0000	1.5000	0.1528	0.0911	-0.1068	0.2331	0.0108	-36.9459
0.0000	2.0000	0.1592	0.0615	-0.0908	0.2135	0.0072	-30.8690
0.0000	2.5000	0.1553	0.0416	-0.0744	0.1920	0.0048	-26.3027
0.0000	3.0000	0.1469	0.0287	-0.0604	0.1723	0.0033	-22.8015
0.0000	3.5000	0.1371	0.0203	-0.0492	0.1551	0.0023	-20.0608
0.0000	4.0000	0.1273	0.0147	-0.0405	0.1404	0.0017	-17.8725
0.2500	0.2500	0.1327	0.1716	-0.0817	0.2362	0.0681	38.2995
0.2500	0.5000	0.1540	0.1737	-0.1157	0.2800	0.0477	42.5666

(*Contd.*)

0.2500	1.0000	0.1887	0.1279	-0.1280	0.2899	0.0267	-38.3224
0.2500	1.5000	0.1982	0.0822	-0.1104	0.2649	0.0154	-31.1461
0.2500	2.0000	0.1913	0.0518	-0.0881	0.2339	0.0092	-25.8206
0.2500	2.5000	0.1778	0.0334	-0.0690	0.2055	0.0057	-21.8589
0.2500	3.0000	0.1629	0.0223	-0.0543	0.1814	0.0038	-18.8567
0.2500	3.5000	0.1486	0.0154	-0.0433	0.1615	0.0026	-16.5302
0.2500	4.0000	0.1358	0.0110	-0.0351	0.1450	0.0018	-14.6874
0.5000	0.2500	0.2516	0.2280	-0.0914	0.3319	0.1476	-41.3211
0.5000	0.5000	0.2574	0.1945	-0.1301	0.3597	0.0921	-38.2022
0.5000	1.0000	0.2620	0.1203	-0.1323	0.3412	0.0411	-30.9077
0.5000	1.5000	0.2473	0.0693	-0.1052	0.2960	0.0205	-24.8796
0.5000	2.0000	0.2235	0.0407	-0.0791	0.2530	0.0112	-20.4344
0.5000	2.5000	0.1991	0.0250	-0.0595	0.2175	0.0066	-17.1784
0.5000	3.0000	0.1774	0.0162	-0.0456	0.1894	0.0042	-14.7453
0.5000	3.5000	0.1588	0.0109	-0.0357	0.1669	0.0028	-12.8802
0.5000	4.0000	0.1431	0.0077	-0.0285	0.1489	0.0019	-11.4149
0.7500	0.2500	0.3742	0.2898	-0.0921	0.4334	0.2307	-32.6905
0.7500	0.5000	0.3692	0.2135	-0.1290	0.4420	0.1406	-29.4491
0.7500	1.0000	0.3390	0.1072	-0.1208	0.3906	0.0557	-23.0933
0.7500	1.5000	0.2948	0.0544	-0.0887	0.3240	0.0252	-18.2070
0.7500	2.0000	0.2526	0.0295	-0.0631	0.2692	0.0129	-14.7531
0.7500	2.5000	0.2175	0.0173	-0.0458	0.2274	0.0073	-12.2954
0.7500	3.0000	0.1894	0.0108	-0.0342	0.1957	0.0045	-10.4942
0.7500	3.5000	0.1670	0.0072	-0.0264	0.1712	0.0029	-9.1317
0.7500	4.0000	0.1489	0.0050	-0.0208	0.1519	0.0020	-8.0712
1.0000	0.2500	0.4969	0.3471	-0.0868	0.5367	0.3074	-24.6021
1.0000	0.5000	0.4797	0.2251	-0.1125	0.5224	0.1825	-20.7373
1.0000	1.0000	0.4092	0.0908	-0.0908	0.4333	0.0667	-14.8588
1.0000	1.5000	0.3341	0.0403	-0.0604	0.3460	0.0283	-11.1715
1.0000	2.0000	0.2749	0.0203	-0.0405	0.2812	0.0140	-8.8265
1.0000	2.5000	0.2309	0.0114	-0.0284	0.2345	0.0077	-7.2514
1.0000	3.0000	0.1979	0.0069	-0.0208	0.2001	0.0047	-6.1355
1.0000	3.5000	0.1727	0.0045	-0.0158	0.1741	0.0030	-5.3090
1.0000	4.0000	0.1529	0.0031	-0.0123	0.1539	0.0021	-4.6746
1.2500	0.2500	0.6172	0.3882	-0.0725	0.6382	0.3672	-16.1728
1.2500	0.5000	0.5792	0.2216	-0.0750	0.5943	0.2065	-11.3743
1.2500	1.0000	0.4591	0.0748	-0.0413	0.4635	0.0704	-6.0664
1.2500	1.5000	0.3578	0.0305	-0.0219	0.3592	0.0290	-3.8108
1.2500	2.0000	0.2872	0.0148	-0.0129	0.2878	0.0142	-2.7142
1.2500	2.5000	0.2378	0.0081	-0.0084	0.2381	0.0078	-2.0962
1.2500	3.0000	0.2022	0.0049	-0.0059	0.2023	0.0047	-1.7062
1.2500	3.5000	0.1755	0.0032	-0.0043	0.1756	0.0031	-1.4392
1.2500	4.0000	0.1548	0.0022	-0.0033	0.1549	0.0021	-1.2452

(Contd.)

Table 3.4. (*Contd.*)

x/b	z/b	σ_z/q	σ_x/q	τ_{xz}/q	σ_1/q	σ_3/q	θ
1.5000	0.2500	0.7272	0.3929	-0.0363	0.7311	0.3890	-6.1297
1.5000	0.5000	0.6448	0.1985	-0.0027	0.6449	0.1984	-0.3518
1.5000	1.0000	0.4726	0.0659	0.0244	0.4741	0.0645	3.4264
1.5000	1.5000	0.3598	0.0285	0.0222	0.3613	0.0270	3.8108
1.5000	2.0000	0.2870	0.0145	0.0168	0.2880	0.0134	3.5085
1.5000	2.5000	0.2374	0.0082	0.0125	0.2380	0.0075	3.1129
1.5000	3.0000	0.2017	0.0050	0.0095	0.2022	0.0046	2.7526
1.5000	3.5000	0.1751	0.0033	0.0074	0.1754	0.0030	2.4483
1.5000	4.0000	0.1545	0.0023	0.0058	0.1547	0.0020	2.1955
1.7500	0.2500	0.7759	0.3295	0.0711	0.7869	0.3185	8.8291
1.7500	0.5000	0.6164	0.1739	0.1149	0.6445	0.1459	13.7213
1.7500	1.0000	0.4360	0.0728	0.0932	0.4586	0.0503	13.5849
1.7500	1.5000	0.3377	0.0360	0.0645	0.3509	0.0228	11.5643
1.7500	2.0000	0.2740	0.0196	0.0450	0.2818	0.0119	9.7457
1.7500	2.5000	0.2294	0.0116	0.0325	0.2341	0.0069	8.3110
1.7500	3.0000	0.1966	0.0074	0.0243	0.1996	0.0043	7.1969
1.7500	3.5000	0.1716	0.0049	0.0187	0.1737	0.0028	6.3228
1.7500	4.0000	0.1521	0.0034	0.0148	0.1535	0.0020	5.6257
2.0000	0.2500	0.4604	0.2943	0.2608	0.6510	0.1037	36.1671
2.0000	0.5000	0.4220	0.1966	0.2128	0.5501	0.0685	31.0441
2.0000	1.0000	0.3524	0.0963	0.1421	0.4156	0.0330	23.9860
2.0000	1.5000	0.2952	0.0513	0.0969	0.3290	0.0174	19.2400
2.0000	2.0000	0.2500	0.0294	0.0683	0.2694	0.0099	15.8831
2.0000	2.5000	0.2148	0.0179	0.0498	0.2267	0.0060	13.4290
2.0000	3.0000	0.1872	0.0116	0.0376	0.1949	0.0039	11.5816
2.0000	3.5000	0.1652	0.0078	0.0291	0.1705	0.0026	10.1536
2.0000	4.0000	0.1476	0.0055	0.0231	0.1513	0.0019	9.0233

Table 3.5. Values of stresses for linearly decreasing triangular loading (x is measured from left hand side edge of the footing, Fig. 3.5)

x/b	z/b	σ_z/q	σ_x/q	τ_{xz}/q	σ_1/q	σ_3/q	θ
0.0000	0.2500	0.4604	0.2943	-0.2608	0.6510	0.1037	-36.1671
0.0000	0.5000	0.4220	0.1966	-0.2128	0.5501	0.0685	-31.0441
0.0000	1.0000	0.3524	0.0963	-0.1421	0.4156	0.0330	-23.9860
0.0000	1.5000	0.2952	0.0513	-0.0969	0.3290	0.0174	-19.2400
0.0000	2.0000	0.2500	0.0294	-0.0683	0.2694	0.0099	-15.8831
0.0000	2.5000	0.2148	0.0179	-0.0498	0.2267	0.0060	-13.4290
0.0000	3.0000	0.1872	0.0116	-0.0376	0.1949	0.0039	-11.5816
0.0000	3.5000	0.1652	0.0078	-0.0291	0.1705	0.0026	-10.1536

(*Contd.*)

0.0000	4.0000	0.1476	0.0055	-0.0231	0.1513	0.0019	-9.0233
0.2500	0.2500	0.7759	0.3295	-0.0711	0.7869	0.3185	-8.8291
0.2500	0.5000	0.6164	0.1739	-0.1149	0.6445	0.1459	-13.7213
0.2500	1.0000	0.4360	0.0728	-0.0932	0.4586	0.0503	-13.5849
0.2500	1.5000	0.3377	0.0360	-0.0645	0.3509	0.0228	-11.5643
0.2500	2.0000	0.2740	0.0196	-0.0450	0.2818	0.0119	-9.7457
0.2500	2.5000	0.2294	0.0116	-0.0325	0.2341	0.0069	-8.3110
0.2500	3.0000	0.1966	0.0074	-0.0243	0.1996	0.0043	-7.1969
0.2500	3.5000	0.1716	0.0049	-0.0187	0.1737	0.0028	-6.3228
0.2500	4.0000	0.1521	0.0034	-0.0148	0.1535	0.0020	-5.6257
0.5000	0.2500	0.7272	0.3929	0.0363	0.7311	0.3890	6.1297
0.5000	0.5000	0.6448	0.1985	0.0027	0.6449	0.1984	0.3518
0.5000	1.0000	0.4726	0.0659	-0.0244	0.4741	0.0645	-3.4264
0.5000	1.5000	0.3598	0.0285	-0.0222	0.3613	0.0270	-3.8108
0.5000	2.0000	0.2870	0.0145	-0.0168	0.2880	0.0134	-3.5085
0.5000	2.5000	0.2374	0.0082	-0.0125	0.2380	0.0075	-3.1129
0.5000	3.0000	0.2017	0.0050	-0.0095	0.2022	0.0046	-2.7526
0.5000	3.5000	0.1751	0.0033	-0.0074	0.1754	0.0030	-2.4483
0.5000	4.0000	0.1545	0.0023	-0.0058	0.1547	0.0020	-2.1955
0.7500	0.2500	0.6172	0.3882	0.0725	0.6382	0.3672	16.1728
0.7500	0.5000	0.5792	0.2216	0.0750	0.5943	0.2065	11.3743
0.7500	1.0000	0.4591	0.0748	0.0413	0.4635	0.0704	6.0664
0.7500	1.5000	0.3578	0.0305	0.0219	0.3592	0.0290	3.8108
0.7500	2.0000	0.2872	0.0148	0.0129	0.2878	0.0142	2.7142
0.7500	2.5000	0.2378	0.0081	0.0084	0.2381	0.0078	2.0962
0.7500	3.0000	0.2022	0.0049	0.0059	0.2023	0.0047	1.7062
0.7500	3.5000	0.1755	0.0032	0.0043	0.1756	0.0031	1.4392
0.7500	4.0000	0.1548	0.0022	0.0033	0.1549	0.0021	1.2452
1.0000	0.2500	0.4969	0.3471	0.0868	0.5367	0.3074	24.6021
1.0000	0.5000	0.4797	0.2251	0.1125	0.5224	0.1825	20.7373
1.0000	1.0000	0.4092	0.0908	0.0908	0.4333	0.0667	14.8588
1.0000	1.5000	0.3341	0.0403	0.0604	0.3460	0.0283	11.1715
1.0000	2.0000	0.2749	0.0203	0.0405	0.2812	0.0140	8.8265
1.0000	2.5000	0.2309	0.0114	0.0284	0.2345	0.0077	7.2514
1.0000	3.0000	0.1979	0.0069	0.0208	0.2001	0.0047	6.1355
1.0000	3.5000	0.1727	0.0045	0.0158	0.1741	0.0030	5.3090
1.0000	4.0000	0.1529	0.0031	0.0123	0.1539	0.0021	4.6746
1.2500	0.2500	0.3742	0.2898	0.0921	0.4334	0.2307	32.6905
1.2500	0.5000	0.3692	0.2135	0.1290	0.4420	0.1406	29.4491
1.2500	1.0000	0.3390	0.1072	0.1208	0.3906	0.0557	23.0933
1.2500	1.5000	0.2948	0.0544	0.0887	0.3240	0.0252	18.2070
1.2500	2.0000	0.2526	0.0295	0.0631	0.2692	0.0129	14.7531
1.2500	2.5000	0.2175	0.0173	0.0458	0.2274	0.0073	12.2954

(*Contd.*)

Table 3.5. (*Contd.*)

x/b	z/b	σ_z/q	σ_x/q	τ_{xz}/q	σ_1/q	σ_3/q	θ
1.2500	3.0000	0.1894	0.0108	0.0342	0.1957	0.0045	10.4942
1.2500	3.5000	0.1670	0.0072	0.0264	0.1712	0.0029	9.1317
1.2500	4.0000	0.1489	0.0050	0.0208	0.1519	0.0020	8.0712
1.5000	0.2500	0.2516	0.2280	0.0914	0.3319	0.1476	41.3211
1.5000	0.5000	0.2574	0.1945	0.1301	0.3597	0.0921	38.2022
1.5000	1.0000	0.2620	0.1203	0.1323	0.3412	0.0411	30.9077
1.5000	1.5000	0.2473	0.0693	0.1052	0.2960	0.0205	24.8796
1.5000	2.0000	0.2235	0.0407	0.0791	0.2530	0.0112	20.4344
1.5000	2.5000	0.1991	0.0250	0.0595	0.2175	0.0066	17.1784
1.5000	3.0000	0.1774	0.0162	0.0456	0.1894	0.0042	14.7453
1.5000	3.5000	0.1588	0.0109	0.0357	0.1669	0.0028	12.8802
1.5000	4.0000	0.1431	0.0077	0.0285	0.1489	0.0019	11.4149
1.7500	0.2500	0.1327	0.1716	0.0817	0.2362	0.0681	-38.2995
1.7500	0.5000	0.1540	0.1737	0.1157	0.2800	0.0477	-42.5666
1.7500	1.0000	0.1887	0.1279	0.1280	0.2899	0.0267	38.3224
1.7500	1.5000	0.1982	0.0822	0.1104	0.2649	0.0154	31.1461
1.7500	2.0000	0.1913	0.0518	0.0881	0.2339	0.0092	25.8206
1.7500	2.5000	0.1778	0.0334	0.0690	0.2055	0.0057	21.8589
1.7500	3.0000	0.1629	0.0223	0.0543	0.1814	0.0038	18.8567
1.7500	3.5000	0.1486	0.0154	0.0433	0.1615	0.0026	16.5302
1.7500	4.0000	0.1358	0.0110	0.0351	0.1450	0.0018	14.6874
2.0000	0.2500	0.0392	0.1269	0.0527	0.1516	0.0145	-25.1001
2.0000	0.5000	0.0749	0.1506	0.0868	0.2074	0.0181	-33.2222
2.0000	1.0000	0.1273	0.1288	0.1125	0.2406	0.0155	-44.8088
2.0000	1.5000	0.1528	0.0911	0.1068	0.2331	0.0108	36.9459
2.0000	2.0000	0.1592	0.0615	0.0908	0.2135	0.0072	30.8690
2.0000	2.5000	0.1553	0.0416	0.0744	0.1920	0.0048	26.3027
2.0000	3.0000	0.1469	0.0287	0.0604	0.1723	0.0033	22.8015
2.0000	3.5000	0.1371	0.0203	0.0492	0.1551	0.0023	20.0608
2.0000	4.0000	0.1273	0.0147	0.0405	0.1404	0.0017	17.8725

Table 3.6. Values of stresses for horizontal linearly increasing triangular loading (x is measured from left-hand side edge of the footing, Fig. 3.6)

x/b	z/b	σ_z/q	σ_x/q	τ_{xz}/q	σ_1/q	σ_3/q	θ
0.0000	0.2500	-0.0527	0.5571	0.1269	0.5825	-0.0780	-11.3003
0.0000	0.5000	-0.0868	0.6680	0.1506	0.6969	-0.1157	-10.8756
0.0000	1.0000	-0.1125	0.8003	0.1288	0.8182	-0.1304	-7.8807
0.0000	1.5000	-0.1068	0.8563	0.0911	0.8648	-0.1153	-5.3573
0.0000	2.0000	-0.0908	0.8750	0.0615	0.8789	-0.0947	-3.6276
0.0000	2.5000	-0.0744	0.8779	0.0416	0.8797	-0.0762	-2.4943

(*Contd.*)

0.0000	3.0000	-0.0604	0.8746	0.0287	0.8755	-0.0613	-1.7543
0.0000	3.5000	-0.0492	0.8694	0.0203	0.8698	-0.0497	-1.2643
0.0000	4.0000	-0.0405	0.8638	0.0147	0.8640	-0.0408	-0.9332
0.2500	0.2500	-0.0817	0.5164	0.1716	0.5622	-0.1275	-14.9234
0.2500	0.5000	-0.1157	0.6662	0.1737	0.7030	-0.1526	-11.9759
0.2500	1.0000	-0.1280	0.8139	0.1279	0.8310	-0.1451	-7.5983
0.2500	1.5000	-0.1104	0.8642	0.0822	0.8711	-0.1173	-4.7845
0.2500	2.0000	-0.0881	0.8765	0.0518	0.8792	-0.0909	-3.0668
0.2500	2.5000	-0.0690	0.8752	0.0334	0.8764	-0.0702	-2.0253
0.2500	3.0000	-0.0543	0.8698	0.0223	0.8704	-0.0549	-1.3824
0.2500	3.5000	-0.0433	0.8636	0.0154	0.8639	-0.0436	-0.9742
0.2500	4.0000	-0.0351	0.8577	0.0110	0.8578	-0.0353	-0.7069
0.5000	0.2500	-0.0914	0.5149	0.2280	0.5911	-0.1675	-18.4719
0.5000	0.5000	-0.1301	0.6837	0.1945	0.7278	-0.1742	-12.7728
0.5000	1.0000	-0.1323	0.8322	0.1203	0.8470	-0.1470	-7.0017
0.5000	1.5000	-0.1052	0.8712	0.0693	0.8761	-0.1100	-4.0383
0.5000	2.0000	-0.0791	0.8756	0.0407	0.8773	-0.0808	-2.4352
0.5000	2.5000	-0.0595	0.8705	0.0250	0.8711	-0.0602	-1.5401
0.5000	3.0000	-0.0456	0.8634	0.0162	0.8637	-0.0459	-1.0191
0.5000	3.5000	-0.0357	0.8566	0.0109	0.8568	-0.0358	-0.7022
0.5000	4.0000	-0.0285	0.8507	0.0077	0.8507	-0.0286	-0.5011
0.7500	0.2500	-0.0921	0.5645	0.2898	0.6741	-0.2017	-20.7203
0.7500	0.5000	-0.1290	0.7277	0.2135	0.7780	-0.1793	-13.2449
0.7500	1.0000	-0.1208	0.8536	0.1072	0.8652	-0.1325	-6.2064
0.7500	1.5000	-0.0887	0.8752	0.0544	0.8782	-0.0917	-3.2172
0.7500	2.0000	-0.0631	0.8715	0.0295	0.8725	-0.0640	-1.8075
0.7500	2.5000	-0.0458	0.8633	0.0173	0.8636	-0.0461	-1.0903
0.7500	3.0000	-0.0342	0.8553	0.0108	0.8554	-0.0344	-0.6985
0.7500	3.5000	-0.0264	0.8485	0.0072	0.8485	-0.0264	-0.4705
0.7500	4.0000	-0.0208	0.8428	0.0050	0.8429	-0.0209	-0.3302
1.0000	0.2500	-0.0868	0.6680	0.3471	0.8033	-0.2222	-21.3048
1.0000	0.5000	-0.1125	0.8003	0.2251	0.8528	-0.1650	-13.1252
1.0000	1.0000	-0.0908	0.8750	0.0908	0.8835	-0.0993	-5.3269
1.0000	1.5000	-0.0604	0.8746	0.0403	0.8764	-0.0621	-2.4607
1.0000	2.0000	-0.0405	0.8638	0.0203	0.8643	-0.0410	-1.2828
1.0000	2.5000	-0.0284	0.8537	0.0114	0.8538	-0.0285	-0.7375
1.0000	3.0000	-0.0208	0.8456	0.0069	0.8457	-0.0208	-0.4578

(Contd.)

Table 3.6. (*Contd.*)

x/b	z/b	σ_z/q	σ_x/q	τ_{xz}/q	σ_1/q	σ_3/q	θ
1.0000	3.5000	-0.0158	0.8393	0.0045	0.8393	-0.0158	-0.3018
1.0000	4.0000	-0.0123	0.8343	0.0031	0.8343	-0.0123	-0.2086
1.2500	0.2500	-0.0725	0.8292	0.3882	0.9733	-0.2166	-20.3643
1.2500	0.5000	-0.0750	0.8960	0.2216	0.9442	-0.1232	-12.2646
1.2500	1.0000	-0.0413	0.8914	0.0748	0.8973	-0.0473	-4.5541
1.2500	1.5000	-0.0219	0.8685	0.0305	0.8695	-0.0229	-1.9600
1.2500	2.0000	-0.0129	0.8526	0.0148	0.8529	-0.0132	-0.9775
1.2500	2.5000	-0.0084	0.8421	0.0081	0.8422	-0.0085	-0.5468
1.2500	3.0000	-0.0059	0.8347	0.0049	0.8348	-0.0059	-0.3335
1.2500	3.5000	-0.0043	0.8294	0.0032	0.8294	-0.0043	-0.2172
1.2500	4.0000	-0.0033	0.8253	0.0022	0.8253	-0.0033	-0.1489
1.5000	0.2500	-0.0363	1.0495	0.3929	1.1767	-0.1636	-17.9479
1.5000	0.5000	-0.0027	0.9949	0.1985	1.0329	-0.0408	-10.8475
1.5000	1.0000	0.0244	0.8970	0.0659	0.9020	0.0195	-4.2955
1.5000	1.5000	0.0222	0.8571	0.0285	0.8581	0.0212	-1.9523
1.5000	2.0000	0.0168	0.8391	0.0145	0.8393	0.0165	-1.0066
1.5000	2.5000	0.0125	0.8292	0.0082	0.8293	0.0124	-0.5743
1.5000	3.0000	0.0095	0.8231	0.0050	0.8232	0.0094	-0.3545
1.5000	3.5000	0.0074	0.8190	0.0033	0.8190	0.0073	-0.2327
1.5000	4.0000	0.0058	0.8159	0.0023	0.8160	0.0058	-0.1604
1.7500	0.2500	0.0711	1.2858	0.3295	1.3695	-0.0126	-14.2395
1.7500	0.5000	0.1149	1.0551	0.1739	1.0862	0.0838	-10.1518
1.7500	1.0000	0.0932	0.8912	0.0728	0.8978	0.0866	-5.1694
1.7500	1.5000	0.0645	0.8432	0.0360	0.8448	0.0628	-2.6377
1.7500	2.0000	0.0450	0.8247	0.0196	0.8252	0.0445	-1.4422
1.7500	2.5000	0.0325	0.8161	0.0116	0.8163	0.0323	-0.8505
1.7500	3.0000	0.0243	0.8114	0.0074	0.8115	0.0242	-0.5356
1.7500	3.5000	0.0187	0.8085	0.0049	0.8086	0.0187	-0.3561
1.7500	4.0000	0.0148	0.8066	0.0034	0.8066	0.0148	-0.2475
2.0000	0.2500	0.2608	1.3423	0.2943	1.4172	0.1858	-14.2795
2.0000	0.5000	0.2128	1.0592	0.1966	1.1026	0.1694	-12.4566
2.0000	1.0000	0.1421	0.8818	0.0963	0.8941	0.1298	-7.2951
2.0000	1.5000	0.0969	0.8302	0.0513	0.8337	0.0934	-3.9802
2.0000	2.0000	0.0683	0.8115	0.0294	0.8126	0.0672	-2.2592
2.0000	2.5000	0.0498	0.8037	0.0179	0.8042	0.0494	-1.3627
2.0000	3.0000	0.0376	0.8001	0.0116	0.8003	0.0374	-0.8706
2.0000	3.5000	0.0291	0.7983	0.0078	0.7984	0.0290	-0.5844
2.0000	4.0000	0.0231	0.7974	0.0055	0.7974	0.0231	-0.4089

3.3 Contact Pressure

Contact pressure is the normal vertical pressure on the surface of contact pressure between the base of a structure and the supporting soil. A knowledge of the contact pressure distribution at the soil-structure interface is necessary for a rational design of footings and rafts since the distribution influences the actual value of bending moment and shear force at any section. The distribution of contact pressure at the base of a structure depends upon (i) the flexural rigidity of footing, (ii) the properties of the supporting soil, and (iii) the surface loads.

The settlement of the base of a perfectly flexible footing loaded uniformly, produces a shallow bowl or trough and the distribution of contact pressure is uniform. The settlement of the base of a perfectly rigid footing is by necessity uniform. Therefore, the contact pressure distribution must conform to the requirement of a uniform settlement.

By adopting Boussinesq's method, the following equation for distribution of contact pressure for any axial loading (p) on the rigid footing resting on perfectly elastic and homogeneous subgrade has been developed (Schultze, 1961):

$$\sigma(x) = \frac{p}{\pi a} \frac{1}{\sqrt{1-\left(\frac{x}{a}\right)^2}} \quad \text{where } a = \frac{B}{2} \tag{3.17}$$

The distribution is independent of the modulus of compressibility E and is valid for surface loading. The equation shows that contact pressure increases from centre to the edges from a finite value to infinity at a small surface load as shown in Fig. 3.7(a).

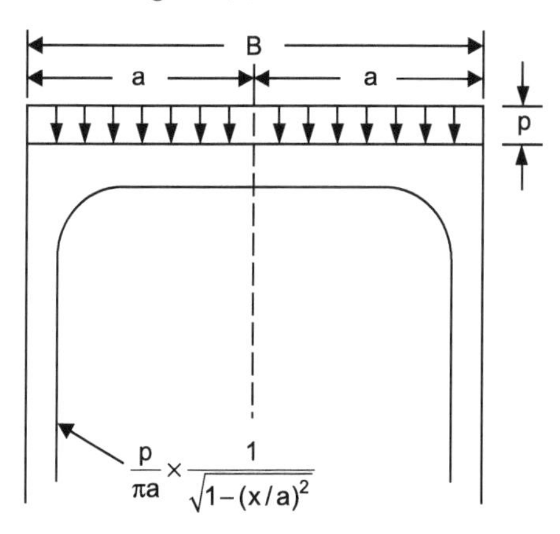

(a) Perfectly elastic and homogeneous subgrade.

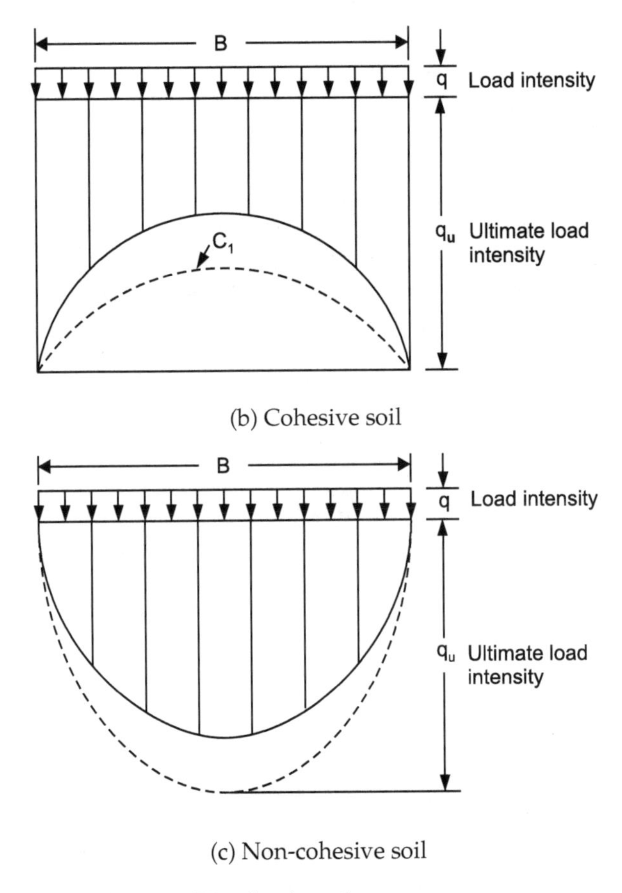

Fig. 3.7. Distribution of contact pressure.

For a real elastic material, the pressure at the edges cannot exceed the ultimate strength (q_u). Therefore, the material at edges passes from elastic to plastic state. The corresponding stress distribution is shown in Fig. 3.7(b) by curve C_1. As the load on the footing is increased, the state of plastic equilibrium spreads from the edges to the centre and the distribution becomes perfectly uniform at failure. The elastic state, therefore, needs modification even at small loads and a different distribution of pressure will apply.

The distribution of contact pressure for a uniformly loaded perfectly rigid footing resting on the surface of a cohesionless material is parabolic (Schultze, 1961). The maximum pressure occurs at the centre and reduces to zero at the edges as there is no resistance to deformation at the edges (Fig. 3.7(c)).

The equations for distribution of contact pressure for various surface loads using non-linear stress-strain relationships are not available.

Illustrative Examples

Example 3.1

A strip footing of width 2.0 m is subjected to a uniformly distributed load of 100 kN/m^2. Considering the soil mass as elastic-homogeneous medium, determine the values of stress σ_z, σ_x and τ_{xz} at depths 1.0 m and 2.0 m for vertical sections passing through the centre and edge of the footing.

Solution

Keeping b = 1.0 m and x = 0.0 m, 1.0 m and –1.0 m, depths 1.0 m and 2.0 m, in Eqs. (3.1a), (3.1b) and (3.1c), values of stresses are given in Table 3.7.

Table 3.7. Values of stresses (Example 3.1)

b (m)	x/b	z/b	σ_z/q	σ_x/q	τ_{xz}/q	σ_z (kN/m^2)	σ_x (kN/m^2)	τ_{xz} (kN/m^2)
1.0	0.0	1.0	0.818	0.182	0.000	81.831	18.169	0.000
1.0	0.0	2.0	0.550	0.041	0.000	54.982	4.052	0.000
1.0	1.0	1.0	0.480	0.225	0.255	47.974	22.509	25.465
1.0	1.0	2.0	0.409	0.091	0.159	40.915	9.085	15.915
1.0	-1.0	1.0	0.480	0.225	-0.255	47.974	22.509	-25.465
1.0	-1.0	2.0	0.409	0.091	-0.159	40.915	9.085	-15.915

Example 3.2

Solve Example 3.1, considering the footing is subjected to a uniform tangential loading of intensity of 100 kN/m^2.

Solution

Keeping b = 1.0 m and x = 0.0 m, 1.0 m and –1.0 m, depths 1.0 m and 2.0 m, in Eqs. (3.2a), (3.2b) and (3.2c), values of stresses are given in Table 3.8.

Table 3.8. Values of stresses (Example 3.2)

b (m)	x/b	z/b	σ_z/q	σ_x/q	τ_{xz}/q	σ_z (kN/m^2)	σ_x (kN/m^2)	τ_{xz} (kN/m^2)
1.0	0.0	1.0	0.000	0.000	0.182	0.000	0.000	18.169
1.0	0.0	2.0	0.000	0.000	0.041	0.000	0.000	4.052
1.0	1.0	1.0	0.255	0.258	0.225	25.465	25.765	22.509
1.0	1.0	2.0	0.159	0.061	0.091	15.915	6.148	9.085
1.0	-1.0	1.0	-0.255	-0.258	0.225	-25.465	-25.765	22.509
1.0	-1.0	2.0	-0.159	-0.061	0.091	-15.915	-6.148	9.085

Example 3.3

Solve Example 3.1, considering the footing is subjected to symmetrical vertical triangular loading with maximum intensity of 100 kN/m^2.

Solution

Keeping b = 1.0 m and x = 0.0 m, 1.0 m and –1.0 m, depths 1.0 m and 2.0 m, in Eqs. (3.3a), (3.3b) and (3.3c), values of stresses are given in Table 3.9.

Table 3.9. Values of stresses (Example 3.3)

b (m)	x/b	z/b	σ_z/q	σ_x/q	τ_{xz}/q	σ_z (kN/m^2)	σ_x (kN/m^2)	τ_{xz} (kN/m^2)
1.0	0.0	1.0	0.500	0.059	0.000	50.000	5.873	0.000
1.0	0.0	2.0	0.295	0.011	0.000	29.517	1.105	0.000
1.0	1.0	1.0	0.205	0.134	0.148	20.483	13.380	14.758
1.0	1.0	2.0	0.205	0.048	0.090	20.483	4.768	9.033
1.0	-1.0	1.0	0.205	0.134	-0.148	20.483	13.380	-14.758
1.0	-1.0	2.0	0.205	0.048	-0.090	20.483	4.768	-9.033

Example 3.4

Solve Example 3.1, considering the footing is subjected to linearly increasing triangular loading with maximum intensity of 100 kN/m^2.

Solution

Keeping b = 1.0 m and x = 0.0 m, 1.0 m and 2.0 m, depths 1.0 m and 2.0 m, in Eqs. (3.4a), (3.4b) and (3.4c), values of stresses are given in Table 3.10.

Table 3.10. Values of stresses (Example 3.4)

b (m)	x/b	z/b	σ_z/q	σ_x/q	τ_{xz}/q	σ_z (kN/m^2)	σ_x (kN/m^2)	τ_{xz} (kN/m^2)
1.0	0.0	1.0	0.127	0.129	-0.113	12.732	12.883	-11.255
1.0	0.0	2.0	0.159	0.061	-0.091	15.915	6.148	-9.085
1.0	1.0	1.0	0.409	0.091	-0.091	40.915	9.085	-9.085
1.0	1.0	2.0	0.275	0.020	-0.041	27.491	2.026	-4.052
1.0	2.0	1.0	0.352	0.096	0.142	35.242	9.627	14.210
1.0	2.0	2.0	0.250	0.029	0.068	25.000	2.936	6.831

Example 3.5

Solve Example 3.1, considering the footing is subjected to linearly decreasing triangular loading with maximum intensity of 100 kN/m^2.

Solution

Keeping $b = 1.0$ m and $x = 0.0$ m, 1.0 m and 2.0 m, depths 1.0 m and 2.0 m, in Eqs. (3.5a), (3.5b) and (3.5c), values of stresses are given in Table 3.11.

Table 3.11. Values of stresses (Example 3.5)

b (m)	x/b	z/b	σ_z/q	σ_x/q	τ_{xz}/q	σ_z (kN/m^2)	σ_x (kN/m^2)	τ_{xz} (kN/m^2)
1.0	0.0	1.0	0.352	0.096	-0.142	35.242	9.627	-14.210
1.0	0.0	2.0	0.250	0.029	-0.068	25.000	2.936	-6.831
1.0	1.0	1.0	0.409	0.091	0.091	40.915	9.085	9.085
1.0	1.0	2.0	0.275	0.020	0.041	27.491	2.026	4.052
1.0	2.0	1.0	0.127	0.129	0.113	12.732	12.883	11.255
1.0	2.0	2.0	0.159	0.061	0.091	15.915	6.148	9.085

Example 3.6

Solve Example 3.1, considering the footing is subjected to horizontal linearly increasing triangular loading with maximum intensity of 100 kN/m^2.

Solution

Keeping $b = 1.0$ m and $x = 0.0$ m, 1.0 m and 2.0 m, depths 1.0 m and 2.0 m, in Eqs. (3.6a), (3.6b) and (3.6c), values of stresses are given in Table 3.12.

Table 3.12. Values of stresses (Example 3.6)

b (m)	x/b	z/b	σ_z/q	σ_x/q	τ_{xz}/q	σ_z (kN/m^2)	σ_x (kN/m^2)	τ_{xz} (kN/m^2)
1.000	0.000	1.000	-0.113	0.800	0.129	-11.255	80.033	12.883
1.000	0.000	2.000	-0.091	0.875	0.061	-9.085	87.500	6.148
1.000	1.000	1.000	-0.091	0.875	0.091	-9.085	87.500	9.085
1.000	1.000	2.000	-0.041	0.864	0.020	-4.052	86.380	2.026
1.000	2.000	1.000	0.142	0.882	0.096	-4.196	80.364	8.687
1.000	2.000	2.000	0.068	0.811	0.029	-6.731	86.545	7.461

Practice Problems

1. A footing of width 1.5 m is subjected to a uniformly distributed load of 75 kN/m^2. Plot σ_z, σ_x and τ_{xz} versus depth (6.0 m) for a section passing through the centre and edge of footing.
2. Solve problem 1, considering the footing is subjected to a uniform tangential loading of intensity 75 kN/m^2.

3. Solve problem 1, considering the footing is subjected to symmetrical vertical triangular loading with maximum intensity 75 kN/m^2.
4. Solve problem 1, considering the footing is subjected to linearly increasing triangular loading with maximum intensity 75 kN/m^2.
5. Solve problem 1, considering the footing is subjected to linearly decreasing triangular loading with maximum intensity 75 kN/m^2.
6. Solve problem 1, considering the footing is subjected to horizontal linearly increasing triangular loading with maximum intensity 75 kN/m^2.

REFERENCES

Boussinesq, J. (1885). Application des Potentials al Etude del Enquilibre et du Movement des Solides Elastiques, Paris, Gauthier-Villard.

Carothers, S.D. (1920). Direct Determination of Stresses. *Proceeding of the Royal Society of London*, Series A, **97(682):** 110-123.

Carrol, W.F. (1963). Dynamic Bearing Capacity of Soil. Technical Report No. 3-599, U.S. Army Engineers, Waterways Experiment Stations, Corps of Engineers, Vicksburg, Mississippi.

Gibson, R.E. (1967). Some Results Concerning Displacement and Stresses in a Non-homogeneous Elastic Half Space. *Geotechnique*, **17(1):** 58-67.

Gray, H. (1936). Stress Distributions in Elastic Solids. *Proc. Int. Conf. Soil Mech. Found. Eng.*, **2:** 157-168.

Harr, M.E. (1966). Foundations of Theoretical Soil Mechanics. McGraw Hill Company, New York.

Jumikis, Alfreds R. (1969). Theoretical Soil Mechanics. Van Nostrand Reinhold.

Kolosov, G.B. (1935). Application of Complex Diagrams and the Theory of Functions of Complex Variables to the Theory of Elasticity. *ONTI.*

Melan, E. (1919). Die Druckverteilungdurcheineelastishe Schieht, *Beton u. Eisen*, **18:** 83-85.

Mindlin, R.D. (1936). Force at a Point in the Interior of a Semi-infinite Solid. *Physics*, **7(5):** 195-202.

Olson, R.E. (1989). Stress Distribution – A Report. Department of Construction Engineering, Chaoyang University of Technology, modified by Jiunnren in 2003.

Poulos, H.G. and Davis, E.H. (1974). Elastic Solutions for Soil and Rock Mechanics. John Wiley and Sons, New York.

Schultze, E. (1961). Distribution of Stress Beneath a Rigid Foundation. *Proc. Fifth Intern. Conf. Soil Mech. and Found. Eng.*, **1:** 807-813.

Westergaard, H.M. (1938). A Problem of Elasticity Suggested by a Problem in Soil Mechanics: Soft Material Reinforced by Numerous Strong Horizontal Sheets. *Contribution to the Mechanics of Solids.* S. Timeshanko, 60th Anniversary Volume. The Macmillan Company, New York.

CHAPTER
4

Strip Footing Subjected to Central Vertical Load

4.1 General

Strip footings are the most common type of foundations used for many structures, specifically buildings. For proportioning of a strip footing, bearing capacity and settlement are the main criteria. Generally the estimation of bearing capacity and settlement are done in two independent steps. Several theories are available for computing bearing capacity (Terzaghi, 1943; Meyerhof, 1951; Balla, 1962; Hansen, 1970; Saran, 1970; Vesic , 1973; Zhu, 2003; Kalinli et al., 2011).

The settlement of a foundation is generally estimated by Terzaghi's theory of one-dimensional consolidation for clay and by performing plate load test for sand. The results of plate load test are extrapolated to the actual size of footing. Terzaghi's one-dimensional consolidation theory is valid only when the lateral yield is negligible as seen in, for example, in the case of a clay layer sandwiched between two sand layers or a large raft resting on a thin clay layer supported by a rigid base. This theory leads to an underestimate of settlement to a large extent for thick layers of clays where the lateral yield is significant. In this theory, the vertical strains are computed from stresses worked out by the elastic theory, which may be true for smaller surface loads but at higher loads, soil becomes plastic and exhibits non-linear stress-strain characteristics.

Pressure-settlement characteristics of a footing were obtained by using the method of characteristics (Kumar and Rao, 2002), discrete element method (Majidi and Mirghasemi, 2008) and finite element method (Basavanna, 1975; Desai, 1971; Shafiee and Jahanandish, 2010).

In this chapter, the general procedure for obtaining the pressure-settlement characteristics of footings using non-linear constitutive laws of soils are described. Two cases have been dealt, namely (i) footings resting on homogeneous isotropic material, and (ii) footings resisting on non-homogeneous an-isotropic material. In the first case, the modulus of elasticity, E or $(1/a)$ may be taken as constant. This may be true to some extent in saturated clays. Sand, on the contrary, is a non-homogeneous anisotropic material. Young's modulus, in this case, is dependent on confining pressure and also varies in both vertical and horizontal directions. In the first case, when E is constant, stress equations based on the theory of elasticity may be used with sufficient accuracy for the analysis. When no stress equations are available where variation of E is there in both vertical and horizontal directions, an empirical analysis has been proposed for getting pressure-settlement characteristics.

In practice, footings are rigid with a rough base. Sharan (1977) studied the effect of flexibility and roughness of the footing on pressure-settlement characteristics of footing subjected to central vertical load. It was found that the average settlement of a flexible footing for the same average pressure intensity (i.e. V/A) is almost same as for the rigid footing; the difference in the two being less than 4 per cent. The effect of roughness on settlement is also negligible. Keeping these facts in view, the analysis is first presented for rigid footing with rough base and resting on clay.

The constitutive relations of soil represented by hyperbola were established from the triaxial compression test. Drainage conditions in the tests were controlled according to the field conditions for obtaining the constitutive relations. Usually in case of footings in clay, the settlement occurs slowly and sufficient time is available for pore pressure dissipation. Therefore, the constitutive relations obtained by performing drained tests are used. In case of footing in sand settlement occurs instantaneously, then constitutive laws for consolidated untrained tests are used. The parameters of constitutive laws obtained from triaxial tests have been suitably modified while using them in two-dimensional stress system.

4.2 Strip Footing with Rough Base on Clay

4.2.1 Assumptions

- The soil mass is assumed to be semi-infinite and an isotropic medium
- The footing base is assumed as being fully rigid with rough base
- The roughness of footing is assumed to generate tangential forces, Fig 4.1(a) and 4.1(b) at the contact surface, which follows the relationship $c_a = c\left(\dfrac{q}{cN_c}\right)$ for cohesive soil, where c_a is the developed

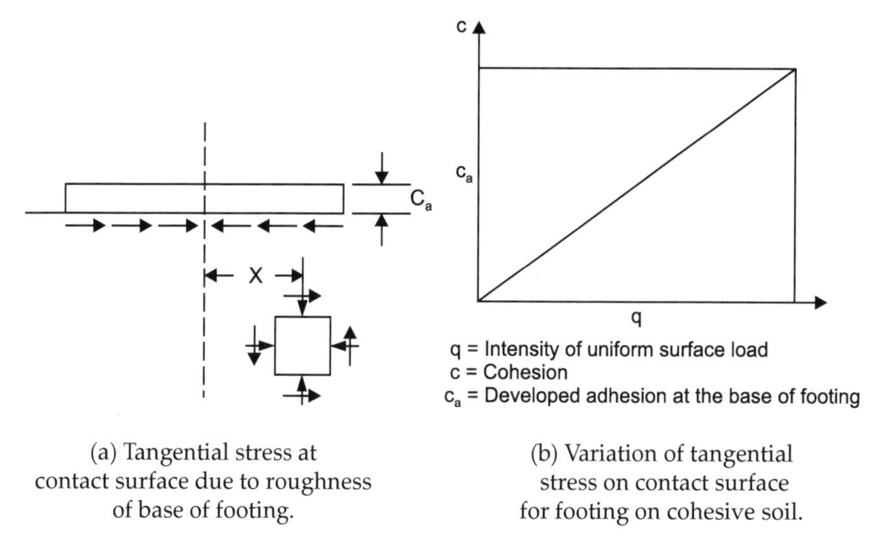

(a) Tangential stress at contact surface due to roughness of base of footing.

(b) Variation of tangential stress on contact surface for footing on cohesive soil.

Fig. 4.1. Tangential stress

adhesion at the contact surface. N_c is the bearing capacity factor for $\phi = 0$ condition, and therefore cN_c represents the ultimate pressure on the footing; q is the applied pressure intensity on the footing

- The contact pressure distribution is assumed as linear from the centre to the edges (Fig. 4.2).
- The soil mass supporting a footing is divided into a large number of thin horizontal strips (Fig. 4.3) in which stresses and strains are assumed to be uniform along any vertical section
  - The stresses in each layer are computed, using Boussinesq's theory as the stress equations for various types of loads that are available.
  - The strains are computed from the known stress condition using constitutive laws
- There is no slippage at the interface of layers of the soil mass.

4.2.2 General Procedure

The general procedure for evaluation of pressure settlement characteristics of footings is described in the following steps:

Step 1

For a given load q_1 on the footing, the contact-pressure distribution at the interface of footing base and supporting soil may be assumed as shown in Figs. 4.1 and 4.2 respectively for roughness of footing and uniformly distributed load acting on the footing. The contact-pressure distributions are the loading patterns for the soil at the surface below the footing and

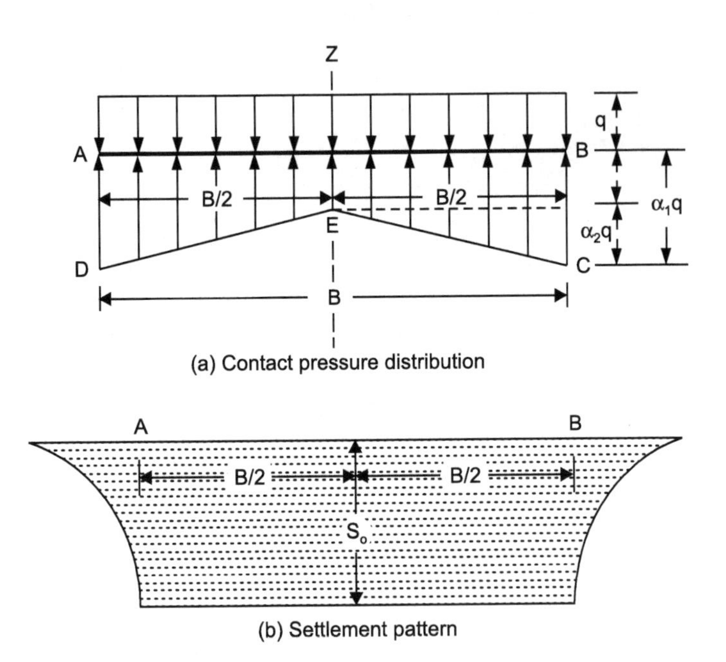

Fig. 4.2. Contact pressure distribution and settlement of centrally loaded footing with rigid base resting on clay.

the stresses in the soil mass induced according to these loading patterns. The correctness of the assumption will be tested in Step 8.

When equating the total vertical load to the area of contact pressure diagram (Fig. 4.2a),

$$q.B = q.\alpha_1.B - q.\alpha_2.\frac{B}{2}$$

or, $$\alpha_2 = 2(\alpha_1 - 1) \tag{4.1}$$

In this step assume a value α_1 and compute α_2 using Eq. (4.1). Thus contact-pressure distribution for vertical loading magnitude is known.

Step 2

The soil mass supporting the footing is divided into n layers as shown in Fig. 4.3.

Step 3

Taking any vertical section, the normal and shear stresses (σ_z, σ_x and τ_{xz}) at the centre of a layer at depth z below the footing were computed using equations of the theory of elasticity given in Chapter 3 for appropriate contact-pressure distribution (Fig. 4.2a) and tangential stress (Fig. 4.1). The

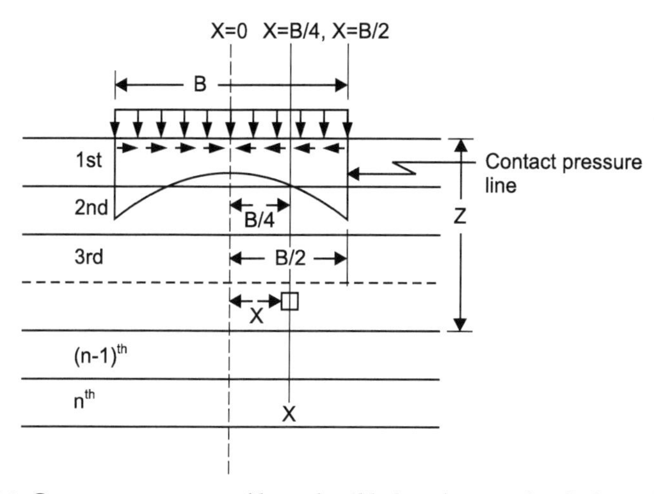

Fig. 4.3. Contact pressure profile and soil below footing divided in *n* layers.

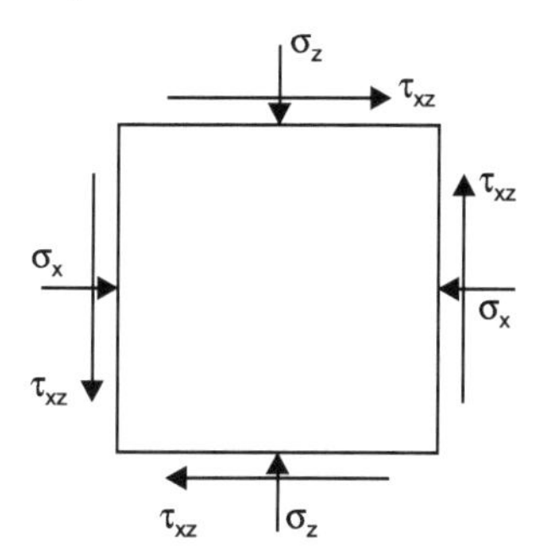

Fig. 4.4. Stresses in soil mass in two-dimensional space.

stress components at a point are shown in Fig. 4.4 for two-dimensional space.

Step 4

The principal stresses and their directions with vertical axis, as shown in Fig. 4.5 are computed by using the following equations:

$$\sigma_1 = \frac{\sigma_Z + \sigma_X}{2} + \sqrt{\left(\frac{\sigma_Z - \sigma_X}{2}\right)^2 + \tau_{XZ}^2} \tag{4.2}$$

$$\sigma_3 = \frac{\sigma_Z + \sigma_X}{2} - \sqrt{\left(\frac{\sigma_Z - \sigma_X}{2}\right)^2 + \tau_{XZ}^2} \tag{4.3}$$

$$\text{Tan } 2\theta = \frac{\tau_{XZ}}{\sigma_Z - \sigma_X} \tag{4.4}$$

Positive value θ of is measured counter clockwise with direction of σ_Z (Fig. 4.5).

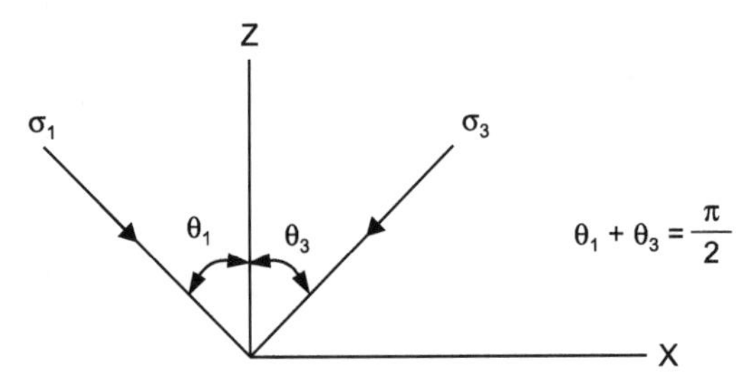

Fig. 4.5. Principal stresses at a point and their directions (two-dimensional).

Step 5 Evaluation of Lateral Strain

Strip footing is a case of plane strain. The strain $\in_2$ in the direction of intermediate principal stress equals zero in plane strain. Therefore, from basic stress-strain equations, we get

$$\in_2 = 0 = \frac{1}{E}\left[\sigma_2 - \mu(\sigma_1 + \sigma_3)\right] \tag{4.5a}$$

or

$$\sigma_2 = \mu(\sigma_1 + \sigma_3) \tag{4.5b}$$

And

$$\in_1 = \frac{1-\mu^2}{E}\left[\sigma_1 - \frac{\mu}{1-\mu}\sigma_3\right] \tag{4.6a}$$

$$\in_3 = \frac{1-\mu^2}{E}\left[\sigma_3 - \frac{\mu}{1-\mu}\sigma_1\right] \tag{4.6b}$$

Assuming, $\mu_1 = \frac{\mu}{1-\mu}$ and $E_1 = \frac{E}{1-\mu^2}$ (4.7)

We have, $\in_1 = \frac{1}{E_1}\left[\sigma_1 - \mu_1\sigma_3\right]$ (4.8a)

$$\in_3 = \frac{1}{E_1}\left[\sigma_3 - \mu_1\sigma_1\right] \tag{4.8b}$$

and

$$\in_3/\in_1 = \frac{\sigma_3 - \mu_1\sigma_1}{\sigma_1 - \mu_1\sigma_3} \tag{4.9}$$

Let $\in_3/\in_1 = -\mu_2$ (4.10)

Then $\mu_2 = \frac{\sigma_3 - \mu_1\sigma_1}{\sigma_1 - \mu_1\sigma_3}$ (4.11)

Step 6 Evaluation of Principal Strains

The strain in the direction of major principal stress is computed from constitutive relations as given below:

$$\epsilon_1 = \frac{a'(\sigma_1 - \sigma_3)}{1 - b'(\sigma_1 - \sigma_3)} \tag{4.12}$$

where $a' = a(1 - \mu^2)$ and $b' = 1.1 \times a$

a' and b' are parameters of Eq. (2.5).

From Eq. 4.10 $\epsilon_3 = -\mu_2 \epsilon_1$

Step 7 Evaluation of Vertical Strain

The strain in the vertical direction (ϵ_z) for each layer is calculated using the following expression:

$$\epsilon_z = \epsilon_1 \cos^2\theta_1 + \epsilon_3 \cos^2\theta_3 \tag{4.13}$$

Since $\epsilon_2 = 0$ is in plain strain condition.

Step 8 Evaluation of Settlement

The evaluation of the total settlement along any vertical section is done by numerically integrating the quantity

$$S = \int_0^n \epsilon_z \, dz \tag{4.14}$$

The total settlement was computed along different vertical sections passing through the base of the footing. Eight vertical sections may be taken for desired accuracy.

Step 9 Value of α_1

Steps 1 to 8 were repeated for different values of α_1. Value of α_1 which gives the settlement of different points on the base of the footing, almost same corresponds to true contact-pressure distribution.

Step 10 Pressure Settlement Curve

The values of total settlement for various pressure intensities are computed according to the method described above and surface load intensity versus settlement curve is then drawn.

It may be noted that if $\alpha_1 = 1.0$, it means the base contact-pressure for vertical load becomes uniform since value of α_2 works out zero. This is possible if the footing base is flexible. The settlement pattern will be non-uniform giving, with maximum settlement at the centre of the footing, while decreasing towards edges. It has been observed that the settlement of a rigid footing is almost the same as the average settlement of flexible footing (Sharan, 1977). The average settlement may be computed by dividing the area of settlement diagram by the width of footing.

4.3 Strip Footing with Rough Base Resting on Sand

4.3.1 Introduction

In Art. 4.2, the methodology has been presented to obtain the pressure-settlement characteristics of footing resting on clay and assuming that this is an isotropic and homogeneous material. The value of Young's modulus E or $(1/a)$ is taken as constant. Sand, on the contrary, is an anisotropic material. The Young's modulus is dependent upon confining pressure and varies in both vertical and horizontal directions. No stress equations are available where variation of E in both vertical and horizontal directions are considered. Keeping this fact in view, an empirical method has been proposed—dividing the stress-strain relationship is various small segments each of which is considered a straight line. In these small stress level range, the soil is considered homogeneous and isotropic.

4.3.2 Assumptions

In the present investigation an empirical approach has been developed which has been found to yield satisfactory results. The following assumptions have been made:

1. The footing base has been assumed to be fully rigid with a rough base.
2. The contact-pressure distribution for vertical loading on the base of footing resting on sand is assumed as shown in Fig. 4.6 equating the vertical load to area of contact pressure.

$$q.B = \alpha_1.q.B + q.\alpha_2.\frac{B}{2}$$

or,

$$\alpha_2 = 2(1 - \alpha_1) \tag{4.15}$$

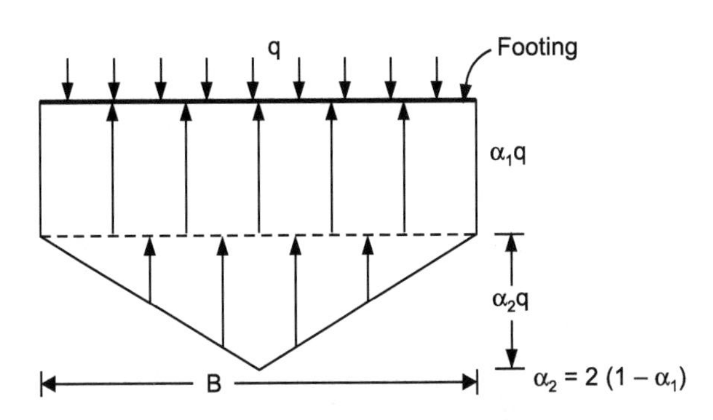

Fig. 4.6. Contact pressure distribution of centrally loaded footing with rigid base resting on sand.

3. The effect of the weight of soil mass has been considered in determination of stresses in sand. Vertical stress due to weight of soil is taken equal to γz where γ is the density of soil and z is the depth. The horizontal stress due to weight of soil is taken as being equal to $K_0 \gamma z$ where K_0 is the coefficient of earth pressure at rest. The soil-state varies from at rest condition at centre to active condition at the edges. However, the pattern of variation of earth pressure coefficient from centre to edges is not known.
4. The ultimate bearing capacity (q_u) has been computed from Terzaghi's equations, i.e.

$$q_u = \frac{1}{2}\gamma.B.N_\gamma \tag{4.16}$$

γ = Unit weight of soil
B = Width of footing
N_γ = Terzaghi's bearing capacity factor (Table A 4.1)

5. A coefficient F has been introduced such that at all stress levels in the following relationship is satisfied:

$$\frac{q_u}{q} = \frac{\sigma_u}{\sigma'_1 - \sigma'_3} = F \tag{4.17}$$

where q is the intensity of surface load,
σ_u is the ultimate stress from hyperbolic relationship and is equal to $(1/b)$,
σ'_1, σ'_3 are the major and minor principal stresses in the soil mass due to surface load q (Fig. 4.6), tangential stress (Fig. 4.7) and weight of soil.

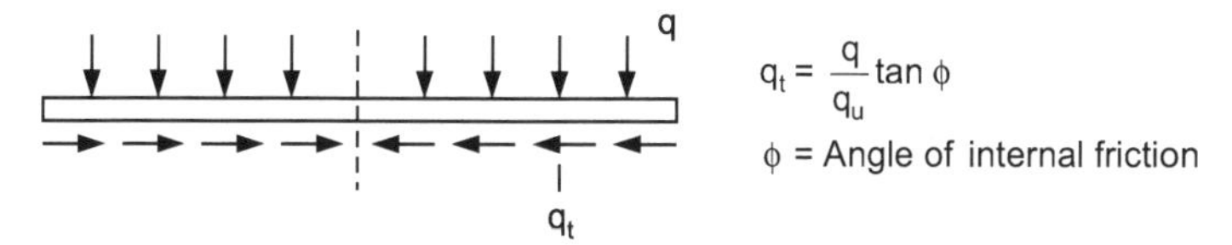

Fig. 4.7. Tangential stress at the base of footing representing roughness.

6. Constitutive law described by hyperbola has been taken. Therefore, σ_u is taken as being equal to $(1/b)$.
7. The modulus of elasticity (E_s) has been calculated from the relevant stress-strain curve at stress level of σ_u/F. The strain at σ_u/F

$$= \frac{a\left(\frac{\sigma_u}{F}\right)}{1-b\left(\frac{\sigma_u}{F}\right)} \tag{4.18a}$$

$$\text{Therefore, } E_s = \frac{\text{Stress}}{\text{Strain}} = \frac{1-b\left(\frac{\sigma_u}{F}\right)}{a} \tag{4.18b}$$

σ_u or $(1/b)$ is a function of the confining pressure (Chapter 2).

4.3.3 General Procedure

The procedure followed in the analysis of uniformly loaded strip footing resting on sand is described in the following steps.

Step 1

Assume a value α_1 for obtaining contact-pressure distribution (Fig. 4.6).

Step 2

The stresses (σ_x, σ_z, and τ_{xz}) at the centre of each layer along different vertical axes passing through the base of the footing as shown in Fig. 4.3 for a given surface load intensity q considering base contact pressure Fig. 4.6 and tangential stress for base roughness (Fig. 4.7) are computed.

Step 3

The effect of the weight of sand has been incorporated by adding γz to σ_z and $K_0\gamma z$ to σ_x where $K_0 = 1 - \sin\varphi$; φ is the angle of internal friction and its value is determined from triaxial tests. The principal stresses σ_1 and σ_3 and their directions with vertical axis are then computed. These principal stresses are designed as σ'_1 and σ'_3.

Step 4

The ultimate shear strength (σ_u) of sand for a given confining pressure σ'_3 is computed from constitutive laws, as discussed in Chapter 2.

Step 5

The ultimate bearing pressure (q_u) is calculated from Terzaghi's (1943) equation (Appendix A4.1).

Step 6

A coefficient F for a given surface load intensity (q) is computed from the following relationship:

$$\frac{q_u}{q} = F \tag{4.19}$$

Step 7

The modulus of elasticity (E_s) is calculated by the following equation:

$$E_s = \frac{1-b\left(\frac{\sigma_u}{F}\right)}{a} \tag{4.20}$$

where *a* and *b* are the constants of hyperbola whose values depend upon the confining pressure (Chapter 2).

Step 8

The strain in each layer in the direction of major principal stress is calculated from the equation

$$\in_1 = \frac{(\sigma_3' - \sigma_3')}{E_s} \tag{4.21}$$

The strain in the direction of minor principal stress is calculated from the following relationship:

$$\in_1 = -\mu_2 \in_1 \tag{4.22}$$

Value of μ_2 is obtained by using Eq. (4.10).

Step 9

The strain in the vertical direction is calculated by using Eq. (4.22)

$$\in_z = \in_1 \cos^2\theta_1 + \in_3 \cos^2\theta_3 \tag{4.23}$$

Step 10

The settlement of each layer and the total settlement along any vertical section are calculated by using Eqs (4.7) and (4.8) respectively.

Step 11

The steps 1 to 9 are repeated for different values of α_1. Values of α_1 which gives settlements at various base points approximately is its true value.

Step 12

The surface load intensity is varied and the Steps 1 to 10 are repeated. The pressure settlement curve is obtained by plotting settlements obtained in Step 10 against corresponding surface load intensity.

4.4 Pressures-Settlement Characteristics of Footings Resting on Buckshot Clay

From the analysis presented in Art 4.2, the following four cases were studied (Sharan, 1977):

- Smooth flexible strip footing by putting $\alpha_1 = 1$ and neglecting the effect of roughness considered as per Fig. 4.1b (assumption No. 3 of Art. 4.2.1).
- Smooth rigid strip footing considering the coefficient α_1 and neglecting the effect of roughness.

- Rough flexible strip footing considering $\alpha_1 = 1$ and roughness at the base of footing as explained in assumption No. 3 of Art. 4.2.1.
- Rough rigid strip analysis as presented in Art. 4.2.

The properties of Buckshot clay has been described in Art. 2.4.1 of Chapter 2. Using the values hyperbolic parameters a and b as 1.9×10^{-4} m^2/kN and 1.13×10^{-4} m^2/kN respectively, and value of cohesion c as 72 kN/m^2, complete procedure has been adopted as described in Art. 4.2 for obtaining the pressure-settlement curves and corresponding contact pressure coefficients. Some typical pressure-settlement curves for rigid strip rough-based footings are shown in Fig. 4.8. Contact pressure distribution curves for the footing of width 250 mm acted upon by different pressure intensities which gave uniform settlement below the base of the footing are shown Fig. 4.9. It is evident from this figure that the tendency of contact pressure distribution is becoming uniform as the pressure intensity reaches near failure. From these curves, values of α_1 were obtained and the following equation is obtained when describing the trend

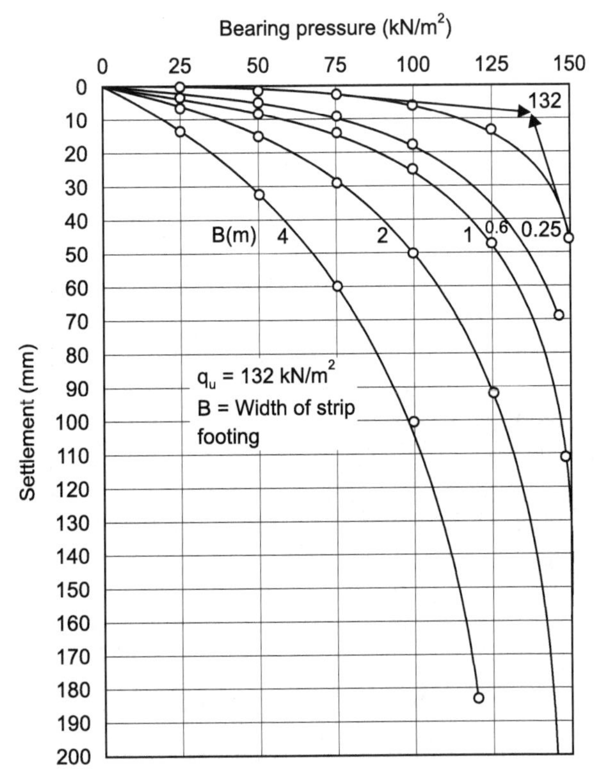

Fig. 4.8. Pressure vs settlement curve for rough rigid strip footing resting on Buckshot clay.

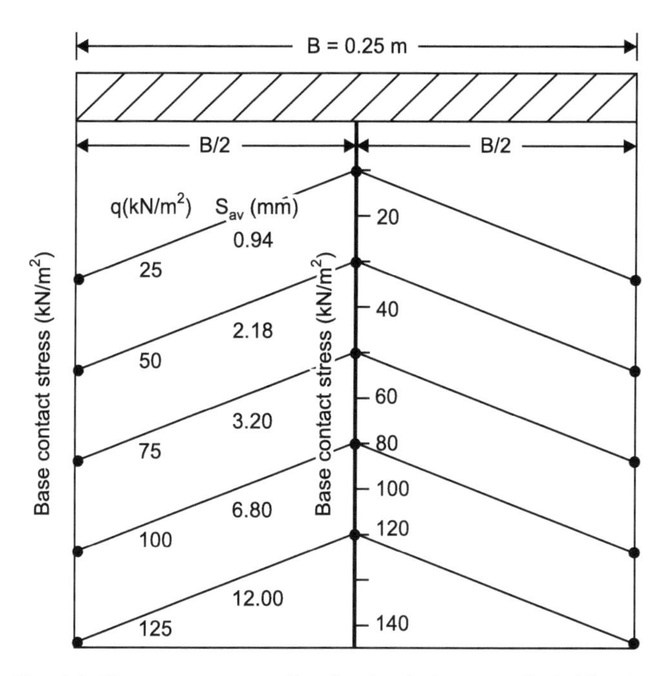

Fig. 4.9. Contact pressure distribution below rough rigid strip footing of width 0.25 m resting on Buckshot clay.

$$\alpha_1 = 1.72 - 0.86\left(\frac{q}{c_u}\right) \tag{4.24}$$

As mentioned earlier, the four cases have been studied independently for obtaining pressure-settlement characteristics of a footing. On studying these cases in detail, the following conclusions were drawn for the footings resting on clay:

- The bearing capacity of flexible strip footings from pressure versus average settlement curves and for rigid footings from pressure versus uniform settlement curves are practically same for both smooth and rough bases
- The bearing capacity of rough rigid strip footing is 5 per cent higher than the bearing capacity of smooth rigid strip footing
- The settlement at failure varies from 4.8-5.5 per cent for strip footings
- The settlement of a footing varies directly in proportion to its width, i.e.

$$\frac{S_f}{S_p} = \frac{B_f}{B_p} \tag{4.25}$$

where S_f and S_p are respectively the settlements of the footings of widths B_f and B_p.

- Contact pressure distribution is dependent on pressure intensity. It is found that as the load reaches the failure load, the contact pressure distribution becomes uniform

Keeping the above facts in view, a parametric study was carried out for the following variables (Biswas *et al.*, 2016 and Biswas, 2017):

Width of footing, B: 0.5 m, 1.0 m, 1.5 m, 2.0 m
Hyperbolic Constitutive Laws Parameters

$1/a$ kN/m^2	5000	7000	9091	12000	15000
$1/b$ kN/m^2	35	50	64	80	100

Pressure intensity, q kN/m^2: 5 to 120 (depending on the values of $1/a$ and $1/b$)

For illustration, $1/a$ = 15000, $1/b$ = 100 and B = 1.0 m, settlements of equally spaced nine points of the base were obtained for different pressure intensities. Typical plots showing the settlement patterns for pressure intensities 25 kN/m^2 and 50 kN/m^2 are given in Figs 4.10 and 4.11 respectively. It is evident that these settlement patterns are parabolic in nature with the maximum settlement at the centre of the footing. Since the settlement of an equivalent rigid footing i.e. of the same width is almost equal to the average settlement of a flexible footing it is obtained by dividing the area of the settlement diagram (i.e. as shown in Figs 4.10 and 4.11) by the width of footing. It is designated as S_{av}. Plots of pressure versus settlement curves were prepared using the average settlement, i.e. considering the footing base as rigid with a smooth base. Effect of roughness is very marginal and therefore neglected.

Typical bearing capacity versus settlement curves for $1/a$ = 9000 kN/m^2 and $1/b$ = 60 kN/m^2 are given in Fig. 4.12 for different widths

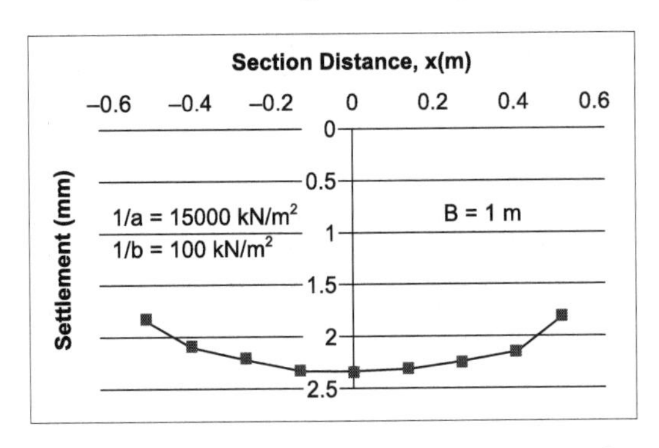

Fig. 4.10. Settlement pattern for q = 25 kN/m^2.

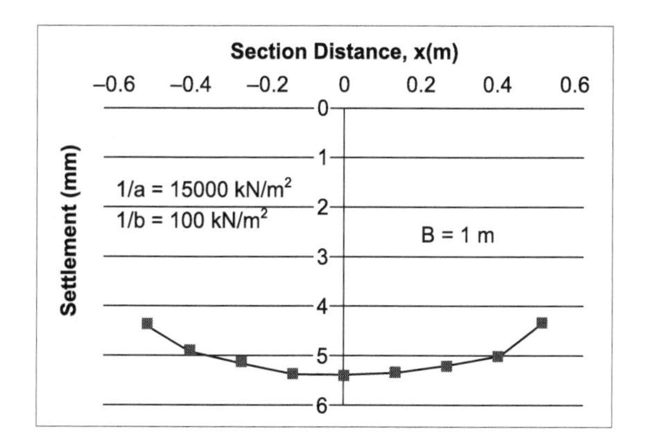

Fig. 4.11. Settlement pattern for q = 50 kN/m^2.

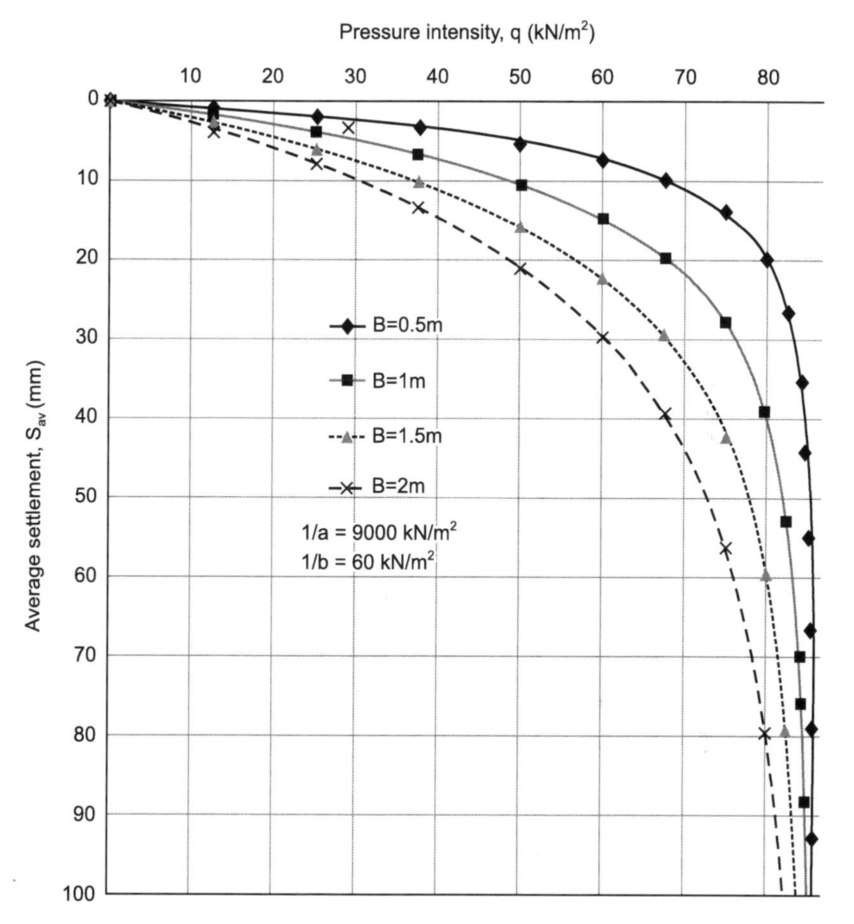

Fig. 4.12. Pressure intensity (q) versus average settlement (S_{av}) curves.

of footing. From these curves, values of settlement were obtained for pressure intensities equal to 12.5 kN/m², 25 kN/m² 37.5 kN/m² and 50 kN/m². Plots were then prepared between (S_{av}/B) ratio and width of footing (B) (Fig. 4.13). As for each pressure intensity, the curve is a straight line parallel to the width axis, indicating that the settlement increases linearly with the width of footing. Hence pressure intensity (q) versus (S_{av}/B) plots have been made for different values of $1/a$ and $1/b$ (Fig. 4.14). From these curves, it is evident that for a given pressure intensity,

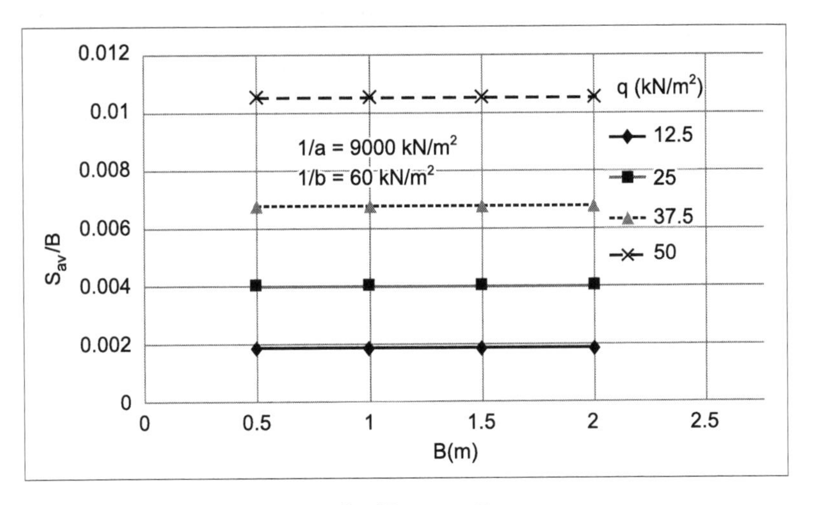

Fig. 4.13. S_{av}/B versus B curves.

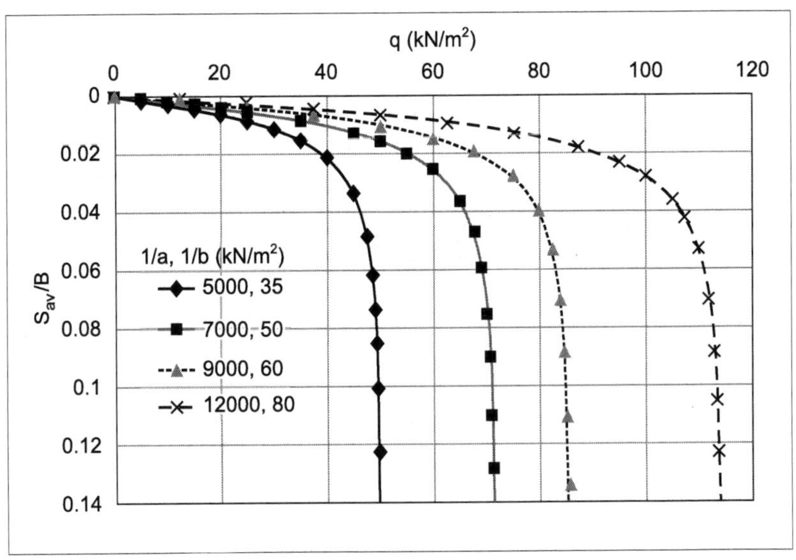

Fig. 4.14. q versus (S_{av}/B) curves.

settlement decreases with the increase in $1/a$ and $1/b$ values. Increase in $1/a$ and $1/b$ values indicate a better soil. In Fig. 4.15 (S_{av}/B) ratio is plotted against $1/a$ for $1/b$ = 60 kN/m^2 for different pressure intensities. These curves indicate that the settlement decreases with increase in $1/a$ ratio. This trend is valid for other values of $1/b$ also. Similar plots have been shown in Fig. 4.16 in which the settlement variation is shown with respect to $1/b$ for a given value of $1/a$. Similar trend is observed in this also.

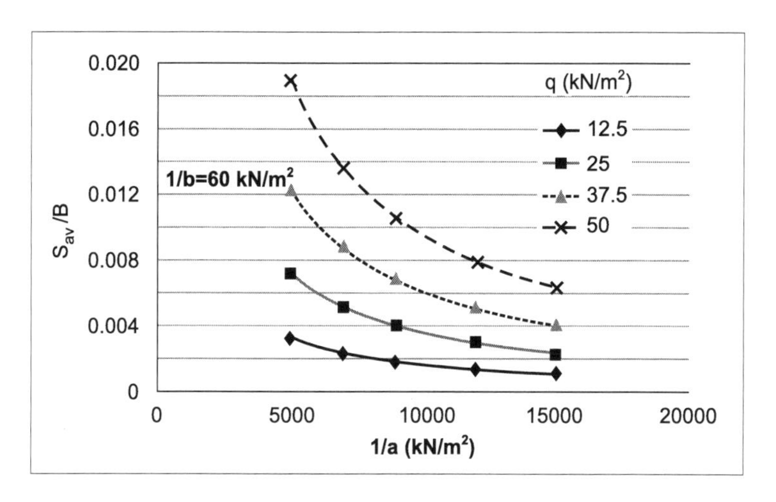

Fig. 4.15. S_{av}/B versus $1/a$ curves.

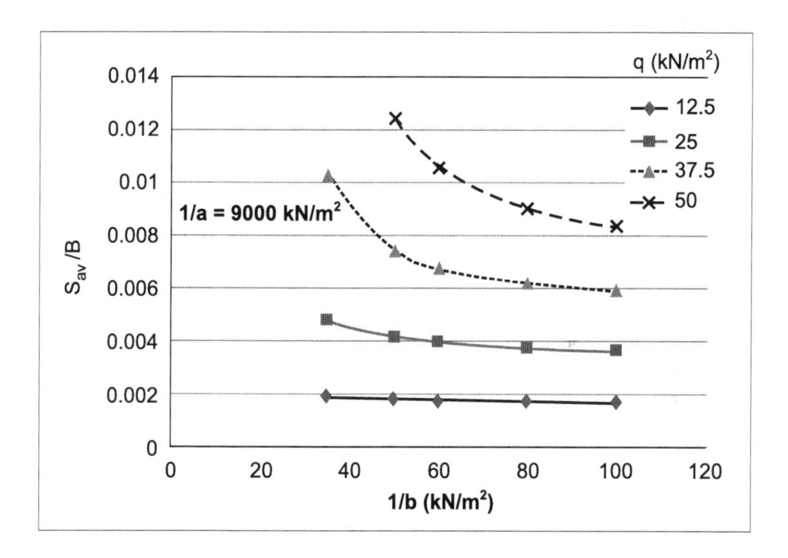

Fig. 4.16. S_{av}/B versus $1/b$ curves.

4.5 Pressure-Settlement Characteristics of Strip Footings Resting on Sand

The pressure settlement characteristics of smooth strip footings resting on Ranipur sand were investigated by Sharan (1977) for footing widths of 50 mm, 100 mm, 150 mm, 600 mm and 2000 mm. The computations were done for sand at two relative densities, viz. 84 per cent and 46 per cent. Settlements at central, quarter and edge sections for a given pressure intensity were computed. The average settlements were calculated by dividing the area of settlement diagram with the width of footing.

Pressure-settlement curves were drawn by plotting the average settlements against bearing pressures. These curves for footings with width of 50 mm 100 mm and 150 mm are plotted in Fig. 4.17, for 600 mm in Fig. 4.18, and for 2000 mm in Fig. 4.19 for sand at relative density of 84 per cent. Similar curves are drawn for footing widths of (50 mm, 100 mm, 150 mm), 600 mm and 2000 mm in Figs 4.20, 4.21 and 4.22 respectively for relative density of 46 per cent. The results show that for the same density and pressure, the settlements increase with footing size. The settlements for same pressure and footing size increase with decrease of relative density.

To study the footing size effect on the pressure settlement curves for footings in sand, the S_f/S_p versus B_f/B_p relation were computed for pressure of 50 kN/m^2, taking B_p = 100 mm, B_f = 50 mm, 100 mm, 150 mm,

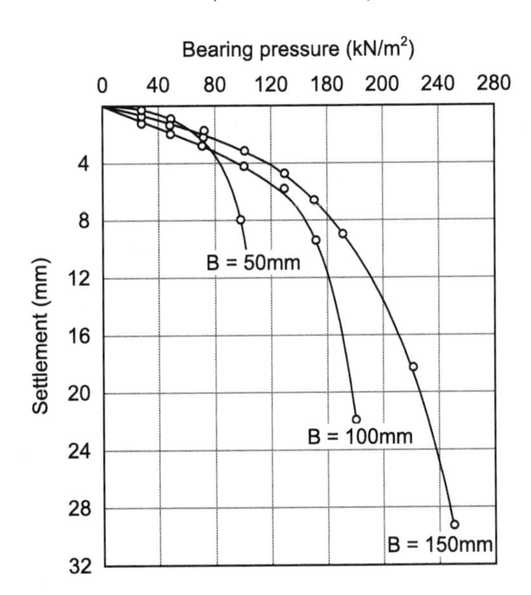

Fig. 4.17. Pressure vs settlement for smooth strip footing resting on Ranipur sand (D_R = 84%).

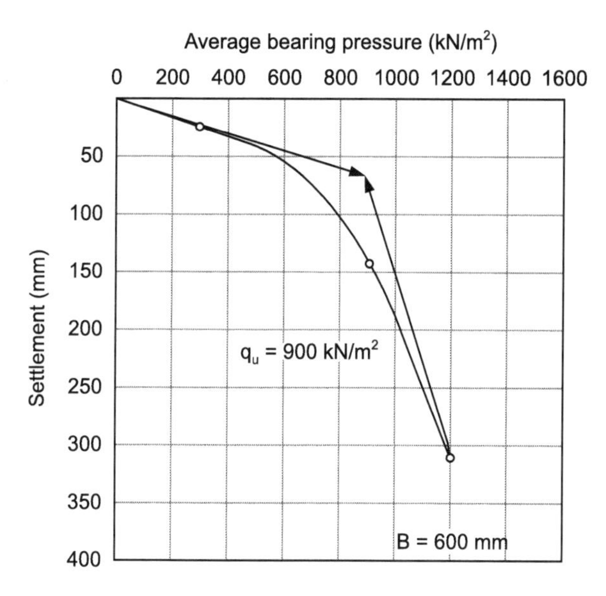

Fig. 4.18. Pressure vs settlement curve for smooth strip footing resting on Ranipur sand (D_R = 84%).

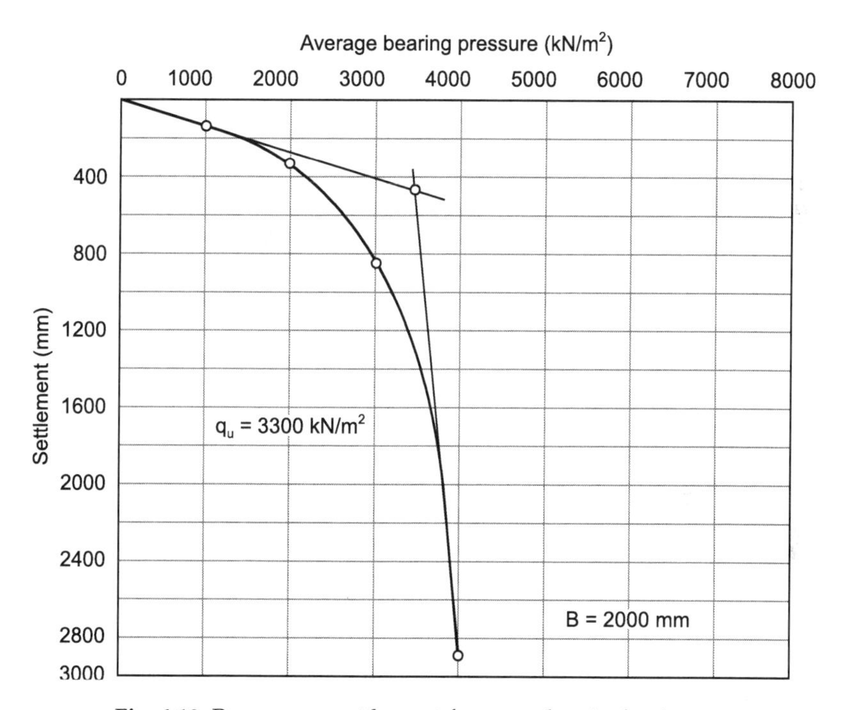

Fig. 4.19. Pressure vs settlement for smooth strip footing on Ranipur sand (D_R = 84%).

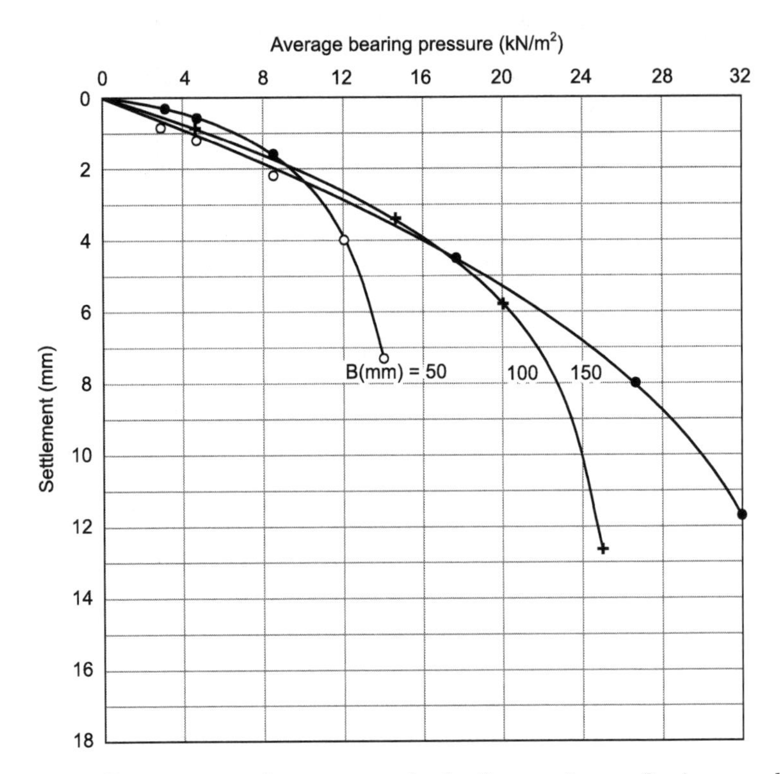

Fig. 4.20. Pressure vs settlement curve for footings resting on Ranipur sand (D_R = 46%, smooth strip).

600 mm and 2000 mm as shown in Fig. 4.23 for relative densities of 84 per cent and 46 per cent. The above relationship computed from conventional formula is also shown in the above figure. It is observed that the curve becomes asymptomatic with increase of footing size (B_f). After B_f/B_p is greater than 6.0 the curve is practically horizontal. Pressure settlement characteristics of these footings were obtained by using the procedure given in Art. 4.3.3.

Illustrative Examples

Example 4.1

A smooth surface strip footing of width 1.5 m rests on clay. It is subjected to a pressure as shown Fig. 4.24. Determine the vertical strain in the soil along the vertical sections passing through the centre and edges of the footing and at depth 1.0 m below the base of the footing. Values of Kondner's hyperbola constants are: $a = 6 \times 10^{-5}$ m²/kN; $b = 0.006$ m²/ kN; value of Poisson ratio is 0.33 and value of cohesion of soil is 75 kN/m².

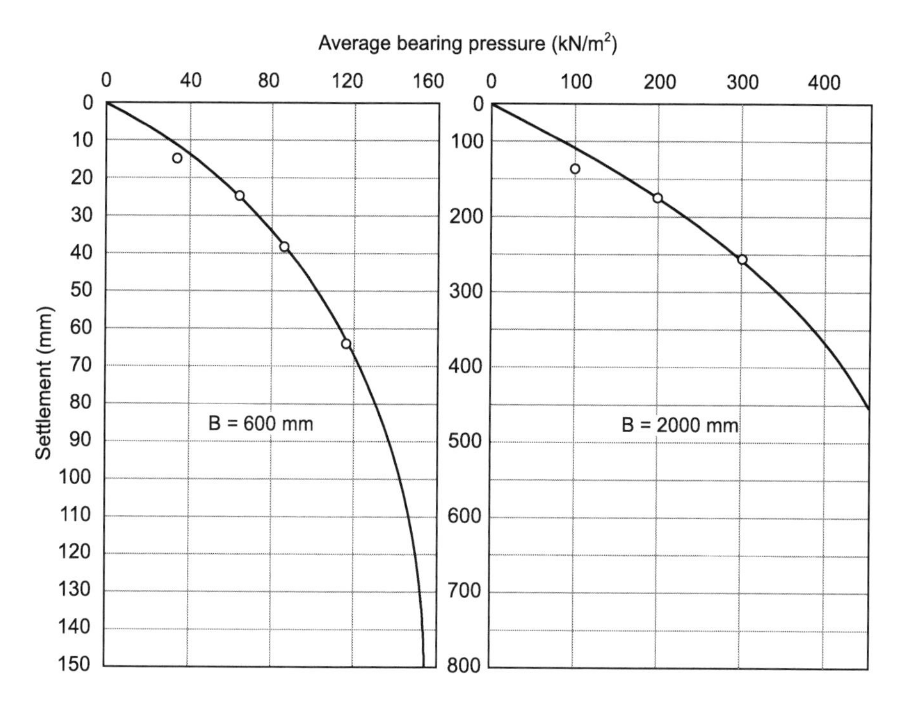

Fig. 4.21. Pressure vs settlement curve for smooth strip footing resting on Ranipur sand (D_R = 46%).

Fig. 4.22. Pressure vs settlement curve for smooth strip footing resting on Ranipur sand (D_R = 46%).

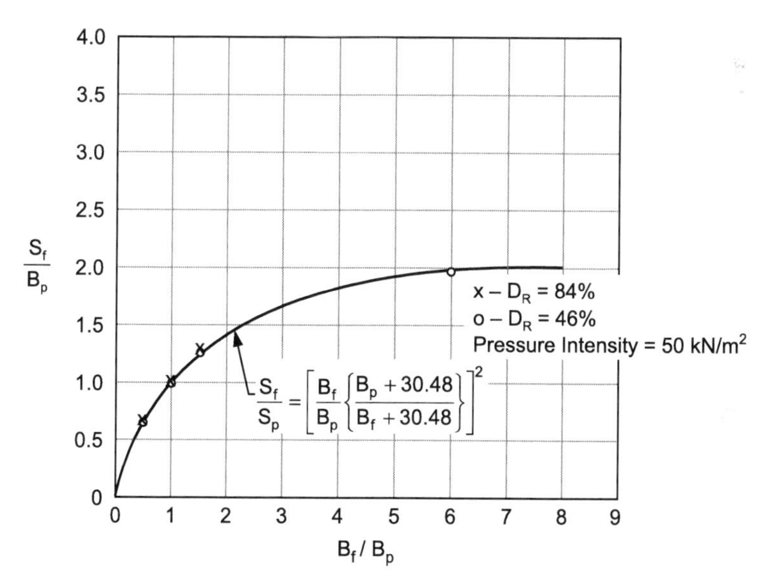

Fig. 4.23. S_f/S_p vs B_f/B_p relationship for Ranipur sand (D_R = 84% and 46%).

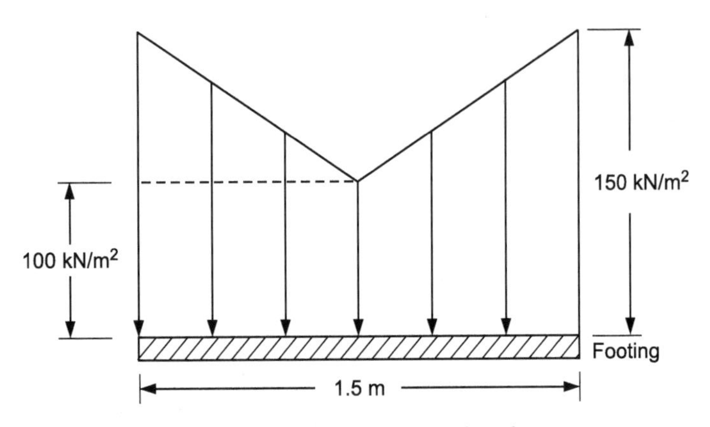

Fig. 4.24. Showing the pressure distribution.

Solution

(i) $2b = 1.5$ m, $b = 0.75$ m

Pressure diagram may be considered equivalent to a uniformly distributed load of 150 kN/m^2 minus a symmetrical triangular load of apex 50 kN/m^2.

For the section passing though the centre:

$$x = 0; \therefore \frac{x}{b} = 0$$

$$z = 2.0 \text{ m}; \therefore \frac{z}{b} = \frac{2.0}{0.75} = 2.66$$

For uniform distributed load:

For $\frac{x}{b} = 1.0$ and $\frac{z}{b} = 2.66$ (using interpolation in Table 3.1)

$$\frac{\sigma_z}{q} = 0.44; \frac{\sigma_x}{q} = 0.02; \frac{\tau_{xz}}{q} = 0.0, \text{ here } q = 150 \text{ kN/m}^2$$

or $\sigma_z = 66.0$ kN/m^2; $\sigma_x = 3.0$ kN/m^2; $\tau_{xz} = 0.0$

For triangular loading:

For $\frac{x}{b} = 0$ and $\frac{z}{b} = 2.66$ (using interpolation in Table 3.3)

$$\frac{\sigma_z}{q} = 0.23; \frac{\sigma_x}{q} = 0.0052; \frac{\tau_{xz}}{q} = 0.0 \text{ here } q = 50 \text{ kN/m}^2$$

or, $\sigma_z = 11.5$ kN/m^2; $\sigma_x = 0.26$ kN/m^2; $\tau_{xz} = 0.0$

Net stresses:

$\sigma_z = 66.0 - 11.5 = 54.5$ kN/m^2; $\sigma_x = 3.0 - 0.26 = 2.74$ kN/m^2; $\tau_{xz} = 0.0$

since $\tau_{xz} = 0.0$, therefore

$$\sigma_1 = \sigma_z = 54.5 \text{ kN/m}^2$$

$$\sigma_3 = \sigma_x = 2.74 \text{ kN/m}^2$$

$$\varepsilon_1 = \frac{a(\sigma_1 - \sigma_3)}{1 - b(\sigma_1 - \sigma_3)} = \varepsilon_z$$

or,
$$\varepsilon_z = \frac{6 \times 10^{-5}\,(54.5 - 2.74)}{1 - 0.006\,(54.5 - 2.74)} = 450 \times 10^{-5}$$

(ii) For the section passing through right-hand side edge of the footing

$$x = 0.75 \quad \therefore \frac{x}{b} = \frac{0.75}{0.75} = 1$$

For uniform distributed load:

For $\frac{x}{b} = 1.0$ and $\frac{z}{b} = 2.66$ (using interpolation in Table 3.1)

$$\frac{\sigma_z}{q} = 0.3650; \frac{\sigma_x}{q} = 0.0531; \frac{\tau_{xz}}{q} = 0.113 \text{ here } q = 150 \text{ kN/m}^2$$

or, $\sigma_z = 54.75 \text{ kN/m}^2$; $\sigma_x = 7.96 \text{ kN/m}^2$; $\tau_{xz} = 16.95 \text{ kN/m}^2$

For triangular loading:

For $\frac{x}{b} = 1$ and $\frac{z}{b} = 2.66$ (using interpolation in Table 3.3)

$$\frac{\sigma_z}{q} = 0.181; \frac{\sigma_x}{q} = 0.0264; \frac{\tau_{xz}}{q} = 0.0644 \text{ here } q = 50 \text{ kN/m}^2$$

or, $\sigma_z = 9.05 \text{ kN/m}^2$; $\sigma_x = 1.32 \text{ kN/m}^2$; $\tau_{xz} = 3.22 \text{ kN/m}^2$;

Total stresses

$\sigma_z = 54.75 - 9.05 = 45.7 \text{ kN/m}^2$; $\sigma_x = 7.96 - 1.32 = 6.64 \text{ kN/m}^2$

$$\tau_{xz} = 16.95 - 3.22 = 13.73 \text{ kN/m}^2$$

$$\sigma_1 = \frac{(45.7 + 6.64)}{2} + \sqrt{\left(\frac{45.7 + 6.64}{2}\right)^2 + 13.73^2}$$

$$= 26.17 + 23.87 = 50.04 \text{ kN/m}^2$$

$$\sigma_3 = 26.17 - 23.87 = 2.3 \text{ kN/m}^2$$

$$\varepsilon_1 = \frac{a(\sigma_1 - \sigma_3)}{1 - b(\sigma_1 - \sigma_3)}$$

$$= \frac{6\times10^{-5}(50.04-2.3)}{1-0.006(50.04-2.3)} = 401 \times 10^{-5}$$

$$\mu_2 = \frac{\sigma_3 - \mu_1\sigma_1}{\sigma_1 - \mu_1\sigma_3};\ \mu_1 = \frac{\mu}{1-\mu} = \frac{0.33}{1-0.33} = 0.492$$

$$= \frac{3.26-0.492\times77.5}{77.5-3.26\times0.492} = \frac{-34.91}{75.89} = -0.46$$

$$\varepsilon_3 = -\mu_2.\varepsilon_1 = 0.46 \times 401 \times 10^{-5}$$

$$= 184 \times 10^{-5}$$

$$\theta = \frac{1}{2}\tan^{-1}\left(\frac{2\tau_{xz}}{\sigma_z - \sigma_x}\right) = \frac{1}{2}\tan-1\left(\frac{2\times13.73}{45.7-6.64}\right) = 17.55°$$

$$[\theta_1 = 17.55°,\ \theta_3 = 107.55°]$$

$$\varepsilon_z = \varepsilon_1 \cos^2\theta_1 + \varepsilon_3 \cos^2\theta_3$$

$$= (401 \cos^2 17.55 + 184 \cos^2 107.55) \times 10^{-5}$$

$$= (364.5 + 16.73) \times 10^{-5} = 381.23 \times 10^{-5}$$

Example 4.2

Solve Example 4.1 assuming that the base of the footing is rough

Solution

(i) $q_u = c.N_c$

$= 75 \times 5.14$ [$N_c = 5.14$ for $\varphi = 0°$, Table A4.1]

$= 385.5\ \text{kN/m}^2$

Average pressure intensity, q

$$q = \frac{150\times1.5-\frac{1}{2}\times50\times1.5}{1.5} = 125\ \text{kN/m}^2$$

$$q_t = \frac{q}{q_u}.c$$

$$= \frac{125}{385.5}\times 75 = 24.3\ \text{kN/m}^2$$

(ii) Due to base roughness, the tangential stress will act as shown in Fig. 4.25. Due to symmetry, the value of vertical strain along the section passing through the point 'c'at depth 1.0 m will be the same as computed in Example 4.1. Effect of tangential stresses along this vertical section will nullify.

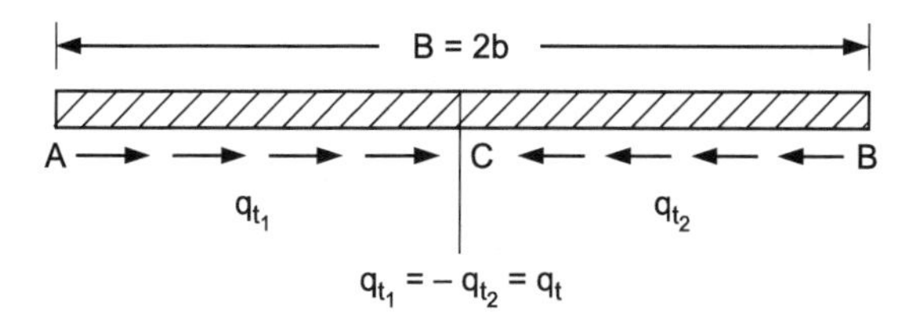

Fig. 4.25. Showing the tangential stress distribution.

Values of vertical strains along the vertical sections passing through points B and C will be the same and are therefore illustrated further for point 'B'.

For tangential load q_{t1}

$$b = \frac{1.5}{4} = 0.375 \text{ m}$$

$$x = \frac{1.5}{2} + \frac{1.5}{4} = 1.125 \text{ m}; z = 1.0 \text{ m}; q_{t1} = 24.3 \text{ kN/m}^2$$

Using Eqs. (3.2a), (3.2b) and (3.2c), we get:

$$\frac{\sigma_z}{q_{t1}} = 0.10578; \frac{\sigma_z}{q_{t1}} = 0.1273; \frac{\tau_{xz}}{q_{t1}} = 0.1138$$

or, $$\sigma_z = 2.57 \text{ kN/m}^2; \sigma_x = 3.09 \text{ kN/m}^2; \tau_{xz} = 2.77 \text{ kN/m}^2$$

For tangential load q_{t2}:

$$b = \frac{1.5}{4} = 0.375 \text{ m}$$

$$x = \frac{1.5}{2} = 0.375 \text{ m}; z = 1.0 \text{ m}; q_{t2} = -24.3 \text{ kN/m}^2$$

Using Eqs. (3.2a), (3.2b) and (3.2c), we get

$$\frac{\sigma_z}{q_{t1}} = 0.11459; \frac{\sigma_z}{q_{t1}} = 0.0274; \frac{\tau_{xz}}{q_{t1}} = 0.0520$$

or, $$\sigma_z = -2.78 \text{ kN/m}^2; \sigma_x = -0.67 \text{ kN/m}^2; \tau_{xz} = -1.26 \text{ kN/m}^2$$

Total stress due to tangential load:

$$\sigma_z = 2.57 - 2.78 = -0.21 \text{ kN/m}^2$$

$$\sigma_x = 3.09 - 0.67 = 2.42 \text{ kN/m}^2$$

$$\sigma_{xz} = 2.76 - 1.26 = 1.50 \text{ kN/m}^2$$

Considering the stresses due to normal loading from Example 4.1, the total stresses will be:

$$\sigma_z = 54.5 - 0.21 = 54.29 \text{ kN/m}^2$$

$$\sigma_x = 2.74 + 2.42 = 5.16 \text{ kN/m}^2$$

$$\sigma_{xz} = 13.73 + 1.50 = 15.23 \text{ kN/m}^2$$

$$\sigma_1 = \frac{(54.29 + 5.16)}{2} + \sqrt{\left(\frac{54.29 - 5.16}{2}\right)^2 + 15.23^2}$$

$$= 29.73 + 28.89 = 56.62 \text{ kN/m}^2$$

$$\sigma_3 = 29.73 - 28.89 = 0.84 \text{ kN/m}^2$$

$$\varepsilon_1 = \frac{a(\sigma_1 - \sigma_3)}{1 - b(\sigma_1 - \sigma_3)}$$

$$= \frac{6 \times 10^{-5}(56.62 - 0.84)}{1 - 0.006(56.62 - 0.84)} = 503 \times 10^{-5}$$

$$\mu_2 = \frac{\sigma_3 - \mu_1\sigma_1}{\sigma_1 - \mu_1\sigma_3}; \mu_1 = \frac{\mu}{1-\mu} = \frac{0.33}{1 - 0.33} = 0.492$$

Therefore,

$$\mu_2 = \frac{0.84 - 0.492 \times 56.62}{56.62 - 0.492 \times 0.84} = \frac{-27.02}{56.21} = -0.48$$

$$\varepsilon_3 = -\mu_2.\varepsilon_1 = -0.48 \times 5.3 \times 10^{-5}$$

$$= 241.8 \times 10^{-5}$$

$$\theta = \frac{1}{2}\tan^{-1}\left(\frac{2\tau_{xz}}{\sigma_z - \sigma_x}\right) = \frac{1}{2}\tan^{-1}\left(\frac{2 \times 15.23}{54.29 - 5.16}\right) = 31.8°$$

$$\theta_1 = 31.8°; \theta_3 = 121.8°$$

$$\varepsilon_z = \varepsilon_1 \cos^2\theta_1 + \varepsilon_3 \cos^2\theta_3$$

$$= (503 \cos^2 31.8 + 241 \cos^2 121.8) \times 10^{-5}$$

$$= (363.3 + 67.1) \times 10^{-5} = 430.4 \times 10^{-4}$$

Example 4.3

A rough-based flexible surface strip footing of width 2.0 m rests on cohesion-less soil and is subjected to a pressure intensity of 300 kN/m^2. The properties of sand are given below:

$$\gamma = 16.5 \text{ kN/m}^3, \varphi = 35°; \text{ value of Poisson ratio is 0.35}$$

Kondner's hyperbola constants: $\frac{1}{a} = 3.89 \times 10^3(\sigma_3)^{0.6}; \frac{1}{b} = 100 + 3.6\ (\sigma_3)$; σ_3 is in kN/m^2. Determine the strain at a point 2.0 m below the right-hand side edge of the footing.

Solution

(i) For $\varphi = 35°$, $N_\gamma = 48.03$ (Table A4.1)

$$q_u = \frac{1}{2}\gamma B N\gamma$$

$$= \frac{1}{2} \times 16.5 \times 2.05 \times 48.03$$

$$= 792.5 \text{ kN/m}^2$$

$$q = 300 \text{ kN/m}^2$$

$$F = \frac{q_u}{q} = \frac{792.5}{300} = 2.64$$

(ii) Stress along a vertical sectioin passing through the edge of footing

$$b = 1.0 \text{ m}, x = 1.0 \text{ m}, \therefore \frac{x}{b} = \frac{1.0}{1.0} = 1.0$$

$$z = 2.0 \text{ m}; \therefore \frac{z}{b} = \frac{2.0}{1.0} = 2.0$$

(a) For uniformly distributed load (as the footing is considered flexible):

For $\frac{x}{b} = 1.0$ and $\frac{z}{b} = 2.0$ (refer to Table 3.1)

$$\frac{\sigma_z}{q} = 0.4092; \frac{\sigma_z}{q} = 0.0908; \frac{\tau_{xz}}{q} = 0.1592$$

$$\sigma_z = 122.8 \text{ kN/m}^2; \sigma_x = 27.2 \text{ kN/m}^2; \tau_{xz} = 47.8 \text{ kN/m}^2$$

(b) For tangential load:

$$\delta = \tan^{-1}\left(\frac{2}{3}\tan\phi\right) = \tan^{-1}\left(\frac{2}{3}\tan 35\right) = \tan^{-1}(0.4668) = 25°$$

$$\delta_m = \tan^{-1}\left(\frac{300}{792.5}\tan\delta\right) = 17.3°$$

$$t_a = q\tan\delta_m = 300 \tan 17.3 = 93.4 \text{ kN/m}^2$$

Refer to Fig. 4.26a

$$b' = \frac{b}{2} = \frac{1.0}{2} = 0.5 \text{ m}$$

$$x = \frac{b}{2} + b = 1.5b = 3b'; \frac{x}{b'} = \frac{3b'}{b'} = 3$$

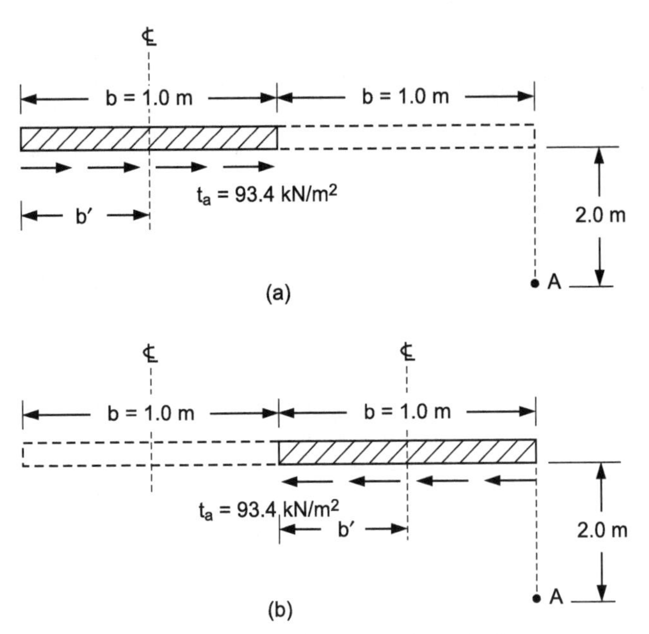

Fig. 4.26. (a) Tangential stress acting in the left hand half portion of the footing, (b) Tangential stress acting in the right hand half portion of the footing.

For $\frac{x}{b'} = \frac{3b'}{b'} = 3; \frac{z}{b'} = \frac{2.0}{0.5} = 4$ (using Eqs.3.2a, 3.2b, 3.2c)

$$\frac{\sigma_z}{t_a} = 0.0954; \ \frac{\sigma_x}{t_a} = 0.0541; \ \frac{\tau_{xz}}{t_a} = 0.07, \ t_a = 93.4 \text{ kN/m}^2$$

$$\sigma_z = 8.31 \text{ kN/m}^2; \ \sigma_x = 5.05 \text{ kN/m}^2; \ \tau_{xz} = 6.54 \text{ kN/m}^2;$$

Refer to Fig. 4.26b

$$x = -b' = -0.5 \text{ m}, \ b' = 0.5 \text{ m}$$

$$\frac{x}{b'} = \frac{-0.5}{0.5} = -1; \frac{z}{b'} = \frac{2.0}{0.5} = 4 \text{ (using Table 3.2)}$$

$$\frac{\sigma_z}{t_a} = -0.06366; \ \frac{\sigma_x}{t_a} = -0.00737; \frac{\tau_{xz}}{t_a} = 0.02026$$

(here $t_a = 93.4 \text{ kN/m}^2$)

$$\sigma_z = -5.94 \text{ kN/m}^2; \ \sigma_x = -0.69 \text{ kN/m}^2; \ \tau_{xz} = 1.89 \text{ kN/m}^2$$

(iii) Total stresses:

$$\sigma_z = 122.8 + 8.91 - (-5.94) = 137.65 \text{ kN/m}^2$$

$$\sigma_x = 27.25 + 5.05 - (-0.69) = 32.94 \text{ kN/m}^2$$

$$\tau_{xz} = 47.8 + 6.54 - (1.89) = 52.45 \text{ kN/m}^2$$

$$\sigma_1 = \frac{(137.65 + 32.94)}{2} + \sqrt{\left(\frac{137.65 - 32.94}{2}\right)^2 + 52.45^2}$$

$$= 85.29 + 74.10 = 159.39 \text{kN/m}^2$$

$$\sigma_3 = 85.29 - 74.10 = 11.19 \text{ kN/m}^2$$

(iv) Computation for evaluating parametars a and b:

$$K_0 = 1 - \sin\phi = 1 - \sin 35 = 0.426$$

$$\sigma_z = 137.65 + 16.5 \times 2.0 = 170.65 \text{ kN/m}^2$$

$$\sigma_x = 32.94 + 0.426 \times 16.5 \times 2.0 = 47.0 \text{ kN/m}^2$$

$$\tau_{xz} = 52.45 \text{ kN/m}^2$$

$$\sigma_1' = \frac{(170.65 + 47.0)}{2} + \sqrt{\left(\frac{170.65 - 47.0}{2}\right)^2 + 52.45^2}$$

$$= 108.82 + 81.07 = 189.89 \text{ kN/m}^2$$

$$\sigma_3' = 108.82 - 81.07 = 27.75 \text{ kN/m}^2$$

$$\frac{1}{a} = 3.89 \text{ x } 10^{-3} (27.75)^{0.6} \text{ kN/m}^2 = 28.57 \times 10^3 \text{ kN/m}^2$$

$$a = 0.035 \times 10^{-3} \text{ m}^2/\text{kN} = 3.5 \times 10^{-5} \text{ m}^2/\text{kN};$$

$$\frac{1}{b} = 100 + 3.6\,\sigma_3 = 100 + 3.6 \times 27.75 = 200 \text{ m}^2/\text{kN}$$

$$b = 0.005 \text{ m}^2/\text{kN}$$

(v) $a = 3.5 \times 10^{-5} \text{ m}^2/\text{kN}$; $b = 0.005 \text{ m}^2/\text{kN}$

$$E_s = \frac{1 - \dfrac{0.005 \times 200}{2.64}}{3.5 \times 10^{-5}} = 0.1775 \times 10^5 \text{ kN/m}^2$$

$$\varepsilon_1 = \frac{(\sigma_1' - \sigma_3')}{E_s} = \frac{189.89 - 27.75}{0.1775 \times 10^5}$$

$$= 913 \times 10^{-5}$$

$$\mu_2 = \frac{\sigma_3' - \mu_1\sigma_1'}{\sigma_1' - \mu_1\sigma_3'}; \mu_1 = \frac{\mu}{1-\mu} = \frac{0.35}{1-0.35} = 0.538$$

$$= \frac{27.75 - 0.538 \times 189.89}{189.89 - 0.538 \times 27.75} = \frac{-74.41}{174.96} = -0.425$$

$$\varepsilon_3 = -\mu_2.\varepsilon_1 = 0.425 \times 913 \times 10^{-5}$$
$$= 388 \times 10^{-5}$$
$$\theta = \frac{1}{2}\tan^{-1}\left(\frac{2\tau_{xz}}{\sigma_z - \sigma_x}\right) = \frac{1}{2}\tan^{-1}\left(\frac{2 \times 52.45}{170.65 - 47.0}\right)$$
$$= \frac{1}{2}\tan^{-1}\left(\frac{104.90}{123.65}\right) = 20.15°;\ \text{so}\ \theta_1 = 20.15°;\ \theta_3 = 110.15°$$
$$\varepsilon_z = \varepsilon_1 \cos^2\theta_1 + \varepsilon_3 \cos^2\theta_3$$
$$= 913 \times 10^{-5} \cos^2 20.15 + 388 \times 10^{-5} \cos^2 110.15$$
$$= 804 \times 10^{-5} + 46 \times 10^{-5} = 820 \times 10^{-5}$$

Practice Problems

1. What are the situations for the suitability of shallow foundations? Explain its design requirements.
2. Give stepwise the procedure of obtaining pressure-settlement characteristics of a smooth flexible footing resting on saturated clay.
3. What will be the modification in the procedure given in Prob. 2 if the footing base is rough.
4. Describe the method of obtaining contact pressure in the case of rigid footing resting on clay.
5. Explain stepwise the procedure of obtaining pressure-settlement characteristics of the footing resting on sand.
6. A surface strip footing of width 2 m rests on clay. It is subjected to pressure as shown in Fig. 4.27. Determine the vertical strain in the soil along the vertical sections passing through the centre and the edges of the footing and at depth 1.0 m below the base of the footing. Values of Kondner's hyperbola constant are $a = 4 \times 10^{-5}$ m^2/kN; $b = 0.003$ m^2/kN; value of Poisson ratio is 0.35.

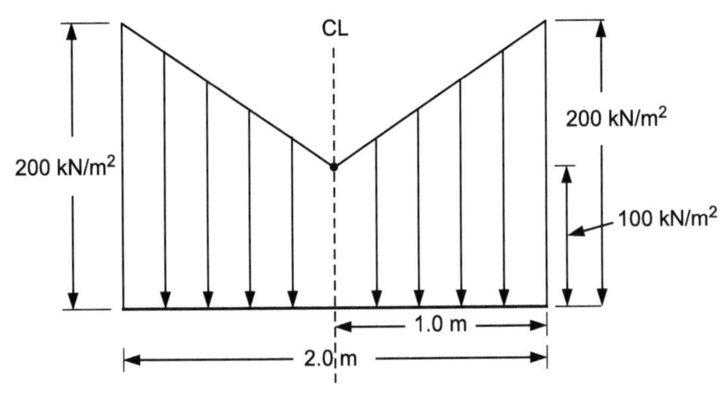

Fig. 4.27. Showing pressure distribution to be adopted in practice problem 6.

7. A surface strip footing of width 1.5 m rests on cohesion-less soil and subjected to a pressure intensity of 200 kN/m^2. The properties of sand are given below:

$$\gamma = 16 \text{ kN/m}^3, \varphi = 30°; \text{ value of Poisson ratio is } 0.33$$

Kondner's hyperbola constants: $\frac{1}{a} = 5 \times 10^3 (\sigma_3)^{0.7}$; $\frac{1}{b} = 75 + 4 (\sigma_3)$; σ_3 is in kN/m^2.

Determine the strain at a point 2.5 m below the right-hand side edge of the footing.

REFERENCES

Balla, A. (1962). Bearing capacity of foundations. *Journal of Soil Mechanics and Foundation Division,* ASCE, **88(5):** 13-36.

Basavanna, B.M. (1975). Bearing Capacity of Soil under Static and Transient Loads. Ph.D. Thesis, IIT Roorkee, Roorkee.

Biswas, T., Saran, S. and Shankar, D. (2016). Analysis of a Strip Footing Using Constitutive Law. *Geosciences,* **6(2):** 41-44.

Biswas, T. (2017). Pseudo-static Analysis of a Strip Footing Using Constitutive Laws of Soils. Ph.D. Thesis, IIT Roorkee, Roorkee.

Desai, C.K. (1971). Nonlinear Analysis Using Spline Functions. *J. Soil Mech. Found. Div.,* ASCE, **97(10):** 1461-1480.

Kalinli, A., Acar, M.C. and Gündüz, Z. (2011). New approaches to determine the ultimate bearing capacity of shallow foundations based on artificial neural networks and ant colony optimization. *Engineering Geology,* **117:** 29-38.

Kumar, J. and Mohan Rao, V.B.K. (2002). Seismic bearing capacity factors for spread foundations. *Géotechnique,* **52(2):** 79-88.

Majidi, A. and Mirghasemi, A. (2008). Seismic 3D-bearing capacity analysis of shallow foundations. *Iranian Journal of Science and Technology,* **32:** 107-124.

Meyerhof, G.G. (1951). The ultimate bearing-capacity of foundations. *Géotechnique,* **2(4):** 301-332.

Saran, S. (1970). Fundamental fallacy in analysis of bearing capacity of soil. *Journal of Institute of Engineers,* **50:** 224-227.

Shafiee, A.H. and Jahanandish, M. (2010). Seismic bearing-capacity factors for stripfootings. *Proceedings of 5th National Congress on Civil Engineering,* Mashhad, 4-6.

Sharan, U.N. (1977). Pressure Settlement Characteristics of Surface Footings Using Constitutive Laws. Ph.D Thesis. University of Roorkee, Roorkee.

Terzaghi, K. (1943). Theoretical Soil Mechanics. John Wiley and Sons Inc., New York.

Vesic, A.S. (1973). Analysis of ultimate loads of shallow foundations. *Journal of the Soil Mechanics and Foundations Division,* ASCE **99(1):** 45-73.

Zhu, D.Y., Lee, C.F. and Law, K.T. (2003). Determination of bearing-capacity of shallow foundations without using superposition approximation. *Canadian Geotechnical Journal*, **40:** 450-459.

Appendix 4.1

Table A4.1: Bearing capacity factors

Φ	N_c	N_q	N_a
0	5.14	1.00	0.00
5	6.49	1.57	0.45
10	8.35	2.47	1.22
15	10.98	3.94	2.65
20	14.83	6.40	5.39
25	20.72	10.66	10.88
30	30.14	18.40	22.40
35	46.12	33.30	48.03
40	75.31	64.20	109.41
45	138.88	134.88	271.76
50	266.89	319.07	762.89

CHAPTER 5

Strip Footing Subjected to Eccentric-inclined Load

5.1 General

A foundation engineer frequently comes across the problem of footings subjected to eccentric-inclined load, e.g. in the case of foundations of retaining walls, abutments, columns, stanchions, tall building, portal framed structures, etc. In these cases, footings in general are acted upon a vertical load, a moment and a horizontal load which result into an eccentric-inclined load. Horizontal loads may be caused by wind pressures, water pressures, seismic forces, etc. Such footings get tilted and as such maximum settlement occurs at the edge of footing. Also these footings get horizontally displaced. A typical settlement pattern is shown in Fig. 5.1. A footing subjected to eccentric-inclined load shows unsymmetrical settlement pattern alongwith horizontal displacement.

Bearing capacity, settlement, tilt and horizontal displacement are the main criteria for proportioning of such footings. All these are obtained in independent steps.

Studies for evaluation of bearing capacity of a footing subjected to eccentric-inclined load can be grouped under: (a) footings subjected to eccentric-vertical loading (Meyerhof, 1953; Prakash and Saran, 1971), (b) footings subjected to central-inclined loading (Schultz, 1952; Meyerhof, 1953; Hjiaj et al., 2004), and (c) footings subjected to eccentric-inclined loading (Saran and Agrawal, 1989; Mahiyar and Patel, 2000; Lau and Boltan, 2011a, 2011b; Patra et al., 2012; Loukidis et al., 2008).

Little attention has been given to obtain the settlement, tilt and horizontal displacement (Agarwal, 1986). They have developed non-

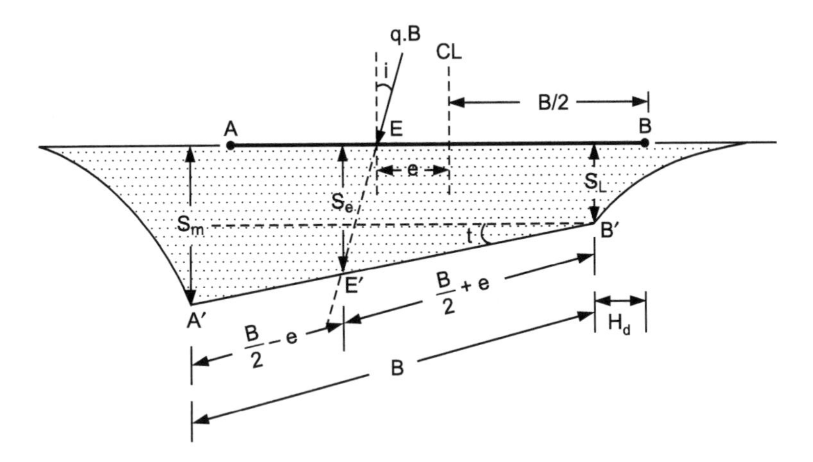

Fig. 5.1. Settlement and horizontal displacement of eccentric inclined loaded rigid strip footing.

dimensional correlation for their evaluation. These correlations are based on a large number of model tests.

In this chapter, procedures have been developed to predict pressure, maximum settlement, pressure tilt and pressure, horizontal displacement curves for the footings subjected to eccentric-inclined load. Both the cases i.e. footings in clays and footings in sand have been considered (Agarwal, 1986).

5.2 Footing on Clay

All the assumption made in the analysis of footing subjected to central-vertical load hold good in this case except that the contact pressure distribution is considered as shown in Fig. 5.2. Salient features of the procedure to evaluate pressure-settlement and pressure tilt characteristics have been described below in the following steps for a typical case (smooth rigid footing).

5.2.1 Vertical Settlement and Tilt

Step 1

Contact pressure distribution as shown in Fig. 5.2(a) is described by three pressure coefficients—$\alpha\alpha_1$, $\alpha\alpha_2$ and $\alpha\alpha_3$. These coefficients are evaluated by satisfying equilibrium conditions as given below:

(i) The total load on a footing equals the area of the contact pressure diagram (Fig. 5.2b).

$$q(a+b)\cos i = \alpha\alpha_1.q.\cos^2 i(a+b) + \tfrac{1}{2}.a.\alpha\alpha_2.q.\cos^2 i + \tfrac{1}{2}.b.\alpha\alpha_3.q.\cos^2 i \quad (5.1)$$

(ii) The pressure distribution diagram is resolved into two parts vertical and horizontal components (Figs 5.2b, 5.2c), such that the vector sum of the area of these two diagrams (Figs 5.2b, 5.2c), should be equal to the area of the contact pressure diagram (Fig. 5.2a).

Taking moment of the area about point E (Fig. 5.2b),

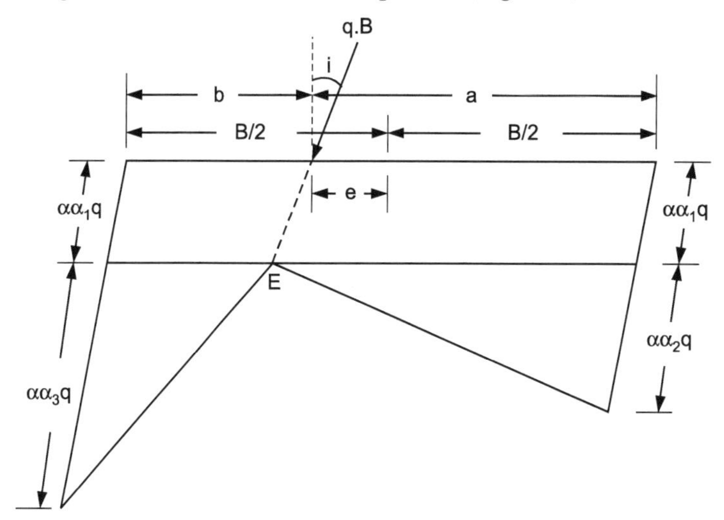

Fig. 5.2(a). Contact-pressure distribution for eccentric inclined load over rigid strip footing

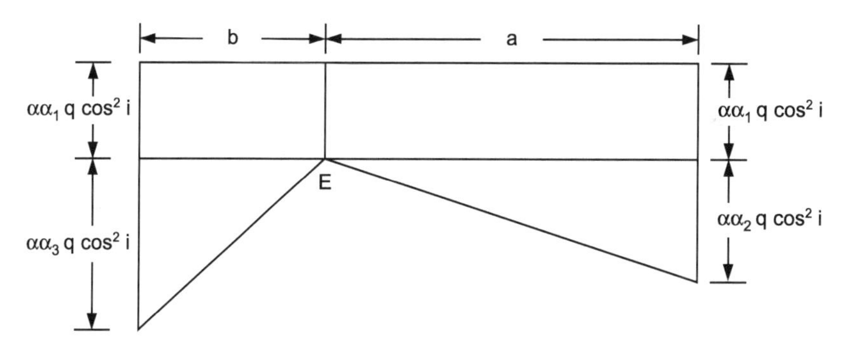

Fig. 5.2(b). Vertical component of pressure diagram

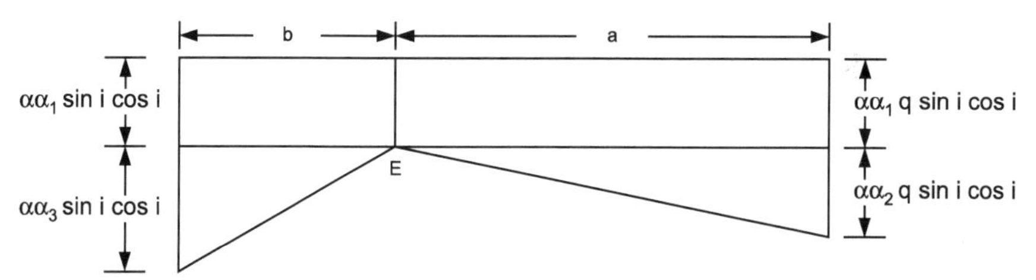

Fig. 5.2(c). Horizontal component of pressure diagram

$$\alpha\alpha_1.q.\cos^2 i.\frac{b^2}{2}+\tfrac{1}{2}.\alpha\alpha_3.q.\cos^2 i+\frac{2b^2}{3}=\alpha\alpha_1.q.\cos^2 i.\frac{a^2}{2}+\alpha\alpha_2.q.\cos^2 i.\frac{a^2}{3} \tag{5.2}$$

Solving the Eqs (5.1) and (5.2) in term of $\alpha\alpha_1$, we get

$$\alpha\alpha_2=\frac{4b-(3a+b)\alpha\alpha_1.cosi}{2acosi} \tag{5.3}$$

$$\alpha\alpha_2=\frac{4a-(a+3b)\alpha\alpha_1.cosi}{2bcosi} \tag{5.4}$$

(iii) The slopes of the lines A′E′ and B′E′ are assumed as being same (Fig. 5.1)

$$\frac{S_m-S_e}{\frac{B}{2}-e}=\frac{S_m-S_L}{\frac{B}{2}+e} \tag{5.5}$$

where e = Eccentricity of the load
S_m = Maximum settlement of the footing
S_e = Settlement of the point of application of load
S_L = Minimum settlement of the footing
B = Width of the footing

(iv) When the eccentricity width ratio (e/B) is equal to zero, i.e. the load is at the centre of the footing, the value of pressure coefficient $\alpha\alpha_1$ was evaluated, considering the settlements at the three different vertical sections ($x = 0, B/2$ and B) being equal.

Step 2

Stresses in each layer of the soil mass (Fig. 5.3) at three vertical sections (at edges and at point E) due to vertical component 'q.cos i' and horizontal component 'q.sin i' of eccentric-inclined load are obtained. Super-position of stresses is done to get the stresses due to the eccentric-inclined load. Equations used to obtain stresses in each layer along any vertical section for different types of loading have already been given in Chapter 3. These stresses are designated as σ_z, σ_x and τ_{xz}.

Step 3

Principal stresses σ_1 and σ_3 and their directions with vertical axis θ_1 and θ_3 are obtained by computing Eqs (4.2) to (4.4).

Step 4

When strip footing is case of a plane strain condition, the strain ε_2 in the direction of intermediate principal stress σ_2 is equal to zero in the plane strain.

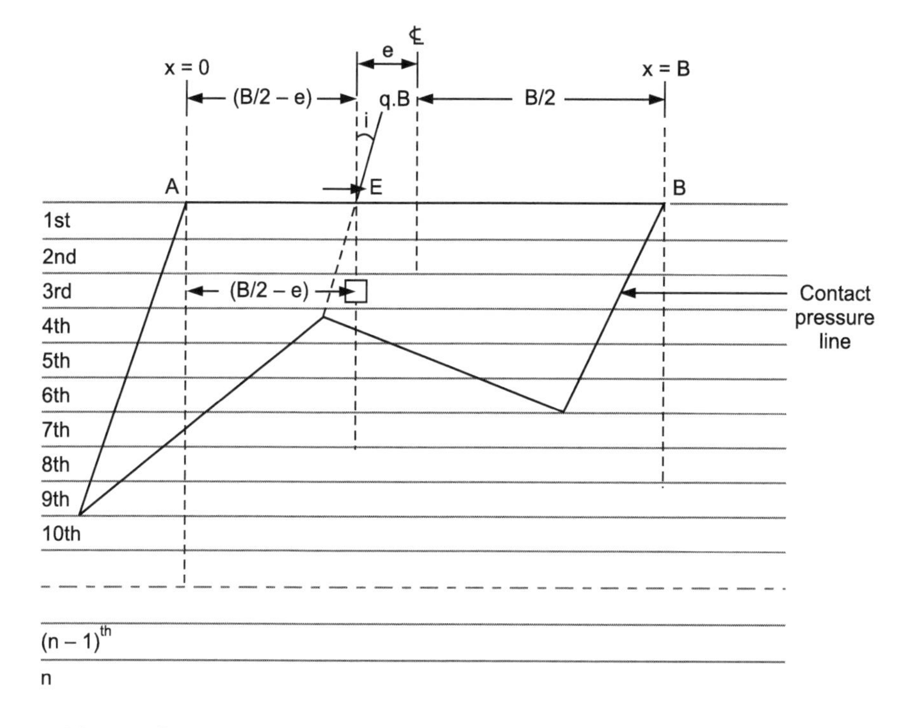

Fig. 5.3. Contact pressure profile and soil below footing divided in *n*-layers.

$$\varepsilon_2 = 0 = \frac{1}{E}\,[\sigma_2 - \mu(\sigma_1 + \sigma_3)] \tag{5.6a}$$

Or

$$\sigma_2 = \mu\,(\sigma_1 + \sigma_3) \tag{5.6b}$$

Step 5

The strains in the direction of major principal and minor principal computed from the constitutive relationship are given below:

$$\varepsilon_2 = \frac{a(\sigma_1 - \sigma_3)}{1 - b(\sigma_1 - \sigma_3)} \tag{5.7a}$$

$$\varepsilon_3 = -\,\mu_2 \varepsilon_1 \tag{5.7b}$$

where a and b are constants of hyperbola (Kondner, 1963) and μ_2 is as given in Eq. (4.11).

Step 6

The strains in the vertical direction (ε_z) in each layer along the vertical sections are computed using the following expression.

$$\varepsilon_z = \varepsilon_1 \cos^2\theta_1 + \varepsilon_3 \cos^2\theta_3 \tag{5.7c}$$

Step 7

The total settlement along any vertical section is obtained by numerically integrating the settlement of each thin strip as follows:

$$S = \int_0^n \varepsilon_z . \Delta z \tag{5.8}$$

where S = Total settlement along any vertical section
ε_z = Strain in the vertical direction (i.e. z-direction)
Δz = thickness of each strip

Step 8

For a given e/B ratio and load inclination (i), the settlement (S_e) at the point of application of load, settlements S_m and S_L (at the edges) are computed for different values of contact pressure coefficients $\alpha\alpha_1$ with a particular load intensity 'q'. The value of $\alpha\alpha_1$ has been taken which satisfied the condition (iii) of Step 1 and the corresponding values of settlements S_m, S_e and S_L are noted.

To get the complete pressure versus settlement curve for a given e/B and i, the load intensity is varied and the set of settlements (S_m, S_e and S_L) are computed. When eccentricity-width ratio (e/B) is equal to zero for a given inclination of load (i), settlements at three different vertical sections—$x = 0$, $x = B/2$ and $x = B$ are obtained for different values of $\alpha\alpha_1$ at a particular load intensity. The value of $\alpha\alpha_1$ is picked up which gives equal settlements of footing at three vertical sections.

Step 9

Tilt (t) of the footing for a known value of e/B ratio, load inclination (i) and load intensity (q) is computed using the following expression (Fig. 5.1):

$$\sin t = \frac{S_m - S_e}{\frac{B}{2} - e} \tag{5.9}$$

5.2.2 Horizontal Displacement

When load is eccentric-inclined on a rigid strip footing, there is horizontal displacement of the footing also alongwith vertical settlement of the footing. The procedure to obtain horizontal displacements of such type of footings is given in the following steps:

Step 1

A thin strip of soil layer just below the footing is taken towards the direction of the horizontal component of the inclined load. The length of this strip of soil has been taken equal to 5 times the width of the footing from the edge of the footing. This thin strip has been divided into n-elements as shown in Fig. 5.4.

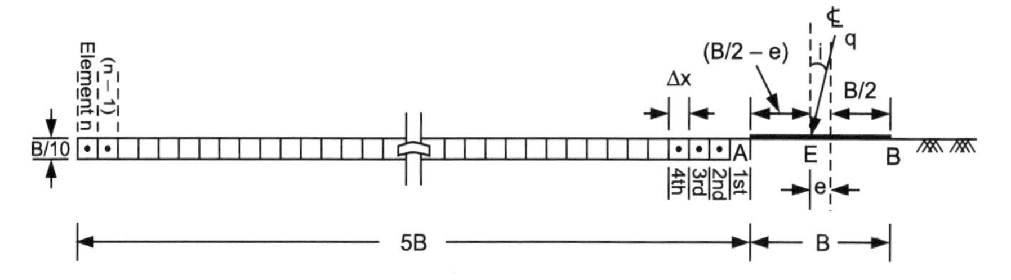

Fig. 5.4. Thin soil layer below rigid strip footing divided into n-number of elements. Length of thin layer 5 times width of footing from the edge of footing.

Step 2

To calculate stresses in the centre of each element for a given e/B ratio, load inclination (i) and load intensity (q), the contact pressure coefficients '$\alpha\alpha_1$' has been taken which satisfied the condition (iii) of Step 1 in section 5.2.1. When eccentricity-width ratio (e/B) is zero, then for a given load inclination (i) and load intensity (q), the value of $\alpha\alpha_1$ which satisfied the condition (iv) of Step 1 in Section 5.2.1 is taken.

Step 3

Stresses σ_z, σ_x, τ_{xz} at the centre of each element are obtained using the stress equations given in Chapter 3.

Step 4

Principal stresses and their directions were then computed in usual way. Principal strains which are obtained as explained in Step 5 of Section 5.2.1.

Step 5

Strain in the direction of the horizontal component of the load (ε_x) has been computed using the following expression:

$$\varepsilon_x = \varepsilon_1 \sin^2 \theta_1 + \varepsilon_3 \sin^2 \theta_3 \tag{5.10}$$

Step 6

Horizontal displacement of the footing (H_d) in the direction of the horizontal component of load is obtained by numerically integrating the displacement of each element in the horizontal domain

$$H_d = \int_0^n \varepsilon_x . \Delta x \tag{5.11}$$

where H_d = Total horizontal displacement of footing

ε_x = Strain in the horizontal direction (i.e. x-direction)

Δx = Width of the element in X-direction

Step 7

The total horizontal displacements in the direction of the horizontal components of the load for the given load inclination (i), e/B ratio for different load intensities are obtained by repeating Steps 2 to 6.

5.2.3 Settlement and Horizontal Displacement for Buckshot-Clay

Pressure versus settlement (S_e), pressure versus maximum settlement (S_m) and pressure versus horizontal displacement curves were obtained for Buckshot clay using the procedures described in the previous sections (Agarwal, 1986). The properties of Buckshot clay has already been described in Chapter 2.

Typically as a pressure-settlement (S_e) curves for eccentricity-width ratios (e/B) equal to 0.1 and 0.2 are shown in Figs 5.5 and 5.6 for various load inclination. Similarly pressure versus maximum settlement curves (S_m) are given in Figs 5.7 and 5.8. Typical pressure versus horizontal displacement curves are shown in Fig. 5.9.

A close observation of Figs 5.5 to 5.9 indicates that obvious findings with the increase in eccentricity e and load inclination i settlement S_e and S_m, and also the horizontal displacement (H_d) increase.

5.3 Footing on Sand

As mentioned in chapter 4 the procedure adopted for the analysis of strip footing resting on clay cannot be applied in sands directly as its modulus of elasticity varies with depth.

5.3.1 Vertical Settlement and Tilt

The procedure adopted for obtaining the pressure-settlement and pressure-tilt characteristics of a footing resting on sand is given in the following steps:

Step 1

In this case, contact pressure is taken as shown in Fig. 5.10a for the vertical component of the eccentric-inclined load and is defined by two pressure coefficients, $\alpha\alpha_1$ and $\alpha\alpha_2$. These coefficients are evaluated by satisfying equilibrium condition: ($\Sigma V = 0$):

$$(a+b)\alpha\alpha_1 q\cos^2 i + \left(\frac{a+b}{2}\right)\alpha\alpha_1 q\cos^2 i = q(a+b).\cos i \tag{5.12}$$

$$\alpha\alpha_2 = \frac{2(1-\alpha\alpha_1\cos i)}{\cos i} \tag{5.13}$$

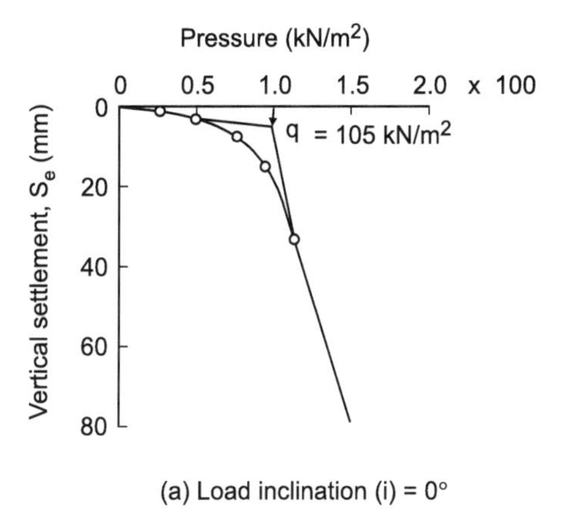

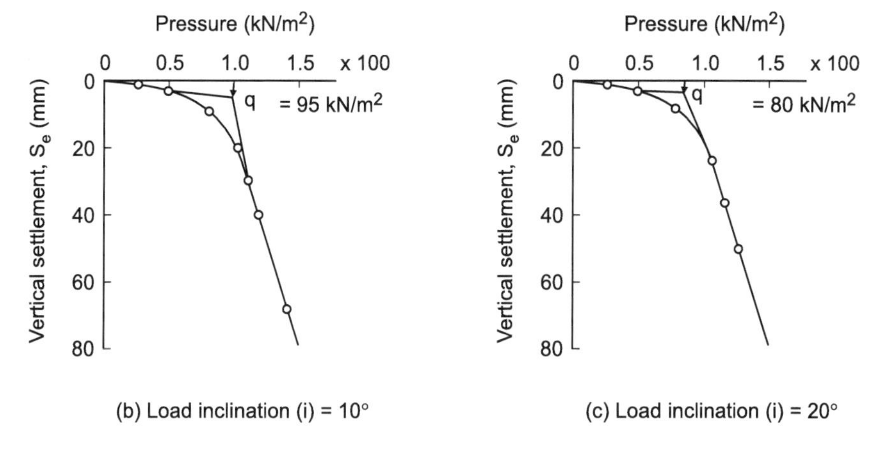

Fig. 5.5. Pressure versus vertical settlement (S_e) curves for rigid strip footing (B = 250 mm) resting on Buckshot clay for e/B = 0.1.

Step 2

Base tangential stress diagram due to horizontal component of the resultant eccentric-inclined load is shown in Fig. 5.10b.

Step 3

Assume a suitable value of $\alpha\alpha_1$ and compute $\alpha\alpha_2$ using Eq. (5.13).

Step 4

Stresses σ_z, σ_x and τ_{xz} were obtained at the desired point due to contact pressure as shown Figs. 10(a) and 10(b) using relevant equations are given in Chapter 3.

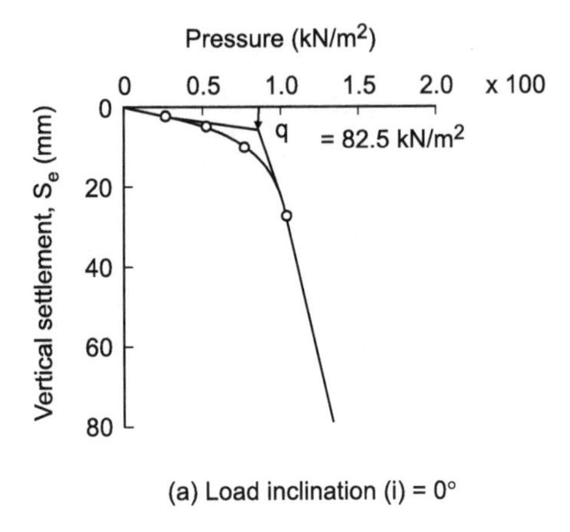

(a) Load inclination (i) = 0°

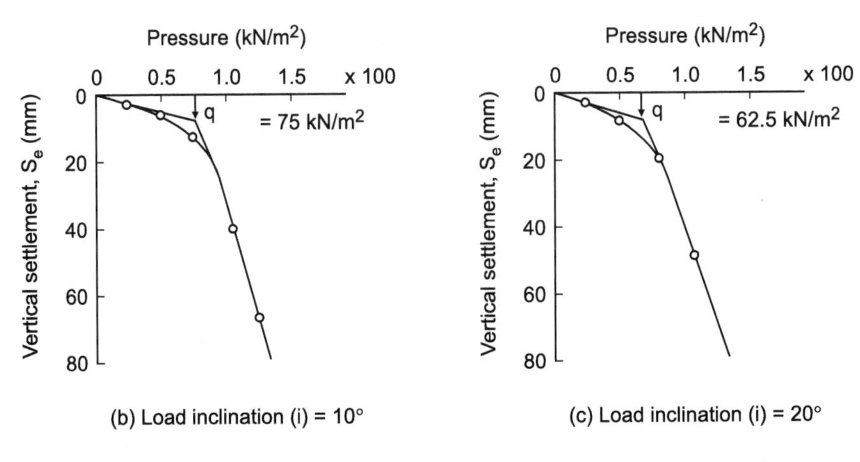

(b) Load inclination (i) = 10°

(c) Load inclination (i) = 20°

Fig. 5.6. Pressure versus vertical settlement (S_e) curves for rigid strip footing (B = 250 mm) resting on Buckshot clay for e/B = 0.2.

Step 5

The value of Poisson's ratio μ has been obtained using the equation:

$$\mu = \frac{K_0}{1+K_0} \tag{5.14}$$

where K_0 = 1-sin φ

φ = angle of internal friction of sand.

Step 6

Stresses due to overburden are added in stresses σ_z and σ_x as given below:

$$\sigma'_z = \sigma_z + \gamma z$$

$$\sigma'_x = \sigma_x + K_0 \gamma z$$

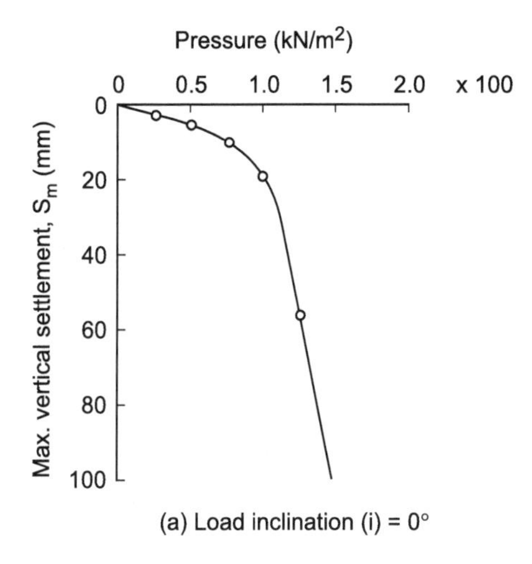

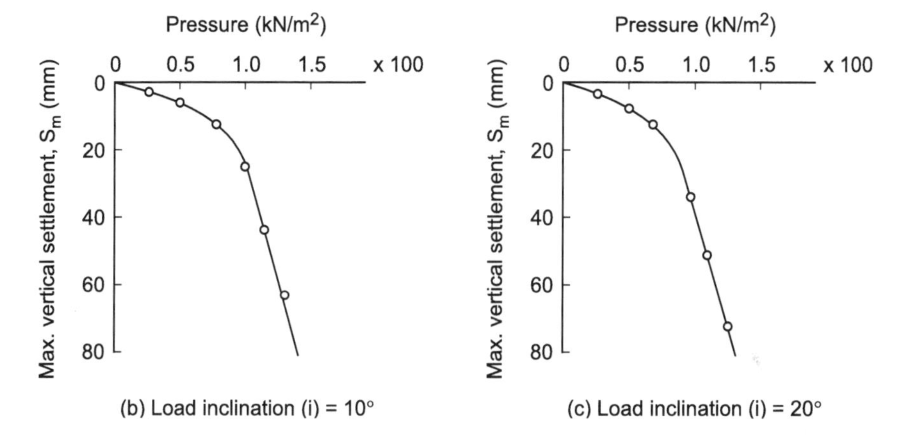

Fig. 5.7. Pressure versus maximum settlement (S_m) curves for rigid strip footing (B = 250 mm) resting on Buckshot clay for e/B = 0.1.

Using stresses σ'_x, σ'_z and τ_{xz}, corresponding values of principal stresses σ_1' and σ_3', along with their directions with vertical (θ_1' and θ_3') were obtained in the usual way.

Step 7

The ultimate pressure (q_u) has been computed from the following equation:

$$q_u = \frac{1}{2}\gamma B N_\gamma + \gamma . D_f . N_q \tag{5.15}$$

where γ = unit weight of sand

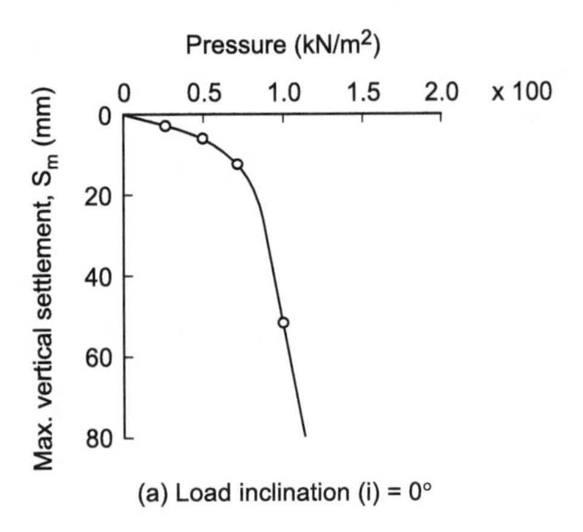

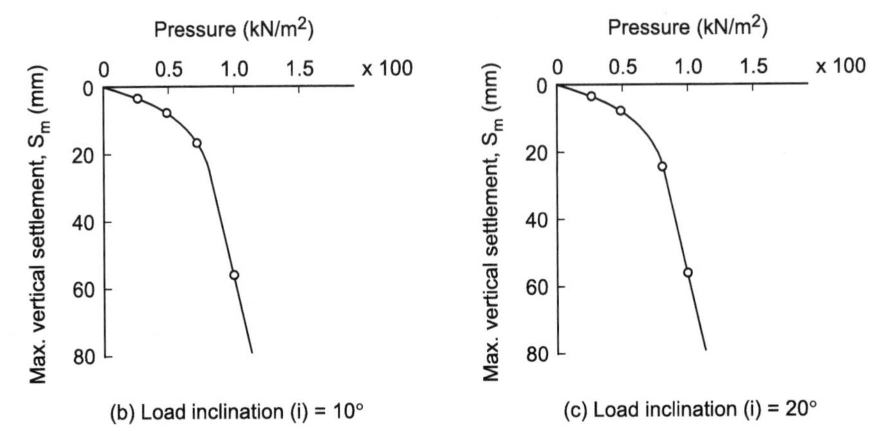

Fig. 5.8. Pressure versus maximum vertical settlement (S_m) curves for rigid strip footing (B = 250 mm) resting on Buckshot clay for e/B = 0.2.

D_f = depth of the footing and

B = width of the footing

N_γ and N_q = Bearing-capacity factors depending on e/B ratio and load inclination i (Appendix A5.1).

Step 8

The factor F for the given surface load intensity (q) has been obtained from the following relationship:

The ultimate bearing pressure (q_u) is calculated from Terzaghi's (1943) equations.

$$\frac{q_u}{q} = F \tag{5.16}$$

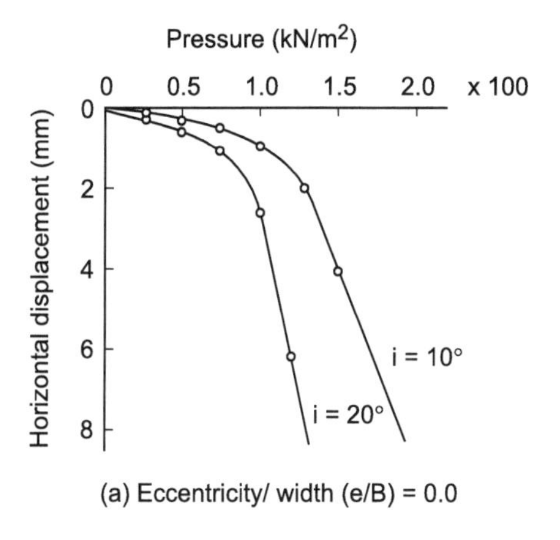

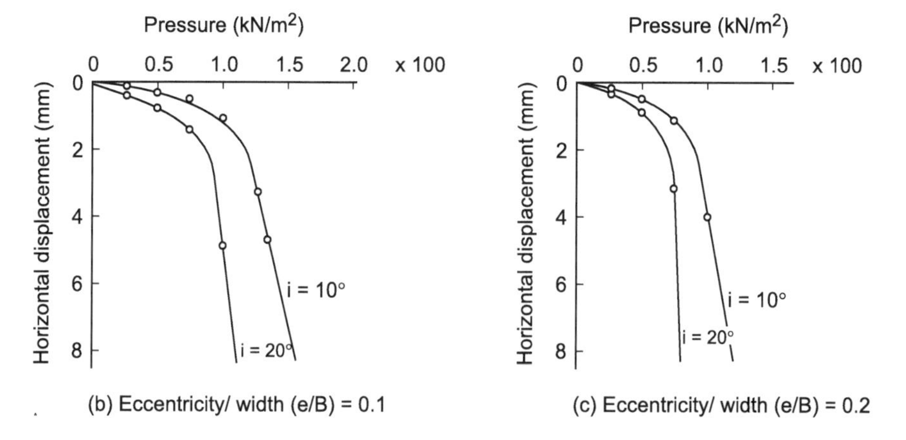

Fig. 5.9. Pressure versus horizontal displacement curves for rigid strip footing (B = 250 mm) resting on Buckshot clay.

Step 9

Modulus of elasticity (E_s) has been calculated by the following equation

$$E_s = \frac{1 - b(\sigma_u / F)}{a} \tag{5.17}$$

where a and b are the constants of a hyperbola whose values depend upon confining pressure (Chapter 2).

Step 10

The strain in each layer in the direction of major principal stress is calculated from the equation

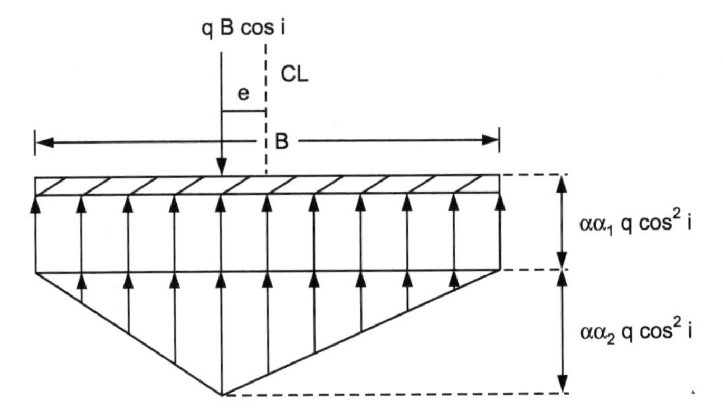

(a) Base contact pressure for vertical component of eccentric inclined load

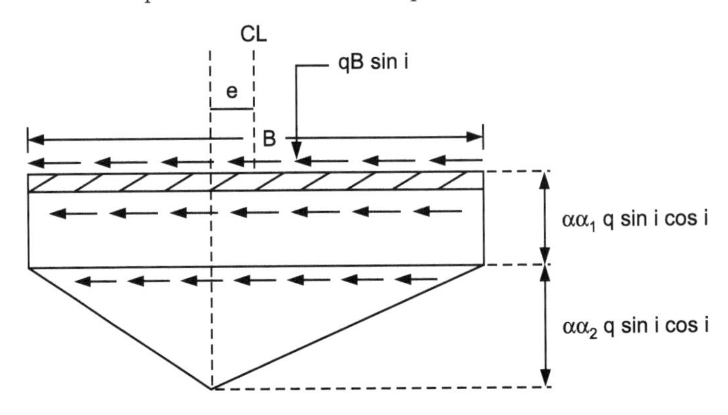

(b) Base contact pressure for horizontal component of eccentric inclined load

Fig. 5.10a, b. Contact pressure distributions.

$$\in_1 = \frac{\sigma_1' - \sigma_3'}{E_s} \tag{5.18}$$

The strain in the direction of minor principal stress is calculated from the following relationship:

$$\in_3 = -\mu_2 \in_1 \tag{5.19}$$

μ_2 is obtained using Eq. (4.11) of Chapter 4.

Step 11

The strain in the vertical direction is calculated using the following relations:

$$\varepsilon_z = \varepsilon_1 \cos^2 \theta_1' + \in_3 \cos^2 \theta_3' \tag{5.20}$$

Step 12

The total settlements and tilt of the footing for the given *e/B* ratio and load inclination (*i*) for different load intensities are obtained using the Steps 7 to 9 as described in Section 5.2.1 for clays.

5.3.2 Horizontal Displacement

The procedure adopted to compute pressure-horizontal displacement characteristics is discussed below.

Contact pressure coefficients ($\alpha\alpha_1$ and $\alpha\alpha_2$, Fig. 5.10) have been used as obtained in Section 5.3.1. Rest of the procedure is same as discussed in Section 5.2.2.

5.3.3 Settlement and Horizontal Displacement of a Footing Subjected to Eccentric-inclined Load and Resting on Sand

Pressure-settlement (S_e), pressure versus maximum settlement (S_m) and pressure versus horizontal displacement (H_d) characteristics were obtained for footing widths 300 mm and 600 mm resting on Ranipur sand for different eccentricities and inclinations of load using the procedures described in Sections 5.3.1 and 5.3.2. Properties of Ranipur sand already given in Chapter 2 for relative density 84% have been used. Typical set of curves are given in Figs 5.11 to 5.16. These figures indicated that all S_e, S_m and H_d increase with the increase in *e/B* ratio and load inclination *i*. Agarwal (1986) on the basis of extensive experimental study showed that the analytical procedures predict settlement (S_e), maximum settlement (S_m) and horizontal displacement (H_d) upto about half the ultimate pressure of the footing under satisfactory.

Illustrative Examples

Example 5.1

A column footing of width 1.75 m resting on clay is subjected to a total vertical load of 1000 kN, moment of 175 kN-m and a shear load of 180 kN. Values of Kondner's hyperbola constants are: $a = 10^{-5}$ m^2/kN; $b = 0.002$ m^2/kN

Adopt Poisson ratio as 0.46.

Determine the values of vertical strains at depth 2.0 m at the edges and at the point where the resultant inclined load acts on the footing.

Solution

(i) $B = 1.75$ m

$M = 175$ kN-m

$V = 1000$ kN

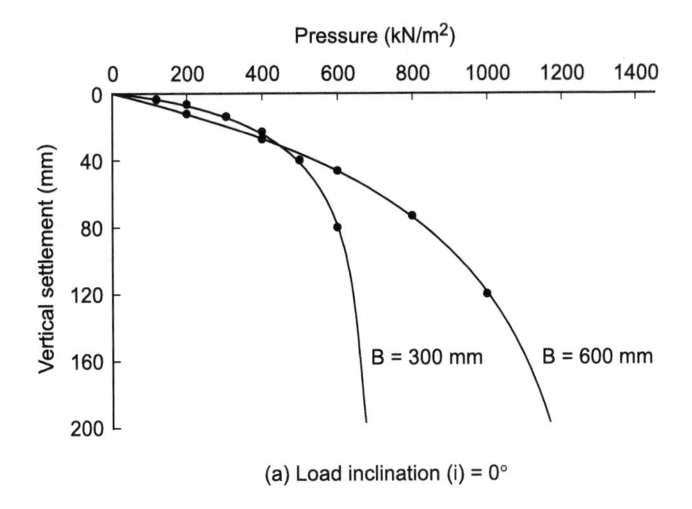

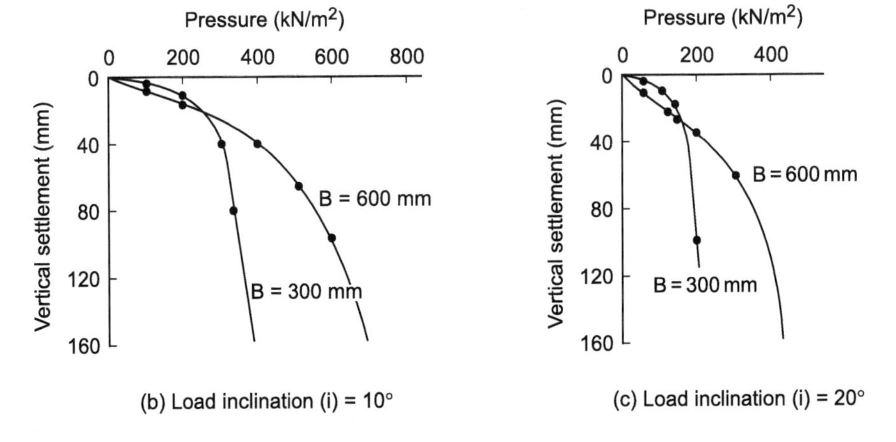

Fig. 5.11. Pressure versus vertical settlement curves for rigid strip footing (B = 300 mm) resting on Ranipur sand (D_R = 84%) for e/B = 0.0.

$$q = \frac{V}{B} = \frac{1000}{1.75} = 571.4 \text{ kN/m}^2$$

$$e = \frac{M}{V} = \frac{175}{1000} = 0.175 \text{ m}$$

$$e/B = \frac{0.175}{1.75} = 0.1$$

$$H = 180 \text{ kN}$$

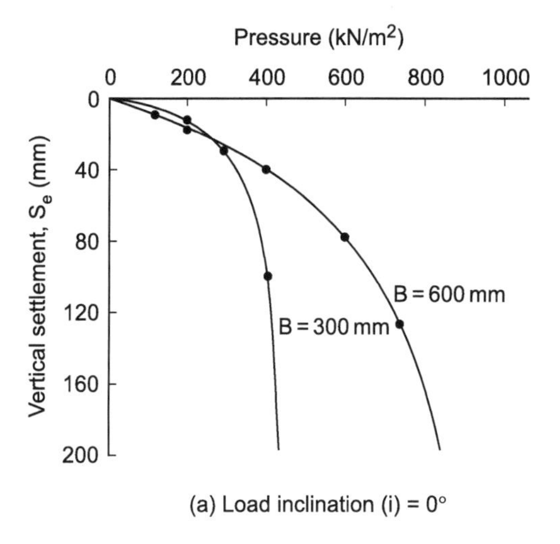

(a) Load inclination (i) = 0°

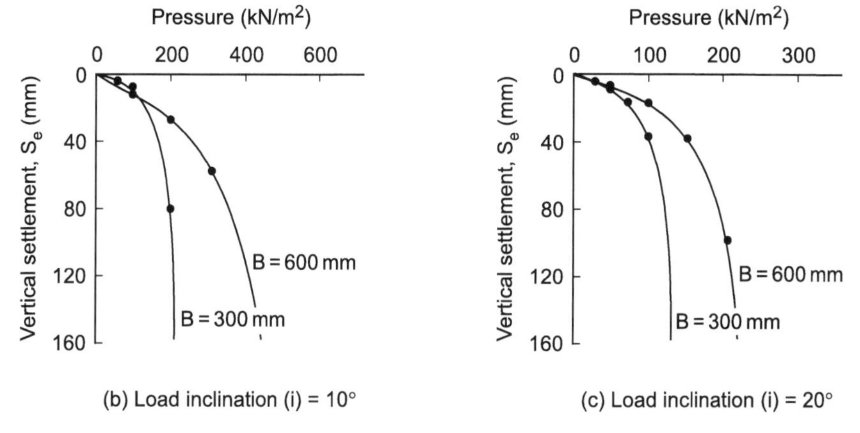

(b) Load inclination (i) = 10° (c) Load inclination (i) = 20°

Fig. 5.12. Pressure versus vertical settlement (S_e) curves for rigid strip footings (B = 300 mm, 600 mm) resting on Ranipur sand (D_R = 84%) for e/B = 0.1.

$$i = \tan^{-1}\left(\frac{H}{V}\right) = \tan^{-1}\left(\frac{180}{10^3}\right) = 10.2^\circ$$

$$b = \frac{B}{2} - e = \frac{1.75}{2} - 0.175 = 0.7 \text{ m}$$

$$a = \frac{B}{2} + e = \frac{1.75}{2} + 0.175 = 1.05 \text{ m}$$

Assuming $\alpha\alpha_1 = 0.5$

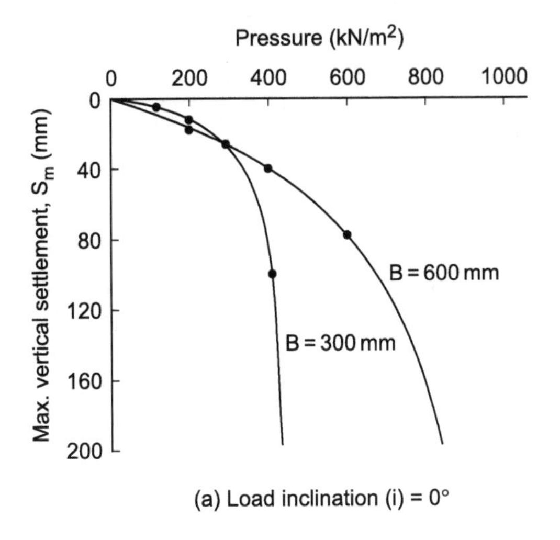

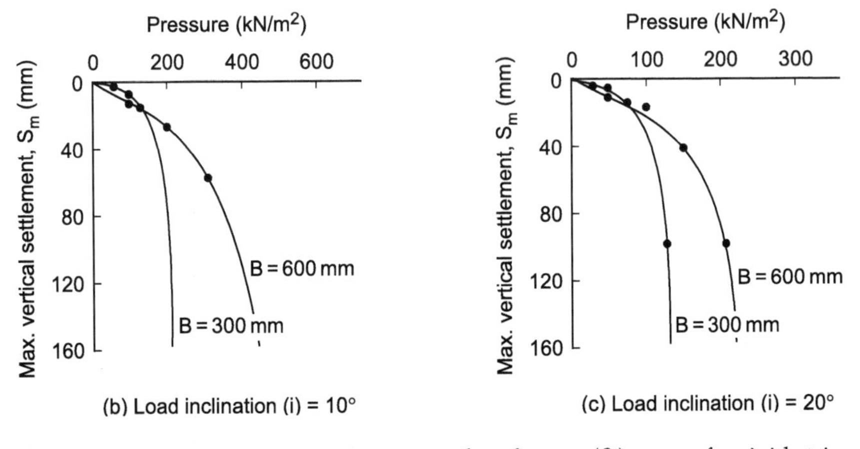

Fig. 5.13. Pressure versus maximum vertical settlement (S_e) curves for rigid strip footing, (B = 300 mm, 600 mm) resting on Ranipur sand (D_R = 84%) for e/B = 0.1.

$$\alpha\alpha_2 = \frac{4\times0.7-\{3\times1.05+0.7\}\cos(10.2)\times0.5}{2\times1.05\times\cos(10.2)}$$

$$\alpha\alpha_3 = \frac{4\times0.7-\{3\times1.05+0.7\}\cos(10.2)\times0.5}{2\times1.05\times\cos(10.2)}$$

$$= \frac{2.65}{1.38} = 1.92$$

Refer to Fig. 5.2b

Vertical loading (q_1, q_2, q_3)

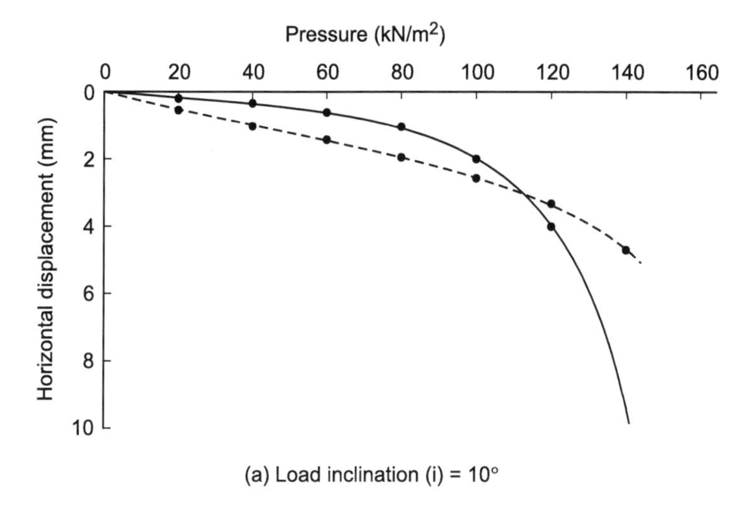

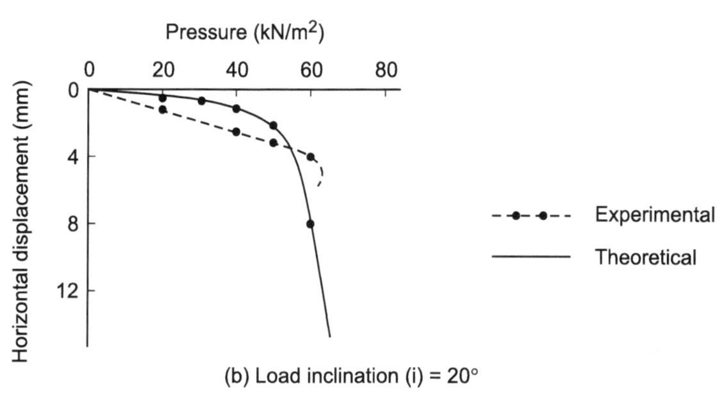

Fig. 5.14. Pressure versus horizontal displacement curves for rigid strip footing (B = 100 mm) resting on Ranipur sand (D_R = 84%) for e/B = 0.0.

$$\cos^2(10.2) = 0.968$$

$$q_1 = \alpha\alpha_1 q \cos^2 i = 0.5 \times 571.4 \times 0.968$$

$$= 0.5 \times 553 = 276.5 \text{ kN/m}^2$$

$$q_2 = \alpha\alpha_2 q\cos^2 i = 0.4378 \times 553 = 242 \text{ kN/m}^2$$

$$q_3 = \alpha\alpha_3 q\cos^2 i = 1.92 \times 553 = 1062 \text{ kN/m}^2$$

Refer to Fig. 5.2c

Tangential loading (q_{t1}, q_{t2}, q_{t3})

$$\sin i \cos i = \frac{\sin(2i)}{2} = \frac{\sin(2\times 10.2)}{2} = 0.1743$$

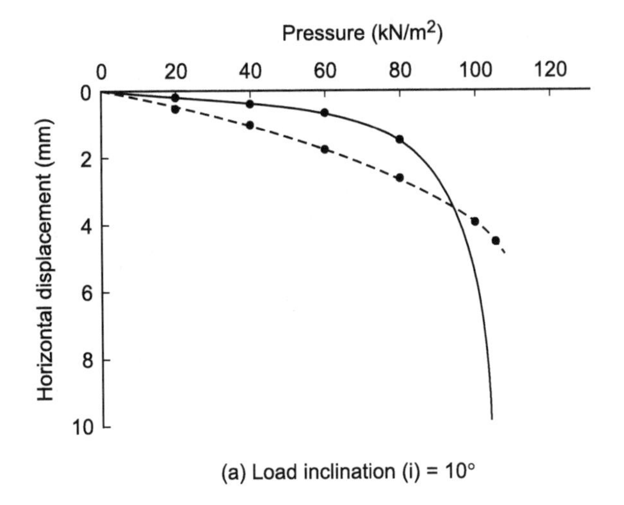

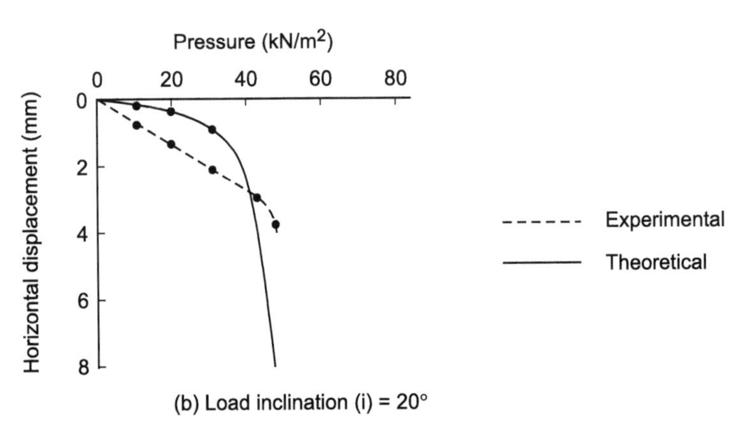

Fig. 5.15. Pressure versus horizontal displacement curves for rigid strip footing (B = 100 mm) resting on Ranipur sand (D_R = 84%) for e/B = 0.1.

$$q \sin i \cos i = 571.4 \times 0.1743 = 99.6 \text{ kN/m}^2$$

$$q_{t1} = \alpha\alpha_1.q \sin i \cos i = 0.5 \times 99.6 = 49.8 \text{ kN/m}^2$$

$$q_{t2} = \alpha\alpha_2.q \sin i \cos i = 0.438 \times 99.6 = 43.6 \text{ kN/m}^2$$

$$q_{t3} = \alpha\alpha_3.q \sin i \cos i = 1.92 \times 99.6 = 191.2 \text{ kN/m}^2$$

Above intensities of pressure were acting on the footing as shown in Figs 5.17(a) and 5.17(b). Location of origins are shown in these figures. For points A, B and C, it is evident from the figures that the values x and b are as follows (Table 5.1).

Table 5.1.

Description of loading	*Origin/s*	*x for point A (m)*	*x for point B (m)*	*x for point C (m)*	*Value of b (m)*	*Value of z/B*
Uniform vertical and horizontal loading	O, Fig. 5.17a, b	0.875	–0.175	–0.875	0.875	2.285
Triangular linearly decreasing vertical loading	C, Fig. 5.17a	1.75	0.7	0	0.70	5.714
Triangular linearly increasing vertical and horizontal loading	B, Fig. 5.17a, b	1.05	0	–0.7	0.525	3.81
Triangular linearly increasing horizontal loading	B, Fig. 5.17b	–1.05	0	0.7	0.35	5.714

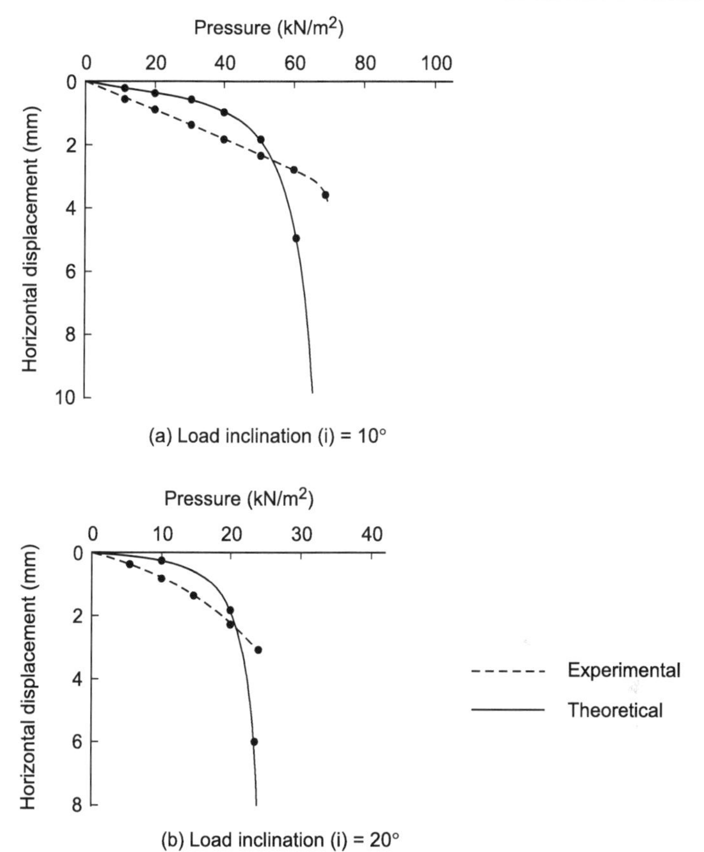

Fig. 5.16. Pressure versus horizontal displacement curves for rigid strip footing (B = 100 mm) resting on Ranipur sand (D_R = 84%) for e/B = 0.2.

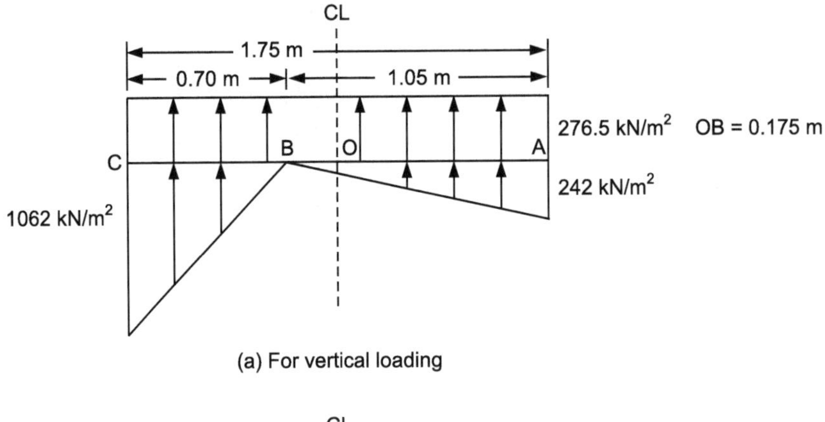

(a) For vertical loading

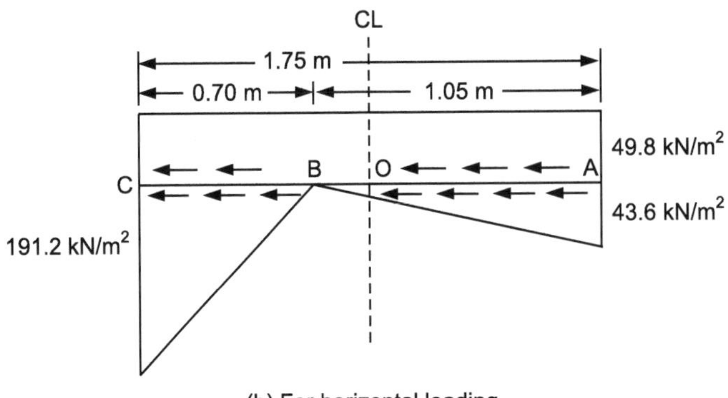

(b) For horizontal loading

Fig. 5.17. Contact pressure diagrams.

Considering the value of z as 2.0 m, z/b were computed and are given in the last column of Table 5.1.

(ii) For each point, i.e. A, B and C, the values of σ_z, σ_x and τ_{xz} were obtained using the equation given in Chapter 3 for the corresponding values of x, z and b. These values are given in Table 5.2. Values at sum of the stresses are given at the bottom row of the Table 5.2.

(iii) Knowing the total values of the stress (σ_z, σ_x and τ_{xz}), the rest procedure is same as discussed in the earlier chapter, i.e. evaluation of principal stresses (σ_1, σ_3), ε_1, μ_2, ε_3 and ε_z. Values of these are given in Table 5.3.

Example 5.2

In Example 5.1, consider that the footing rest on sand with the values of a and b are as given below:

$$\frac{1}{a} = 4.2 \times 10^3(\sigma_3)^{0.5}; \ \frac{1}{b} = 90 + 4(\sigma_3); \ \sigma_3 \text{ is in kN/m}^2$$

Table 5.2. Values of σ_z, σ_x and τ_{xz} (Example 5.1)

		Section A			*Section B*			*Section C*		
	$q (kN/m^2)$	$\sigma_z (kN/m^2)$	$\sigma_x (kN/m^2)$	$\tau_{xz} (kN/m^2)$	$\sigma_z (kN/m^2)$	$\sigma_x (kN/m^2)$	$\tau_{xz} (kN/m^2)$	$\sigma_z (kN/m^2)$	$\sigma_x (kN/m^2)$	$\tau_{xz} (kN/m^2)$
$Uniform_{(vertical)}$	276.5	106.883	19.649	38.165	135.847	8.516	-9.412	106.883	19.649	–38.165
Decrease $Linear_{(vertical)}$	106.2	48.172	27.083	35.884	105.404	6.207	24.021	113.811	2.200	–12.870
Increase $Linear_{(vertical)}$	242	37.240	1.522	6.097	31.703	4.015	–10.547	18.549	8.659	–12.444
$Uniform_{(horizontal)}$	49.8	6.874	2.138	3.539	–1.695	–0.332	1.534	–6.874	–2.138	3.539
Increase $inear_{(horizontal)}$	191.2	4.817	152.517	1.203	–8.333	165.563	3.172	–9.832	169.117	6.841
Increase $inear_{(horizontal)}$	43.6	–1.473	38.834	1.112	–0.986	36.961	0.255	0.528	34.714	0.090
Total		**202.512**	**241.743**	**85.990**	**261.940**	**220.930**	**9.023**	**223.065**	**232.201**	**-53.009**

Table 5.3. Showing the Details of Computations of σ_z

Section	σ_z (kN/m^2)	σ_x (kN/m^2)	τ_{xz} (kN/m^2)	σ_1	σ_3	θ_1 *(Deg.)*	θ_3 *(Deg.)*	a (m^2/kN)	b (m^2/kN)	μ	μ_1	a' $(\sigma_1–\sigma_3)$	$1\text{-}b'^*$ $(\sigma_1–\sigma_3)$	ε_1	μ_2	ε_3	ε_z
A	202.512	241.743	85.990	310.336	133.921	–38.576	51.424	0.000010	0.002	0.460	0.852	0.001	0.612	0.00227	0.665	–0.00151	0.00080
B	261.940	220.930	9.023	263.836	219.033	11.876	101.876	0.000010	0.002	0.460	0.852	0.000	0.901	0.0004	0.074	–0.00003	0.00037
C	223.065	232.201	–53.009	280.839	174.430	42.538	132.538	0.000010	0.002	0.460	0.852	0.001	0.766	0.0011	0.490	–0.00054	0.00035

Determine the vertical strain at depth 2.0 m below the base of footing and lying on vertical section passing through the right-hand edge of footing, i.e. point A. The value of angle of internal friction is 35°.

Solution

(i) Determination of Stresses

Total values of stresses σ_x, σ_z, τ_{xz}, will be same as give in Table 5.2, i.e. for point '*A*';

$$\sigma_x = 241.7 \text{ kN/m}^2, \sigma_z = 202.5 \text{ kN/m}^2, \tau_{xz} = 86.0 \text{ kN/m}^2$$

In case sand, overburden stresses are added i.e.

$$\sigma_z = 202.5 + 16.5 \times 2.0 = 235 \text{ kN/m}^2, [\gamma = 16.5 \text{ kN/m}^3]$$

$$\sigma_x = 241.7 + 0.4 \times 16.5 \times 2.0 = 255 \text{ kN/m}^2, [K_0 = 0.4]$$

$$\sigma_1 = \frac{(235+255)}{2} - \sqrt{\left(\frac{235-255}{2}\right)^2 + 86^2} = 331.6 \text{ kN/m}^2$$

$$\sigma_3 = \frac{(235+255)}{2} - \sqrt{\left(\frac{235-255}{2}\right)^2 + 86^2} = 158.4 \text{ kN/m}^2$$

$$\tan 2\theta = \left(\frac{2\times 86}{235-255}\right), \theta_1 = -41.68, \theta_3 = 48.31$$

(ii) Evaluation of hyperbola parameters a and b

$$\frac{1}{a} = 4.0 \times 10^3(158.4)^{0.5} = 50.3 \times 10^3 \text{ kN/m}^2$$

$$\frac{1}{b} = 90 + 4(158.4) = 724 \text{ kN/m}^2$$

$$a = 20 \times 10^{-6} \text{ m}^2/\text{kN}$$

$$b = 0.00138 \text{ m}^2/\text{kN}$$

(iii) For $e/B = 0.0$ and $i = 0°$

$$\text{For } \phi = 35°, N_\gamma = 48 \text{ [Table A4.1]}$$

$$q_u = \frac{1}{2}\gamma (B - 2e) N_\gamma (1 - i/\phi)^2$$

$$\frac{1}{2} \times 16.5 \times (1.75 - 2 \times 0.175) \times 48 \times (1 - 10/35)^2$$

$$= 8.25 \times 1.4 \times 48 \times 0.51 = 283 \text{ kN/m}^2$$

$$F = \frac{q_u}{q} = \frac{283}{100} = 2.83$$

$$\varepsilon_s = \frac{(1-1/F)}{a} = \frac{(1-1/2.83)}{20\times10^{-6}} = 0.0323 \times 10^6$$

$$\varepsilon_1 = \frac{(\sigma_1 - \sigma_3)}{E_s} = \frac{331.6-158.4}{20\times10^{-6}\times10^6} = 5362 \times 10^{-6}$$

$$\mu_1 = \frac{\mu}{1-\mu} = \frac{0.46}{1-0.46} = 0.85$$

$$\mu_2 = \frac{\sigma_3 - \mu_1\sigma_1}{\sigma_1 - \mu_1\sigma_3}$$

$$= \frac{158.4-0.85\times331.6}{331.6-0.85\times158.4} = \frac{-123.46}{196.96} = -0.627$$

$$\varepsilon_3 = -\mu_2\varepsilon_1 = -(-0.627) \times 5362 \times 10^{-6}$$

$$= 338.8 \times 10^{-6}$$

$$\theta = \frac{1}{2}\tan^{-1}\left(\frac{2\tau_{xz}}{\sigma_z - \sigma_x}\right) = \frac{1}{2}\tan^{-1}\left(\frac{2\times52.45}{137.65-32.94}\right)$$

$$= \frac{1}{2}\tan^{-1}\left(\frac{104.90}{104.71}\right) = 22.5°;\ \text{so}\ \theta_1 = 22.5°;\ \theta_3 = 112.5°$$

$$\varepsilon_z = \varepsilon_1\cos^2\theta_1 + \varepsilon_3\cos^2\theta_3$$

$$= 5362 \times 10^{-6}\cos^2(-41.68) + 3388 \times 10^{-6}\cos^2(48.31)$$

$$= 2994 \times 10^{-6} + 1498 \times 10^{-6} = 4492 \times 10^{-6}$$

Practice Problems

1. Explain stepwise the procedure for obtaining pressure versus tilt-curve for a footing resting on clay and subjected to eccentric-inclined load.
2. Explain stepwise the procedure for obtaining pressure versus maximum settlement curve for a footing resting on sand and subjected to eccentric-inclined load.
3. A column footing of width 1.5 m resting on clay is subjected to a total vertical load of 750 kN, moment of 125 kN-m and a shear load of 150 kN. Values of Kondner's hyperbola constants are: $a = 0.8 \times 10^{-5}$ m²/kN; $b = 0.003$ m²/kN. Adopt Poisson ratio as 0.35.

   Determine the values of vertical strains at depth 1.0 m at the edges and at the point where the resultant inclined load acts on the footing.
4. In Example 3, consider that the footing rests on sand with the values of a and b as given below:

$$\frac{1}{a} = 3.0 \times 10^3 (\sigma_3)^{0.4} ; \frac{1}{b} = 100 + 3.2(\sigma_3);$$

σ_3 is in kN/m^2

Determine the vertical strain at depth 1.5 m below the base of footing and lying on vertical section passing through the right-hand edge of footing, i.e. point *A*. Value of angle of internal friction is 30°.

REFERENCES

Agarwal, R.K. (1986). Behaviour of Shallow Foundations Subjected to Eccentric-Inclined Loads. Ph.D thesis, University of Roorkee, Roorkee.

Hjiaj, M., Lyamin, A.V. and Sloan, S.W. (2004). Bearing capacity of a cohesive frictional soil under non-eccentric inclined loading. *Computers and Geotechnics*, **31:** 491-516.

Lau, C. and Bolton, M. (2011a). The bearing capacity of footings on granular soils I: Numerical analysis. *Geotechnique*, **61(8):** 627-638.

Lau, C. and Bolton, M. (2011b). The bearing capacity of footings on granular soils II: Experimental evidence. *Geotechnique*, **61(8):** 639-650.

Loukidis, D., Chakraborty, T. and Salgado, R. (2008). Bearing capacity of strip footings on purely frictional soil under eccentric and inclined loads. *Canadian Geotechnical Journal*, **45:** 768-787.

Mahiyar, H. and Patel, A.N. (2000). Analysis of angle-shaped footing under eccentric loading. *Journal of Geotechnical and Geo-environmental* Engineering, **126(12):** 1151-1156.

Meyerhof, G.G. (1953). The bearing capacity of foundations under eccentric-inclined loads. *Proc. 3rd International Conference on Soil Mechanics and Foundation Engineering*, Zurich, Switzerland, **1:** 440-445.

Patra, C.R., Behara, R.N., Sivakugan, N. and Das, B.M. (2012). Ultimate bearing capacity of shallow strip foundation under eccentrically inclined load. Part I. *International Journal of Geotechnical Engineering*, **6(3):** 343-352.

Prakash, S. and Saran, S. (1971). Bearing capacity of eccentrically loaded footings. *Journal of the Soil Mechanics and Foundations Division*, ASCE, **97(1):** 95-117.

Saran, S. and Agarwal, R.K. (1989). Eccentrically obliquely loaded footing. *Journal of Geotechnical Engineering*, ASCE, **115(11):** 1673-1680.

Schultz, E. (1952). Der Widerstand des Baugrundes gegen scharge sohlpressungen (in German). *Die Bautechnik*, **29(12):** 336-342.

Appendix 5.1

Charts for obtaining bearing capacity factors for a footing subjected to eccentric incline load

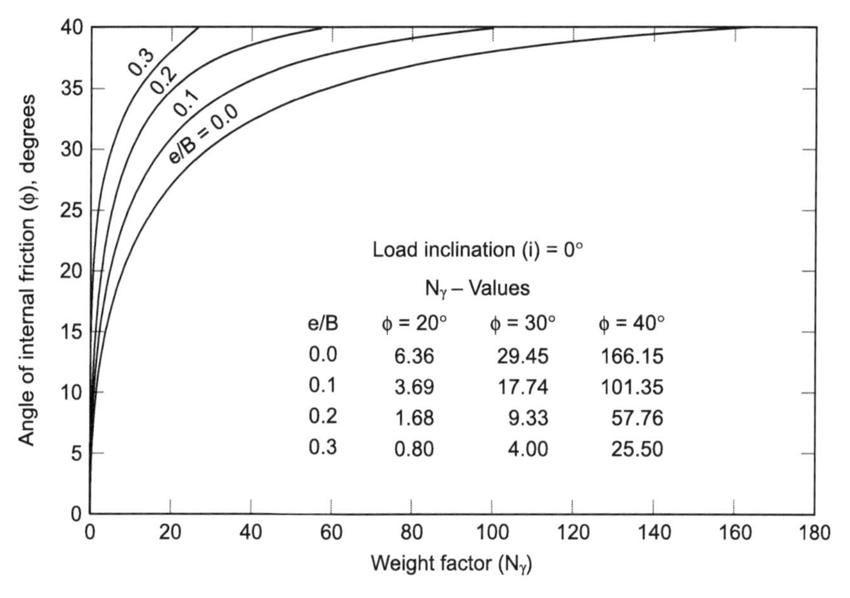

Fig. A5.1. N_γ versus φ for different e/B at $i = 0°$

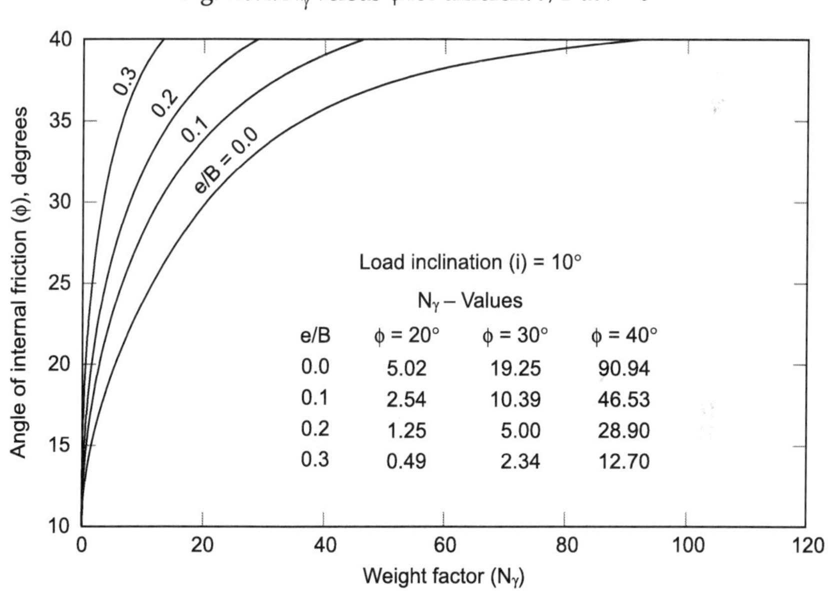

Fig. A5.2. N_γ versus φ for different e/B at $i = 10°$

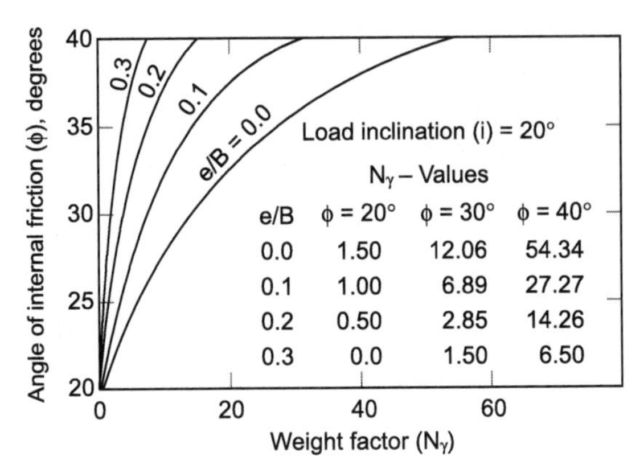

e/B	φ = 20°	φ = 30°	φ = 40°
0.0	1.50	12.06	54.34
0.1	1.00	6.89	27.27
0.2	0.50	2.85	14.26
0.3	0.0	1.50	6.50

Fig. A5.3. N_γ versus φ for different e/B at $i = 20°$

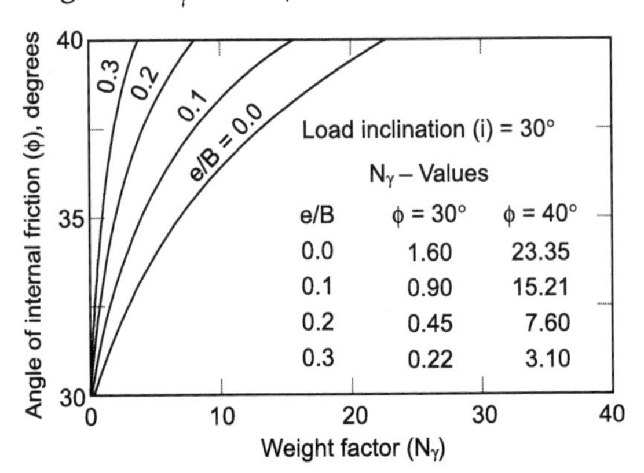

e/B	φ = 30°	φ = 40°
0.0	1.60	23.35
0.1	0.90	15.21
0.2	0.45	7.60
0.3	0.22	3.10

Fig. A5.4. N_γ versus φ for different e/B at $i = 30°$

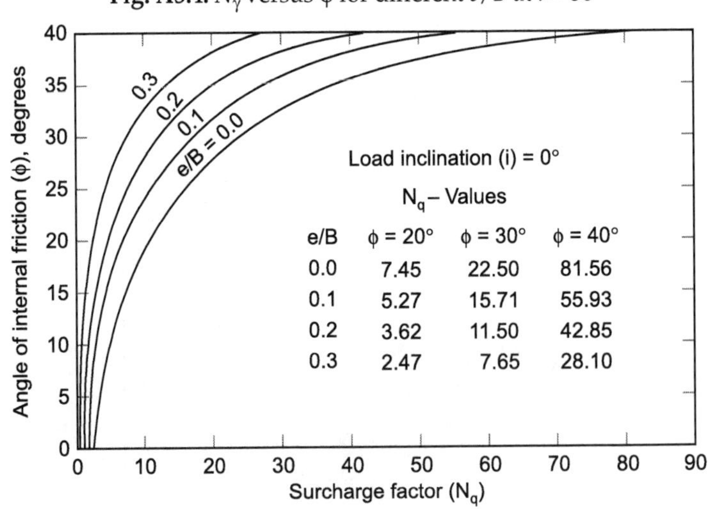

e/B	φ = 20°	φ = 30°	φ = 40°
0.0	7.45	22.50	81.56
0.1	5.27	15.71	55.93
0.2	3.62	11.50	42.85
0.3	2.47	7.65	28.10

Fig. A5.5. N_q versus φ for different e/B at $i = 0°$

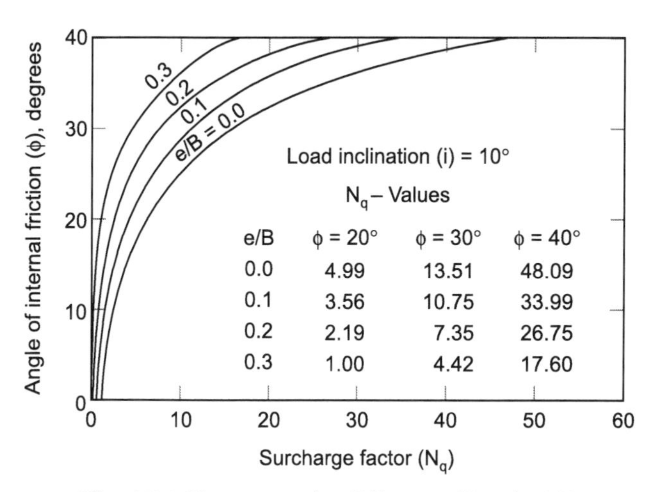

Fig. A5.6. N_q versus φ for different e/B at $i = 10°$

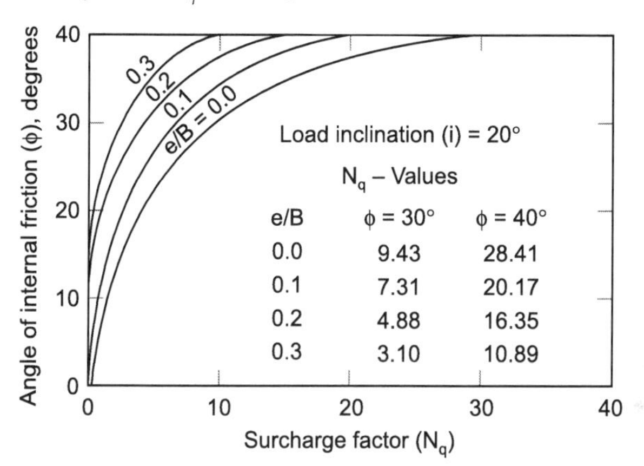

Fig. A5.7. N_q versus φ for different e/B at $i = 20°$

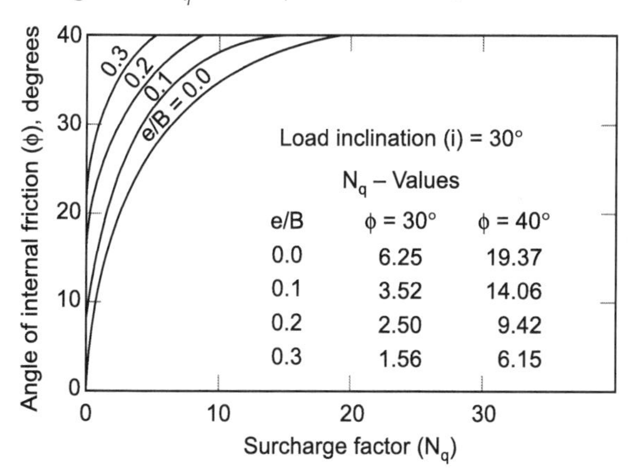

Fig. A5.8. N_q versus φ for different e/B at $i = 30°$

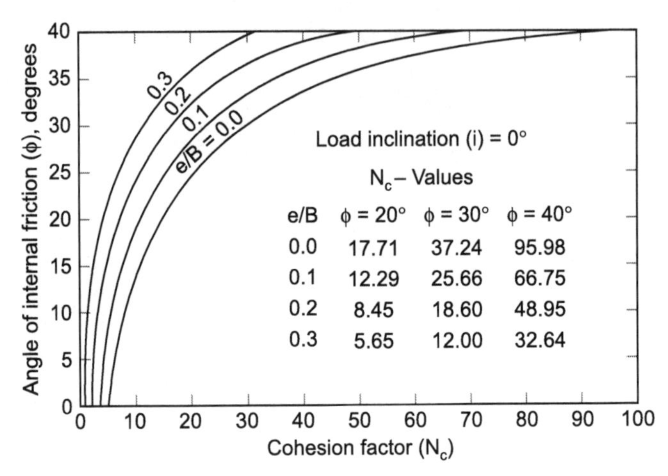

e/B	ϕ = 20°	ϕ = 30°	ϕ = 40°
0.0	17.71	37.24	95.98
0.1	12.29	25.66	66.75
0.2	8.45	18.60	48.95
0.3	5.65	12.00	32.64

Fig. A5.9. N_c versus φ for different e/B at $i = 0°$

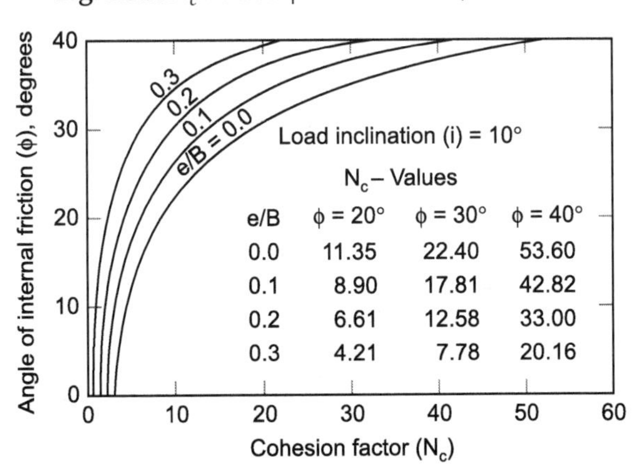

e/B	ϕ = 20°	ϕ = 30°	ϕ = 40°
0.0	11.35	22.40	53.60
0.1	8.90	17.81	42.82
0.2	6.61	12.58	33.00
0.3	4.21	7.78	20.16

Fig. A5.10. N_c versus φ for different e/B at $i = 10°$

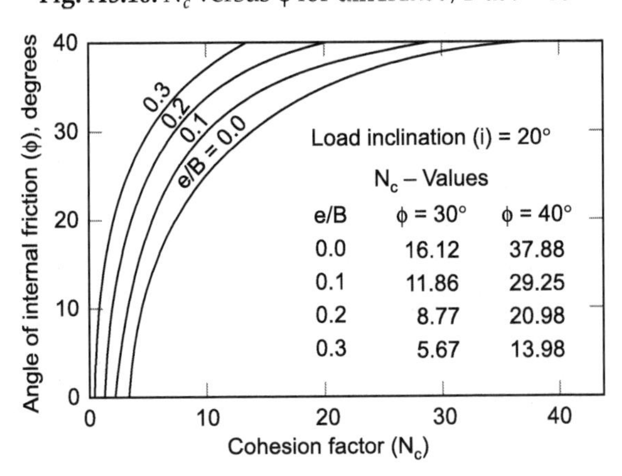

e/B	ϕ = 30°	ϕ = 40°
0.0	16.12	37.88
0.1	11.86	29.25
0.2	8.77	20.98
0.3	5.67	13.98

Fig. A5.11. N_c versus φ for different e/B at $i = 20°$

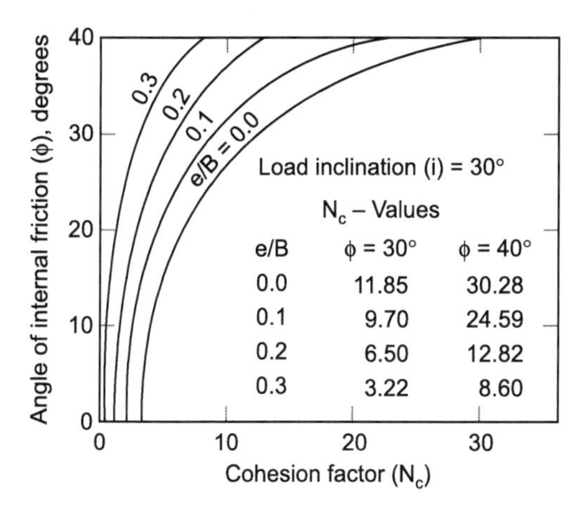

Fig. A5.12. N_c versus φ for different e/B at $i = 30°$

CHAPTER
6

Strip Footing Adjacent to a Slope

6.1 General

Foundations are sometimes placed on slopes, or adjacent to slopes, or near a proposed excavation. Presently in the case of bridges, footings are usually not placed within the fill; instead, pile or other foundations are considered. These alterations may not be most economical.

Foundations are also sometimes situated near the open section of underground railways. In such a situation, the problem becomes that of obtaining the minimum value of the bearing capacity: (1) from foundation failure, and (2) from overall stability of the slope. In case of non-cohesive soil, the bearing capacity is always governed by foundation failure, while in cohesive material, the bearing capacity of the foundation may be dictated by stability.

In general the problem may be described pictorially as shown in Fig. 6.1. Where D_e represents the edge distance, and if it is zero footing, it may

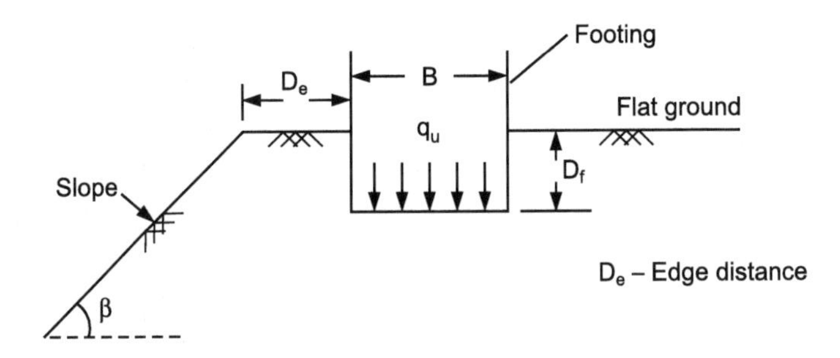

Fig. 6.1. Footing adjacent to a slope.

be considered resting completely on the slope. It will lead the results little on to a conservative side.

Proportioning of a footing resting adjacent to a slope also requires the evaluation of its bearing capacity and settlement and are determined in two independent steps. Earlier, the problem of bearing capacity was solved by three different approaches, namely: (1) slip line analysis (Sokolovski, 1960; Siva Reddy and Mogaliah, 1975); (2) limit equilibrium analysis (Meyerhof, 1957; Mizuno et al., 1960; Siva Reddy and Mogliah, 1976; Bowles, 1977; Myslivec and Kysela, 1978; Bauer et al., 1981; Sud, 1984; Saran, et al., 1989); and (3) limit analysis (Chen, 1975; Shields et al., 1977, 1990; Kusakabe et al., 1981; Saran et al., 1989; Narita and Yamaguchi, 1990). Some investigators have solved this problem by using numerical analysis (Castelli and Motta, 2010; Castelli and Lentini, 2012; Mohammadreza and Adel, 2015).

On the basis of analyzing the large number of model tests data, Sud (1984) developed a non-dimensional correlation for the estimation of settlement of a footing resting adjacent to a slope. No other method is available which gives the closed form solution except a method based on advanced numerical techniques (Acharyya and Dey, 2016).

In this chapter, a general procedure based on constitutive laws of soil is evolved to predict the settlement of a footing placed near the edge of a slope.

Soil, in general, is an anisotropic material and Young's modulus E is dependent upon the confining pressure. No method is available where variation of E with confining pressure is considered for computing the stresses in the soil medium. Further, no method is available to get stresses due to foundation load on soil mass restricted by a slope even if E and μ are considered independent to confining pressure. Hence, a semi-empirical method is formulated to estimate the pressure settlement characteristics of footings adjacent to slopes and consisting of cohesive-frictional soils.

Since the confining pressures are the major criteria to evaluate the settlement of soil mass, they have been assumed to be provided by the passive earth pressure developed on the sides of the slope. Maximum shearing resistance has been assumed to develop at the base of the footing and minimum at the depth where the stresses become zero.

This assumption is due to the observation that the degree to which the strength can be mobilized is directly dependent on the amount of movement of the soil mass (Terzaghi, 1943). The movement of the soil will be maximum at the base of the foundation and it decreases with depth. Using constitutive relation of the soil, Sud (1984) developed a procedure for obtaining pressure-settlement characteristic of a footing resting adjacent to a slope, and the same has been included in this chapter in concise form.

6.2 Analysis

Assumptions

The following assumptions have been made in the analysis:

- The footing base is assumed to be flexible and as such uniform contact pressure distribution has been assumed for obtaining pressure settlement curves. Flexible footings arc generally not encountered in practice. However, the average settlement of flexible footing is almost same as the settlement of a rigid footing for average pressure intensities (Sharan, 1977). The average settlements have been computed for various pressure intensities. The pressure settlement curve computed by the present empirical approach may be taken as for rigid strip footing.
- The whole soil mass supporting a footing is assumed to be a vertical column of soil as shown in Fig. 6.2 and this column of soil is divided into a large number of thin horizontal strips in which stresses and strains are assumed to be uniform along any vertical section. In this figure, B, D_f, D_e, H_s and β respectively represent width of footing, depth of footing, edge distance, height of slope and slope angle with horizontal. The settlements can be expected to be approximately the same even if the column of soil, which is stressed due to footing load is taken as spreading in its area as the depth increases (as the general situation), because as the load intensity decreases due to enlarged area, the confining pressure on the side of the slope decrease.

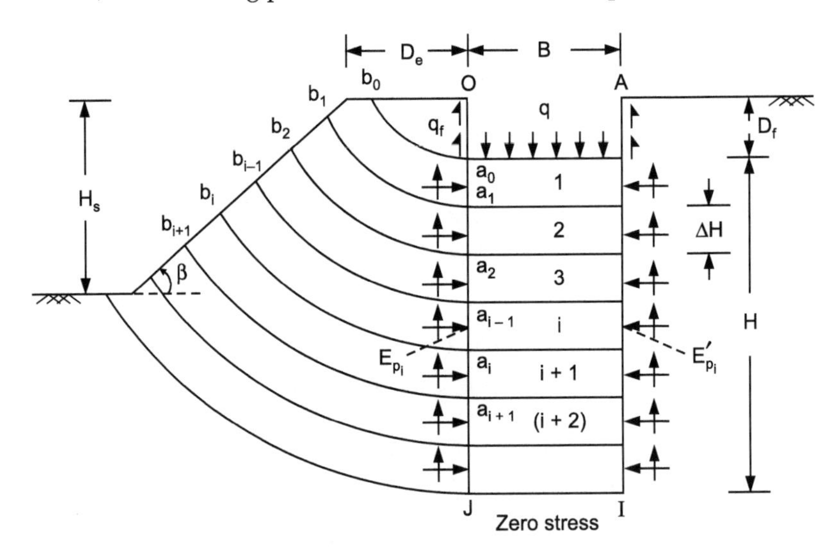

Fig. 6.2. Soil mass below the footing.

- The passive earth pressure has been evaluated by taking the failure surface as a log spiral with centre of rotation at the edge of the footing.
- The effect of the weight of soil mass is considered in determination of stresses in soil mass. Vertical stresses due to weight of soil have been taken equal to γz, where γ is the density of soil and z is the depth.
- Shear stress has been assumed to vary linearly along the length of a strip.
- The ultimate bearing capacity q_u is computed from limit equilibrium analysis (Sud, 1984; Saran, Sud and Handa, 1989). A brief description is given in Appendix 6.1.
- A coefficient F has been introduced such that at all the stress levels in the following relationship are satisfied

$$\frac{q_u}{q} = \frac{\sigma_u}{\sigma_1 - \sigma_3} = F \tag{6.1}$$

  where q is intensity of load, σ_u is the ultimate stress from hyperbola relationship of Kondner (1963) and is equal to $(1/b)$. σ_1, σ_3 are the major and minor principal stresses in the soil mass due to load q and weight of soil.
- There is no slippage at the interface of layers of the soil mass.

6.3 Procedure for Analysis

The procedure for evaluation of settlement of uniformly loaded strip footing near the edge of the slope on soils is described in the following steps:

Step 1

For a given load, the depth H of the soil mass under the footing, at which the stresses become zero is assumed (Fig. 6.2).

Step 2

This column of soil is divided into n number of thin horizontal strips.

Step 3

Angle of shearing resistance is varied from full value of ϕ at the base of the footing to zero at depth H, where the stresses become zero.

$$\varphi_{mi} = \phi - \frac{\phi i \Delta H}{H} \tag{6.2}$$

where φ_{mi} is the mobilized value of ϕ at depth $i\Delta$H. This variation of ϕ is due to the observation that the degree to which strength mobilizes is directly dependent on the amount of movement of the soil mass. Similarly, cohesion '*c*' has been varied with the full value of c at the base of the footing to zero at depth H.

$$c_{mi} = c - \frac{ci\Delta H}{H}; \; c_{ai} = \frac{c_{mi} + c_{m(i-1)}}{2}$$

where c_{mi} = mobilized value at depth $i\Delta H$

$c_{m(i-1)}$ = mobilized value at depth $(i-1)\,\Delta H$

c_{ai} = average cohesion for the ith strip

Step 4

The confining pressure (σ_3) acting at the centre of each strip is taken as

$$\sigma_3 = E_{pi} \cos \phi_{ai}$$

where $$\phi_{ai} = \frac{\phi_{mi} + \phi_{m(i-1)}}{2} \tag{6.3}$$

and E_{pi} is the passive resistance offered to each strip and is evaluated by calculating the moments of resistance due to the soil mass as shown in Fig. 6.2. ϕ_{ai} is the average ϕ for i^{th} strip.

$$E_{pi} = \frac{M_{\gamma(i)} + M_{c(i)} + M_{\gamma(i-1)} + M_{c(i-1)}}{\Delta H\left(\frac{i+(i-1)}{2}\right) + D_f} \tag{6.4}$$

where E_{pi} = passive earth pressure for i^{th} strip

$M_{c(i)}$ = moment of resistance due to cohesion at i^{th} strip

$M_{\gamma(i)}$ = moment of resistance due to soil mass at i^{th} strip

$M_{c(i-1)}$ = moment of resistance due to cohesion at $(i-1)^{th}$ strip

$M_{\gamma(i-1)}$ = moment of resistance due to soil mass at $(i-1)^{th}$ strip

ΔH = thickness of the strip

To evaluate passive earth pressure E_p for any particular depth, the failure surface has been taken as a log spiral having centre of rotation at 0 (Fig. 6.3) at the edge of the footing at ground surface.

$$R_0 = OD = i\Delta H + D_f$$

$$R_1 = OE = R_0 e^{\theta \tan \phi_{mi}} \tag{6.5}$$

where θ is the angle of log spiral. There can be two cases: Case I, where the rupture surface meets the slope (Fig. 6.3a) and Case II the rupture surface meets the base of slope (Fig. 6.3b).

Case I

Rupture surface meeting the slope

$$R_1 = \frac{D_e \sin\beta}{\sin(\beta + \theta - 90)} \tag{6.6}$$

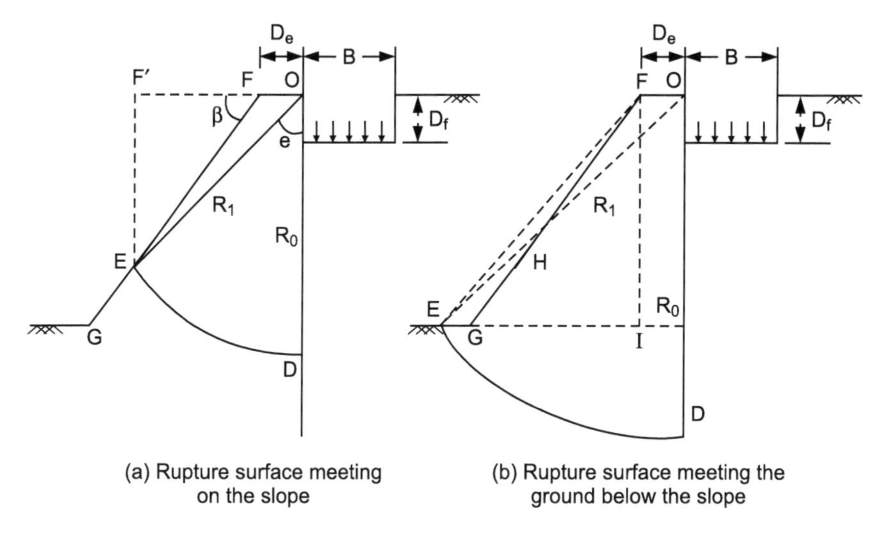

Fig. 6.3. Rupture surfaces for evaluating E_p.

From Eqs. (6.5) and (6.6)

$$R_0\, e^{\theta \tan \phi_{mi}} = \frac{D_e \sin\beta}{\sin(\beta+\theta-90)} \tag{6.7}$$

From the above transcendental equation, the value of angle of log spira θ is obtained by trial and error.

Taking moments about the point 0, the centre of rotation

M_γ = Moment of mass of area OED + Moment of mass of area EF′O – Moment of mass of area F′FE

$$= \frac{\gamma R_0^3}{3(\tan^2\phi+1)}\left[e^{3\theta \tan\phi_{mi}}(3\tan\phi\sin\theta - \cos\theta) + 1\right]$$

$$+\frac{1}{3}\gamma R_0^3 e^{3\theta \tan\phi_{mi}} \cos\theta \sin^2\theta \tag{6.8}$$

$$-\frac{1}{2}\gamma R_0 e^{\theta \tan\phi_{mi}}(R_0 e^{\theta\tan\phi} - D_e)\cos\theta$$

$$\left[\frac{2}{3}R_0 e^{\theta\tan\phi_{mi}}\sin\theta + \frac{1}{3}D_e\right]$$

$$M_c = \frac{c_{mi}R_0^2}{2\tan\phi_{mi}}\left[e^{3\theta\tan\phi}{}_{mi-1}\right] \tag{6.9}$$

Case II

Rupture surface meeting the base of slope

$$R_1 = \frac{H_s}{\cos\theta} \tag{6.10}$$

From Eqs (6.5) and (6.10)

$$R_0\, e^{\theta \tan \phi_{mi}} = \frac{H_s}{\cos\theta}$$

From the above equation, the value of θ is obtained by trial and error.

Taking moments about the point 0 (Fig. 6.3b)

M_γ = Moment of mass of area OED – Moment of mass of area EFI + Moment of mass of area IGF + Moment of mass of area HFO.

$$M_\gamma = \frac{\gamma R_0^{\;3\theta\tan\phi_{mi}}}{3\left(9\tan^2\phi+1\right)}[e^{3\theta \tan \phi_{mi}}(3 \tan \phi_{mi} \sin \theta - \cos \theta) + 1]$$

$$-\frac{1}{6}\gamma H_s(\sin \theta R_0 e^{3\theta\tan \phi_{mi}} - D_e)(\sin \theta R_0\, e^{\theta \tan \phi_{mi}} - 2D_e) \tag{6.12}$$

$$+\frac{1}{2}\gamma\frac{H_s^2}{\tan\beta}\left(\frac{1}{3}\frac{H_s}{\tan\beta}+D_e\right)+\frac{1}{3}\gamma H_s \sin^2\theta R^2{}_0\, e^{2\theta\tan\phi_{mi}}$$

$$-\frac{1}{3}\gamma H_s(R_1 \sin \theta - D_e)^2$$

From the Eqs (6.8) or (6.12) and (6.9), the value of M_γ and M_c are obtained for a particular depth $i\Delta H + D_f$ and substituted *m* Eq. (6.4) to get the value of E_{pi}.

Step 5

Force on the sides of the wall of the footing is calculated by the following equation:

$$q_f = \frac{1}{2} D^2{}_f K_0 \tan \delta + c_a D_f$$

where γ = angle of wall friction and may been taken equal to 2/3 ϕ

K_0 = coefficient of earth pressure at rest and is calculated by the equation

$K_0 = (1 - \sin \phi)$

c_a = adhesion between the side walls of the foundation and soil

Step 6

The vertical load for any particular height of soil mass *H* can be computed by considering the overall equilibrium of soil column. As the value of ϕ

and cohesion c has been assumed to vary with depth, the earth pressure developed on the sides of the column will also vary with depth.

$$qB = \Sigma E_{pi} \cdot \sin \varphi_{ai} + \Sigma E'_{pi} \sin \varphi'_{ai} - \gamma BH + 2q_f + 2\Sigma c_{ai} \Delta H \tag{6.13}$$

where c_{ai} = average value of cohesion for the ith strip

φ_{ai} = average value of φ for the ith strip

q = intensity of loading

E_{pi} = earth pressure on the side of slope at i^{th} strip inclined at an mobilized angle φ_{ai}

E'_{pi} = earth pressure on the side away from slope at the ith strip inclined at an angle φ'_{ai}

Subsequently, in Step 8, it is proved that $E_{pi} \sin \varphi_{ai}$ is equal to $E'_{pi} \sin \varphi'_{ai}$. So Eq. (6.13) becomes

$$qB = 2\Sigma E_{pi} \sin \varphi_{ai} - \gamma D^2_f K_0 \tan \delta + 2\Sigma c_{ai} \Delta H$$

$$\frac{q}{\gamma H} = \frac{2\sum E_{pi} \sin \phi_{ai}}{\gamma BH} - 1 + \frac{D_f^2 K_0 \tan \delta}{BH} + \frac{2\sum c_{ai} \Delta H}{\gamma BH} \tag{6.14}$$

Step 7

A plot is drawn between the values of H/B and $q/\gamma H$ from Eq. (6.15) and the value of H is obtained from this plot for the given load q (Fig. 6.7).

Step 8

The vertical stress on any strip is calculated by considering the static equilibrium of the forces acting on the strip. The forces acting on the strip are shown in Fig. 6.4. Considering $\Sigma V = 0$

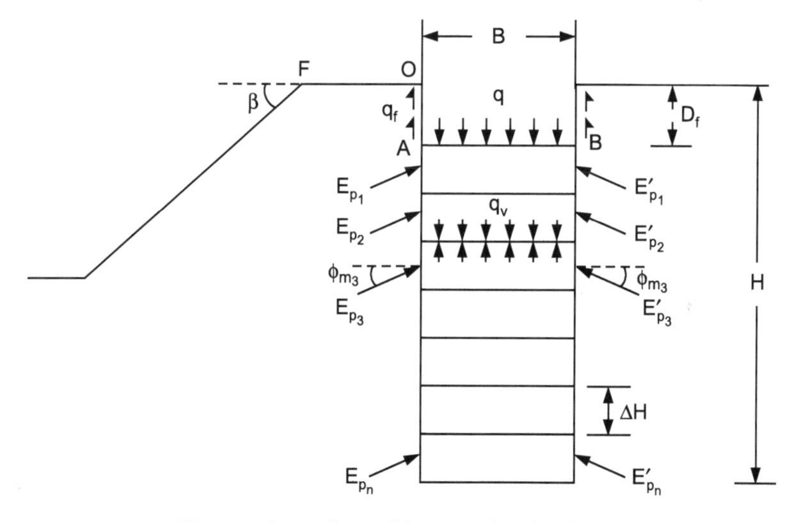

Fig. 6.4. Overall equilibrium of soil column.

$$q_vB - 2q_f - (q_v + dq_v)B - E_{pi}\sin\varphi_{ai}dH + E'_{pi}\sin\varphi'_{ai}dH - \gamma BdH - 2c_{ai}dH = 0$$

$$Bdp_v - \gamma D^2{}_fK_0\tan\delta - 2c_{ai}D_f - E_{pi}\sin\phi_{ai}dH - E'_{pi}\sin\varphi'_{ai}dH + \gamma BdH - 2c_{ai}dH = 0$$

Integrating

$$q_vB - \gamma D^2{}_fK_0\tan\delta - 2\Sigma c_{ai} - \Sigma E_{pi}\sin\varphi_{ai} - \Sigma E'_{pi}\sin\varphi'_{\mathrm{ai}}E'_{pi} + \gamma BH = 0 \quad (6.15)$$

Considering the equilibrium condition of $\Sigma H = 0$

$$E_{pi}\sin\varphi_{ai}\,dH = E'_{pi}\sin\varphi'_{ai}\,dH$$

or

$$\sum E_{pi}\sin\varphi_{ai} = \sum E'_{pi}\sin\varphi'_{ai}$$

The Eq. (6.15) becomes

$$q_vB = 2\Sigma E_{pi}\sin\varphi_{\mathrm{ai}} - \gamma BH + \gamma D^2{}_fK_0\tan\delta + 2\Sigma c_{ai}\Delta H = 0 \quad (6.16)$$

Where q_v = vertical stress on the strip

From Eq. (6.16), vertical stress on a particular strip can be obtained.

Step 9

The confining stress (σ_3) for the ith strip is taken as

$$\sigma_3 = E_{pi}\cos\varphi_{ai}$$

Step 10

The ultimate strength (σ_u) of sand for a given confining pressure σ_3 is computed from the constitutive law of soil obtained by triaxial testing in-the lab.

Step 11

The ultimate bearing capacity (q_u) is calculated from the charts and equation given in Appendix 6.1.

Step 12

The shear stresses on the side of the strip are taken as ($E_{pi}\sin\phi_{ai} + C_{ai}\Delta H$) and are assumed to vary linearly along the width of the strip.

Step 13

The state of stresses are obtained for three points—*C*, *D*, *E*—on the horizontal plane passing through the centre of the strip. *C* is at the centre of the strip. *E* is at the edge and *D* is midpoint between *C* and *E* (Fig. 6.5). The stresses on these elements are shown in Fig. 6.6.

For element at point *E*

$$\sigma_z = q_v$$

$$\sigma_x = E_{\mathrm{p}}\cos\phi_m/\Delta H$$

$$\tau_{xz} = \left\{(E_p\sin\phi_m)/\Delta H\right\} + C_m/\Delta H$$

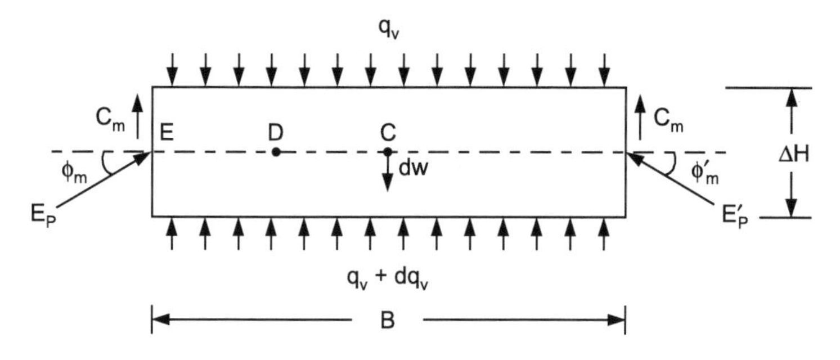

Fig. 6.5. Stresses on the strip.

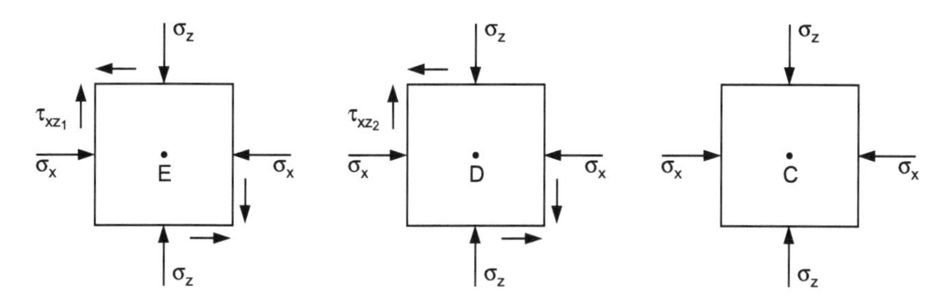

Fig. 6.6. Stresses on the elements.

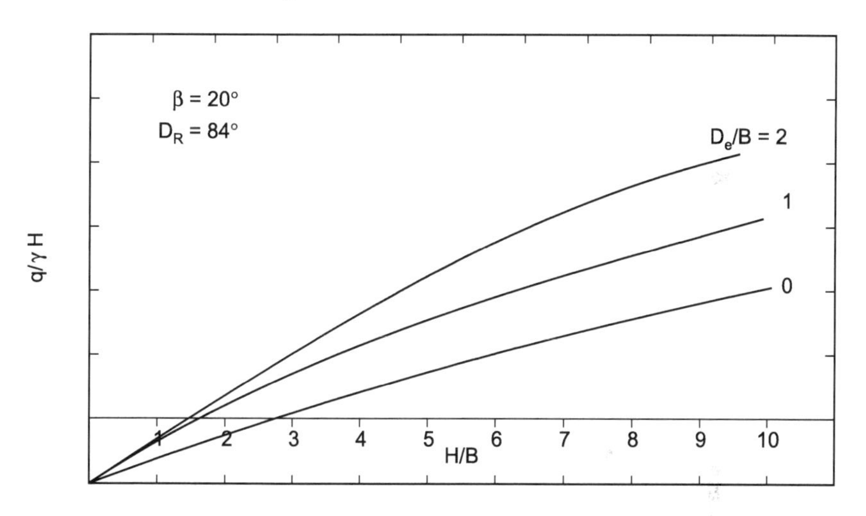

Fig. 6.7. Plot of $q/\gamma H$ versus H/B.

For element at point D

$$\sigma_z = q_v$$

$$\sigma_x = E_p \cos \phi_m / \Delta H$$

$$\tau_{xz} = \left[\left\{E_p \cos\phi_m / \Delta H\right\} + C_m / \Delta H\right] / 2$$

For element at point C

$$\sigma_z = q_v$$

$$\sigma_x = E_p \cos \phi_m / \Delta H$$

$$\tau_{xz} = 0$$

where σ_z = vertical stress

σ_x = stress in x-direction

τ_{xz} = shear stress on xz plane

The principal stresses on the elements and their directions with respect to the vertical z-axis have been computed using the equations of the theory of elasticity as given below.

$$\sigma_1 = \frac{\sigma_z + \sigma_x}{2} + \sqrt{\left(\frac{\sigma_z - \sigma_x}{2}\right)^2 + \tau_{xz}^2} \quad (6.17)$$

$$\sigma_3 = \frac{\sigma_z + \sigma_x}{2} + \sqrt{\left(\frac{\sigma_z - \sigma_x}{2}\right)^2 + \tau_{xz}^2} \quad (6.18)$$

$$\tan 2\theta = \frac{2\tau_{xz}}{\sigma_z - \sigma_x} \quad (6.19)$$

Positive value of θ is measured counter clockwise with direction of σ_z where σ_1 and σ_3 are the major and minor principal stresses.

Step 14

A coefficient F for a given load intensity (q) is computed from the following relationship:

$$\frac{q_u}{q} = F \quad (6.20)$$

Step 15

The modulus of elasticity (E) is calculated from the Eq. (6.21) at stress level of σ_u/F

$$E = \frac{1 - b\left(\sigma_v/F\right)}{a} \quad (6.21)$$

where a and b are the constants of hyperbola whose values depend upon confining pressure.

Rest procedure is same as illustrated in chapter for footings resting on flat ground.

6.4 Pressure Settlement Curves

The pressure settlement curves for a footing of width 300 mm for 30° slope have been plotted in Fig. 6.8 for D_e/B = 0.0, 1.0, 2.0 and 3.0 for the relative density of 84%. Kondner's (1962) hyperbola constants were obtained using triaxial test and the following relationships were developed: $\frac{1}{2}$ 800 + (σ_3); $\frac{1}{b}$ = 220 + 2.2(σ_3); units of (1/a), (1/b) and σ_3 are kN/m^2. It is

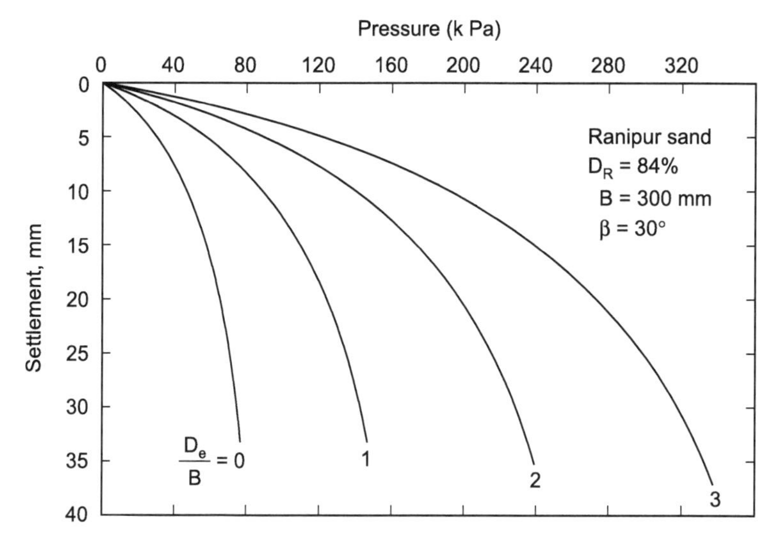

Fig. 6.8. Pressure settlement curves for different D_e/B ratios.

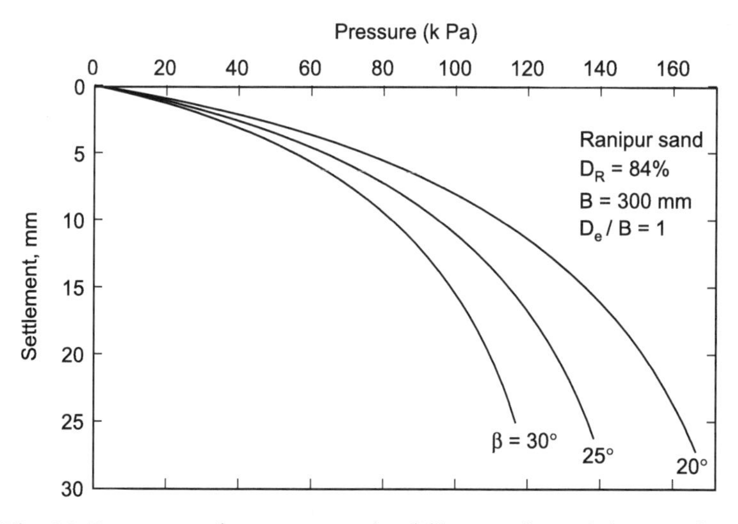

Fig. 6.9. Pressure settlement curves for different values of slope angles.

evident from Fig. 6.8 that the settlement of the footing at a given pressure decreases with increase in the edge distance.

The pressure-settlement curves for different slope angles are plotted in Fig. 6.9. It is evident from this figure that the settlement of a footing at given pressure increases with increase in the slope angle.

To find the effect of height of slope H_s on settlement, the pressure settlement curves for 300 mm footing resting on the shoulder ($D_e/B = 0.0$) of 30° slope for $H_s = B$, $2B$ and $4B$ are plotted in Fig. 6.10. The settlements

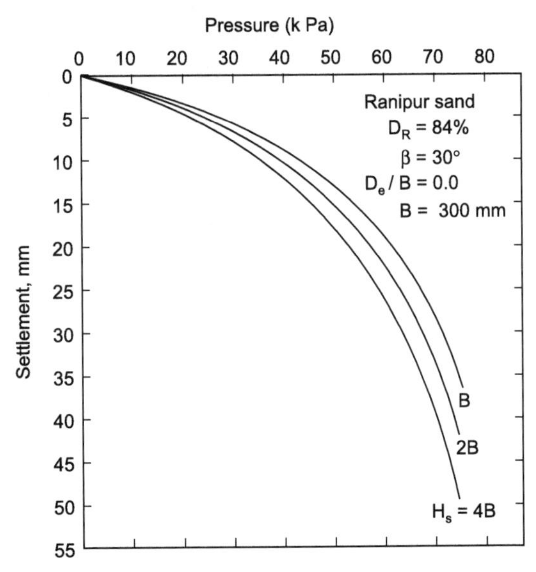

Fig. 6.10. Pressure settlement curves for different heights of slopes, H_s.

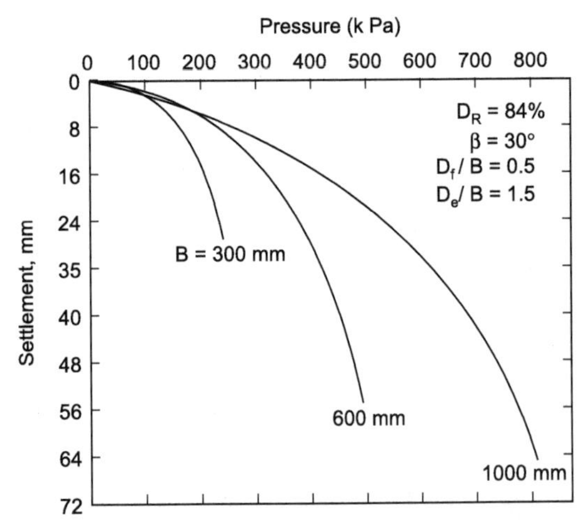

Fig. 6.11. Pressure settlement curves for different sizes of foundation for $D_f/B = 0.5$.

increase with increase in the height of slope. However, as the ultimate bearing capacity is independent of the height of slope, so will be the factor *F* (Eq. 6.1) and in this analysis, the settlements depend on the factor F. The settlement so computed for different heights of slope may not give the actual values.

Figure 6.11 gives the pressure settlement curves for foundations of width 300 mm, 600 mm and 1000 mm respectively for $D_e/B = 0.5$, $\beta = 30°$ and $D_f/B = 0.5$. The trend in behaviour of foundations at shallow depths is similar to that of surface footings. In general, for the same pressure intensity, settlement decreases with increase in the width of footing. However, the trend is reverse when the settlements are obtained for pressure intensities corresponding to the factor of safety.

Illustrative Examples

Example 6.1

Draw a pressure-settlement curve for a footing of width 1.0 m resting on cohesion-less soil slope ($\beta = 30°$) at an edge distance of 1.0 m. The footing was placed at a depth of 0.5 m below the ground surface. The unit weight of soil and angle of internal friction are respectively 16.3 kN/m^{3}and 39°. Using Kondner's hyperbolic parameters as below:

$$\frac{1}{a} = 800 + (\sigma_3);\ \frac{1}{b} = 220 + 2.2\ (\sigma_3);\ \text{proportion the footing.}$$

Solution

1. Using the described Art. 6.3, the pressure-settlement characteristics of 1.0 m wide strip footing was obtained for slope ($\beta = 30°$), $D_e/B = 1.0/1.0 = 1.0$ and $D_f/B = 0.5/1.0 = 0.5$ and is shown in Fig. 6.12. The relation of $1/a$ and $1/b$ as given in Example 6.1 was used.

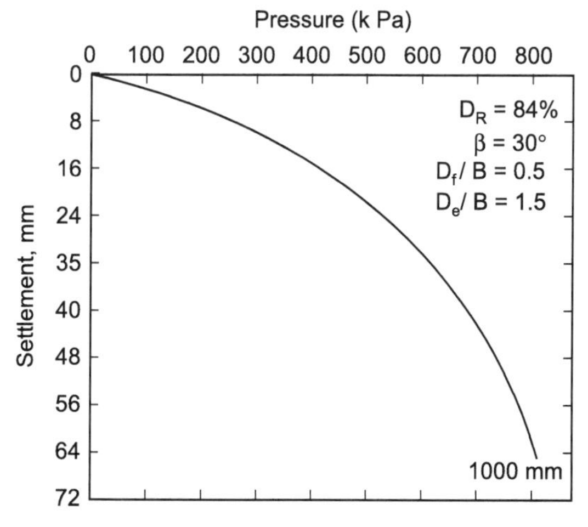

Fig. 6.12. Pressure settlement curves for $D_f/B = 0.5$, Example 6.1.

2. The curves become steep beyond the pressure intensity of about 750 kN/m^2 and therefore, may be taken as ultimate bearing capacity.
3. The ultimate bearing capacity was also obtained using the N_γ and N_q charts developed by Sud (1984). From Figs. A6.6 and A6.14, values of N_γ and N_q were obtained for ($\beta = 30°$), $D_e/B = 1.0/1.0 = 1.0$, $\varphi = 39°$ and $D_f/B = 0.5/1.0 = 0.5$ as below:

$$N_\gamma = 80;\ N_q = 16.5$$

Therefore,

$$q_u = \frac{1}{2}\gamma B N_\gamma + \gamma D_f N_q$$

$$= \frac{1}{2} \times 16.3 \times 1.0 \times 80 + 16.3 \times 0.5 \times 16.5$$

$$= 652 + 134 = 786 \text{ kN/m}^2$$

This value matches with the q_u value obtained by pressure settlement characteristics (Step No. 2).

$$\text{Allowable soil pressure} = \frac{750}{3} = 250 \text{ kN/m}^2.$$

For pressure intensity of 250 kN/m^2 the settlement is 7.5 mm. It is well within the permissible limit usually adopted for strip footing. Therefore the footing may be designed for pressure intensity of 250 kN/m^2 or less.

Practice Problems

1. What important factors are involved in checking the safety of a footing resting near the edge of a slope? Draw neatly a probable rupture surface in shear failure of a footing resting adjacent to a slope.
2. Give stepwise a procedure for obtaining pressure-settlement characteristics of a footing resting near the edge of a slope. Consider the slope material as:
   (i) Clay
   (ii) Sand
   (iii) Cohesive-frictional soil
3. Draw a pressure-settlement curve for a footing of width 1.5 m resting on cohesion-less soil slope ($\beta = 25°$) at an edge distance of 2.0 m. The footing was placed at a depth of 1 m below the ground surface. The unit weight of soil and the angle of internal friction are 16.5 kN/m^3 and 37° respectively. Using Kondner's hyperbolic parameters as below:

   $\frac{1}{a} = 800 + (\sigma_3)$; $\frac{1}{b} = 220 + 2.2(\sigma_3)$, proportion the footing.

REFERENCES

Acharyya, R. and Dey, A. (2016). Square footings on unreinforced sandy slopes: Numerical modelling using plaxis 3d. Geotechnics for Infrastructure Development, March 11-12. Kolkata, pp. 1-8.

Bauer, G.E., Shields, D.H., Scott, J.D. and Gruspier, J.E. (1981). Bearing capacity of footing in granular slope. *Proc. of 11th International Conference on Soil Mechanics and Foundation Engineering*. Balkema, Rotterdam. The Netherlands, **2:** 33-36.

Bowles, J.P. (1997). Foundation Analysis and Design. McGraw Hill, New York.

Castelli, F. and Motta, E. (2010). Bearing capacity of strip footings near slopes. *Geotechnical and Geological Engineering*, **28:** 187-198.

Castelli, F. and Lentini, V. (2012). Evaluation of the bearing capacity of footings on slopes. *International Journal of Physical Modelling in Geotechnics*, **12(3):** 112-118.

Chen, W.F. (1975). Limit Analysis and Soil Plasticity. Elsevier.

Kondner, R.L. and Zelasko, S. (1963). A hyperbolic stress-strain formulation for sands. *Proceedings of 2nd Pan American Conference on Soil Mechanics and Foundation Engineering*, Brazil, **I:** 289-324.

Kondner, R.L. (1963). Hyperbolic stress-strain response of cohesive soils. *Journal of Soil Mechanics and Foundation Div.*, ASCE, **89(3):** 115-143.

Kusakabe, O., Kimura, T. and Yamaguchi, H. (1981). Bearing capacity of slopes under strip loads on the top surface. *Soils Found.*, **21(4):** 29-40.

Meyerhof, G.G. (1957). The ultimate bearing capacity of foundation on slopes. *Proceedings of 4th International Conference on Soil Mechanics and Foundation Engineering*, **I:** 384-386.

Mizuno, T., Youshiharu, T. and Hiroshi, K. (1960). On the bearing capacity of a slope on cohesionless soils. *4th Int. Conf. on Soil Mech. and Found. Engg.*, **Vol. 1:** 384-386.

Mohammadreza, H.A. and Asakereh, A. (2015). Numerical analysis of the bearing capacity of strip footing adjacent to slope. *International Journal of Engineering Trends and Technology*, **29(6):** 313-317.

Myslivec, A. and Kysela, Z. (1978). The Bearing Capacity of Building Foundations. Elsevier, Amsterdam.

Narita, K. and Yamaguchi, H. (1990). Bearing capacity analysis of foundations on slopes by use of log-spiral sliding surfaces. *Soils Found.*, **30(3):** 144-152.

Saran, S., Sud, V.K. and Handa, S.C. (1989). Bearing capacity of footings adjacent to slopes. *J. Geotech. Eng.* ASCE, **115(4):** 553-573.

Sharan, U.N. (1977). Pressure-settlement Characteristics of Surface Footing from Constitutive Laws. Ph.D. Thesis, IIT Roorkee.

Shields, D.H., Scott, J.D., Bauer, G.E., Deschenes, J.H. and Barsvary, A.K. (1977). Bearing capacity of foundation near slopes. *Proc. 10th International Conference on Soil Mechanics and Foundation Engineering*, Tokyo, Japan, **2:** 715-720.

Shields, D.H., Chandler, N. and Garnier, J. (1990). Bearing capacity of foundation in slopes. *J. Geotech. Eng.* ASCE, **116(3):** 528-537.

Siva Reddy and Mogaliah, G. (1976). Bearing capacity of shallow foundations on slopes. *Indian Geotechnical Journal*, **5(4):** 237-253.

Siva Reddy and Mogaliah, G. (1976). Stability of slopes under foundation load. *Indian Geotechnical Journal*, **6(2):** 91-111.

Sud, V.K. (1984). Behaviour of Shallow Foundations Adjacent to Slopes. Ph.D. Thesis, IIT Roorkee.

Terzaghi, K. (1943). Theoretical Soil Mechanics. John Wiley and Sons, Inc. N.Y.

Appendix 6.1

Sud (1984) studied the problem of obtaining ultimate bearing capacity (q_u) of footing resting adjacent to a slope (Fig. A6.1). He developed the solutions using both 'limit equilibrium analysis' and 'limit-analysis' and summarized the results in the form of non-dimensional bearing capacity factors: N_γ, N_q and N_c (Figs A6.2 to A6.26). These bearing capacity factors depend on the angle of internal friction φ, slope angle β, edge distance-width ratio D_e/B. Knowing all these factors, the ultimate bearing capacity may be obtained using the following standard equation:

$$Q_u = q_u.B = B\left(\frac{1}{2}\gamma B N_\gamma + \gamma D_f N_q + cN_c\right) \tag{A6.1}$$

γ and *c* are respectively the unit weight and cohesion of the soil.

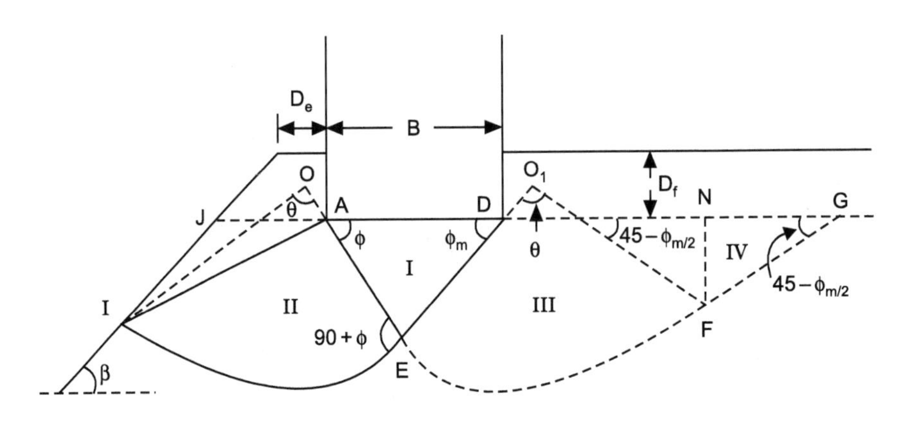

Fig. A6.1. Rupture surface.

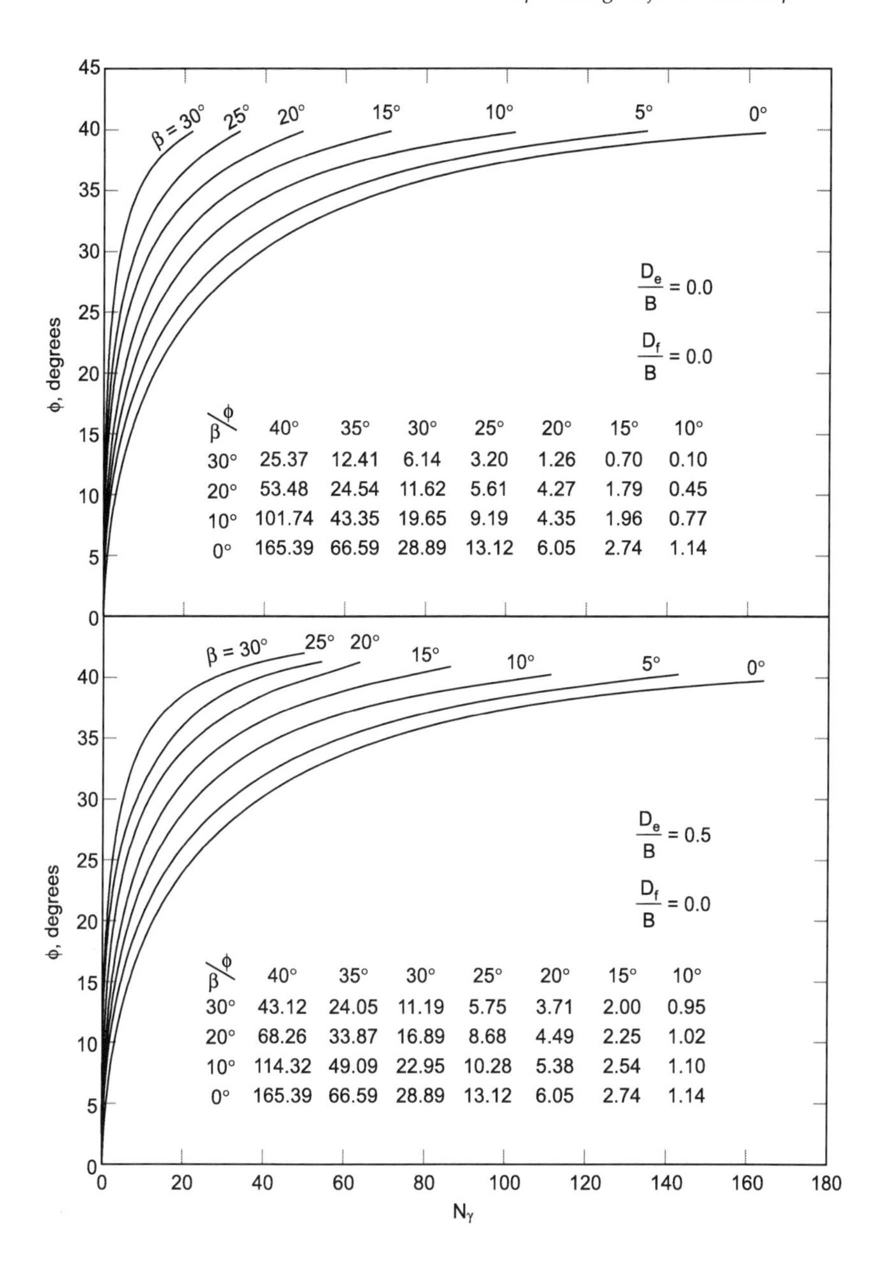

$D_e/B = 0.0$, $D_f/B = 0.0$

β \ ϕ	40°	35°	30°	25°	20°	15°	10°
30°	25.37	12.41	6.14	3.20	1.26	0.70	0.10
20°	53.48	24.54	11.62	5.61	4.27	1.79	0.45
10°	101.74	43.35	19.65	9.19	4.35	1.96	0.77
0°	165.39	66.59	28.89	13.12	6.05	2.74	1.14

$D_e/B = 0.5$, $D_f/B = 0.0$

β \ ϕ	40°	35°	30°	25°	20°	15°	10°
30°	43.12	24.05	11.19	5.75	3.71	2.00	0.95
20°	68.26	33.87	16.89	8.68	4.49	2.25	1.02
10°	114.32	49.09	22.95	10.28	5.38	2.54	1.10
0°	165.39	66.59	28.89	13.12	6.05	2.74	1.14

Fig. A6.2. N_r vs ϕ ($D_f/B = 0.0$ and $D_e/B = 0.0, 0.5$).

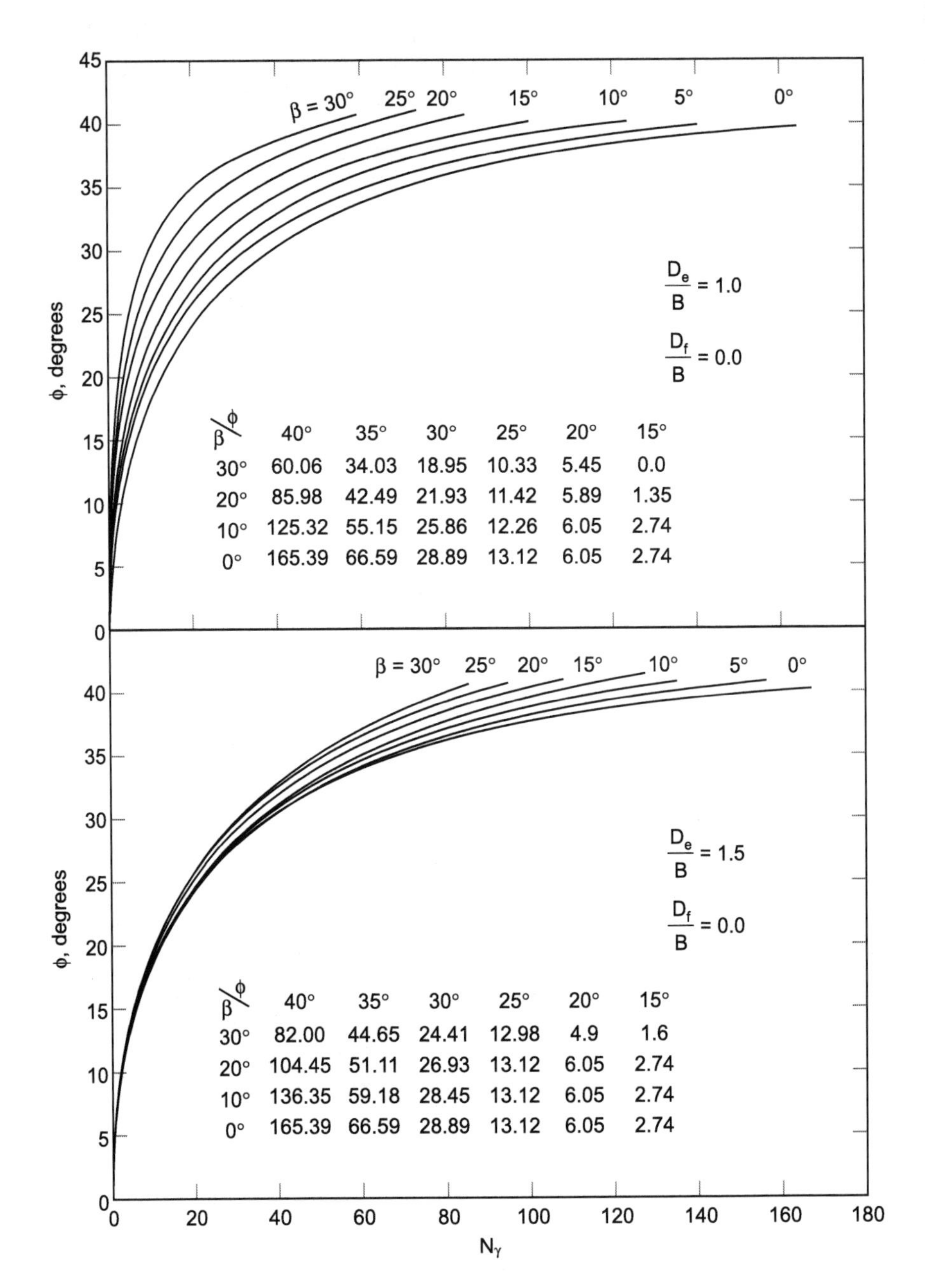

β \ φ	40°	35°	30°	25°	20°	15°
30°	60.06	34.03	18.95	10.33	5.45	0.0
20°	85.98	42.49	21.93	11.42	5.89	1.35
10°	125.32	55.15	25.86	12.26	6.05	2.74
0°	165.39	66.59	28.89	13.12	6.05	2.74

β \ φ	40°	35°	30°	25°	20°	15°
30°	82.00	44.65	24.41	12.98	4.9	1.6
20°	104.45	51.11	26.93	13.12	6.05	2.74
10°	136.35	59.18	28.45	13.12	6.05	2.74
0°	165.39	66.59	28.89	13.12	6.05	2.74

Fig. A6.3. N_r vs ϕ ($D_f/B = 0.0$ and $D_e/B = 1.0, 1.5$).

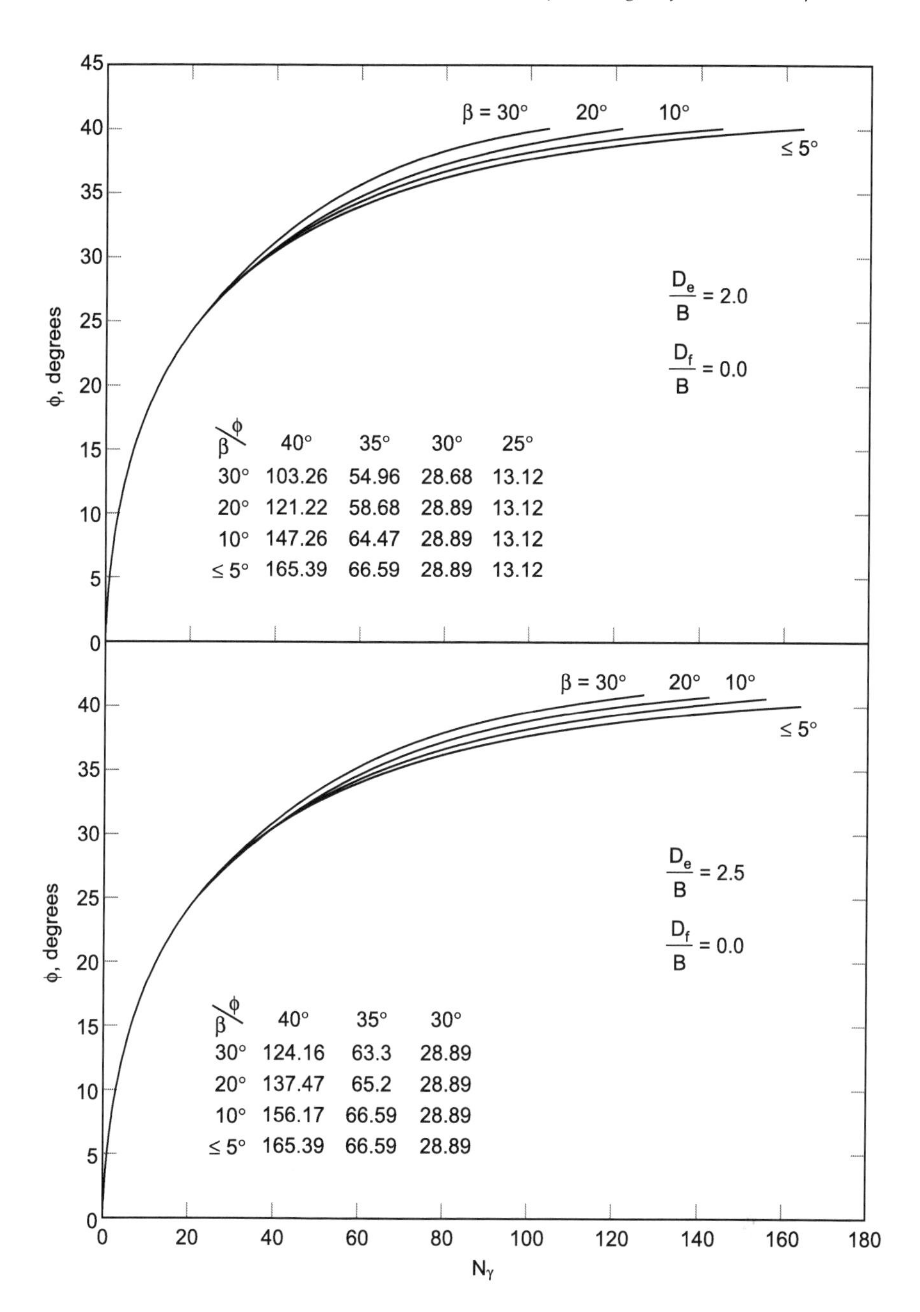

Fig. A6.4. N_r vs φ (D_f/B = 0.0 and D_e/B = 2.0, 2.5).

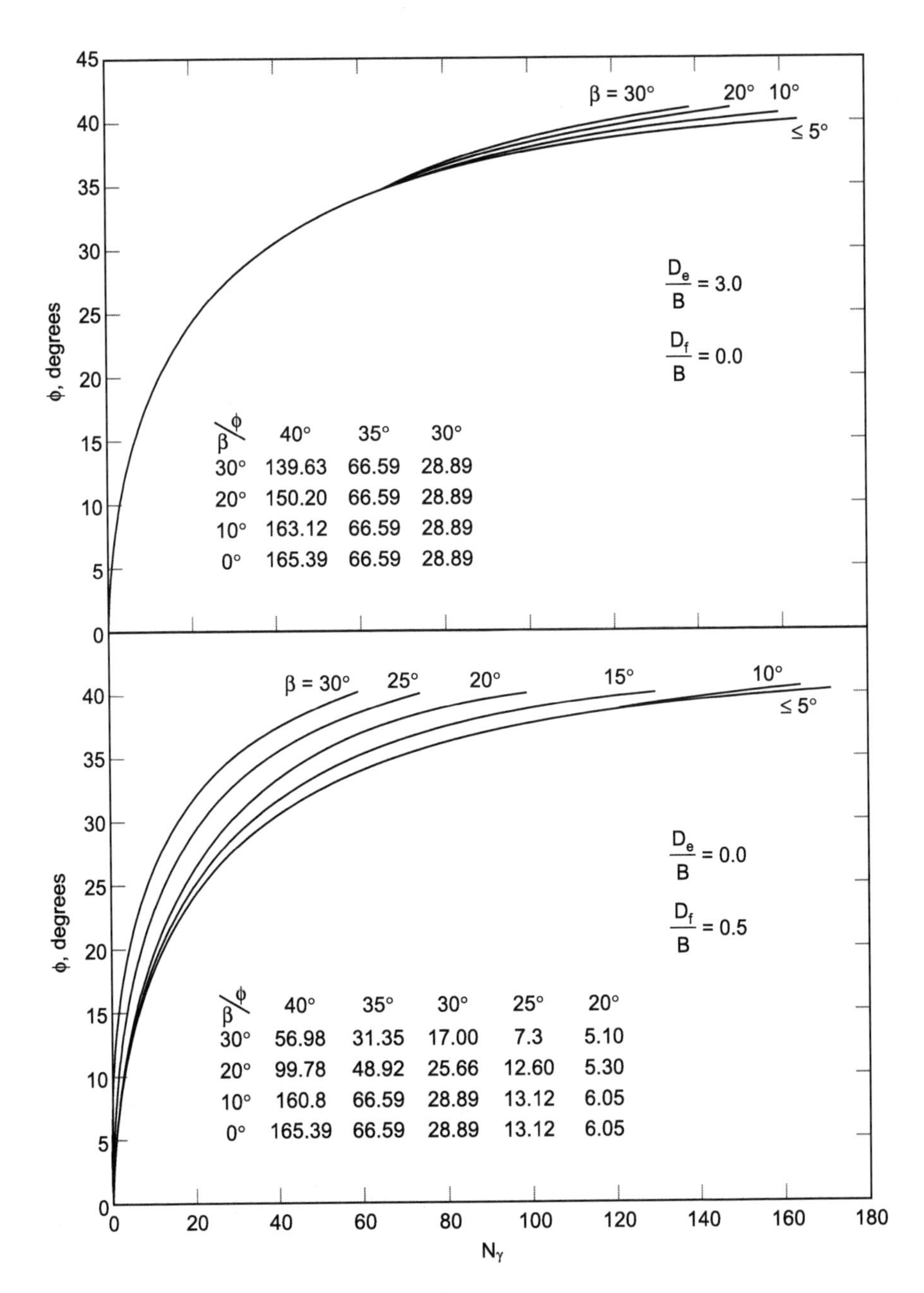

Fig. A6.5. N_r vs ϕ ($D_f/B = 0.0$, $D_e/B = 3.0$ and $D_f/B = 0.5$, $D_e/B = 0.0$).

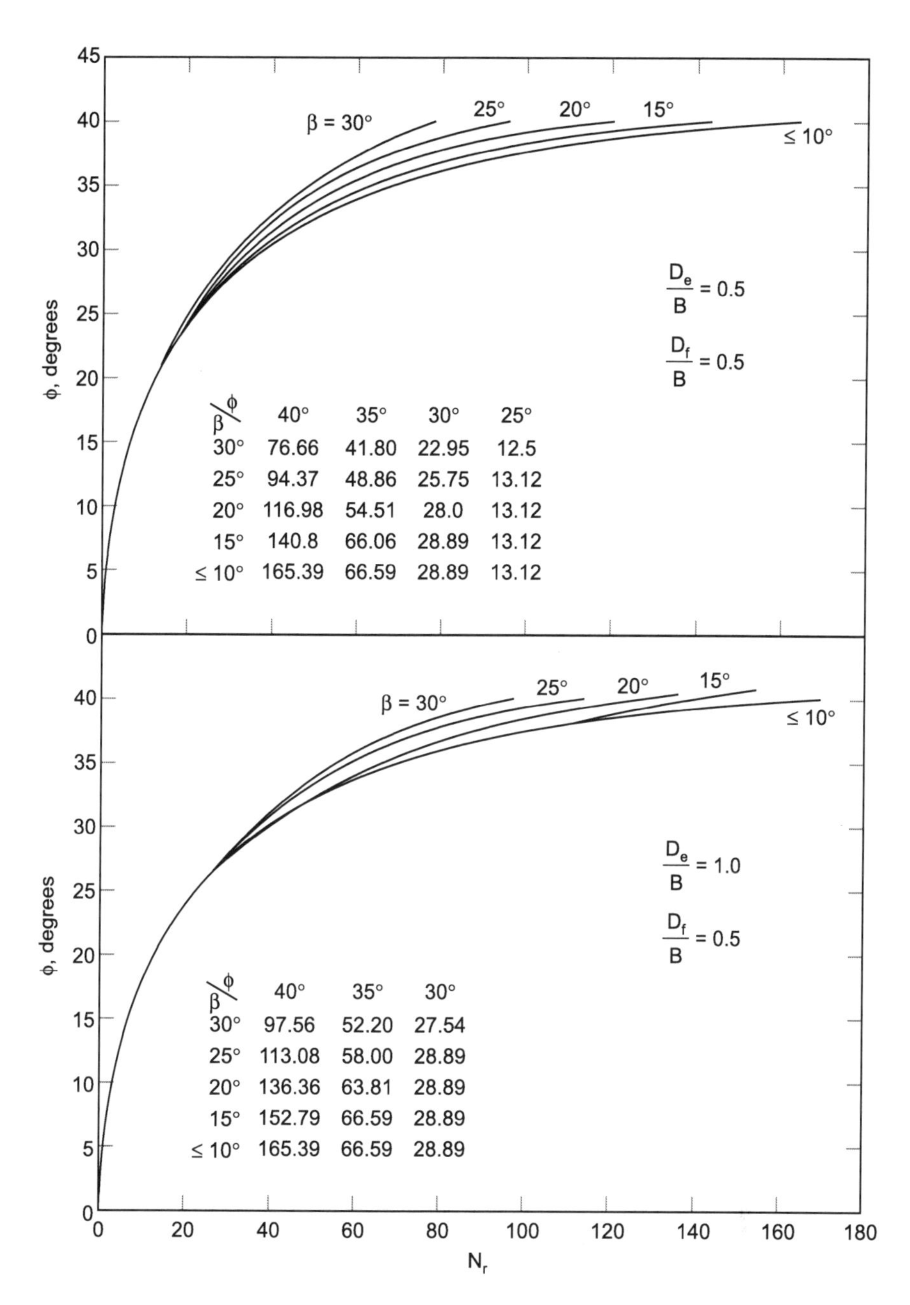

Fig. A6.6. N_r vs ϕ ($D_f/B = 0.5$ and $D_e/B = 0.5, 1.0$).

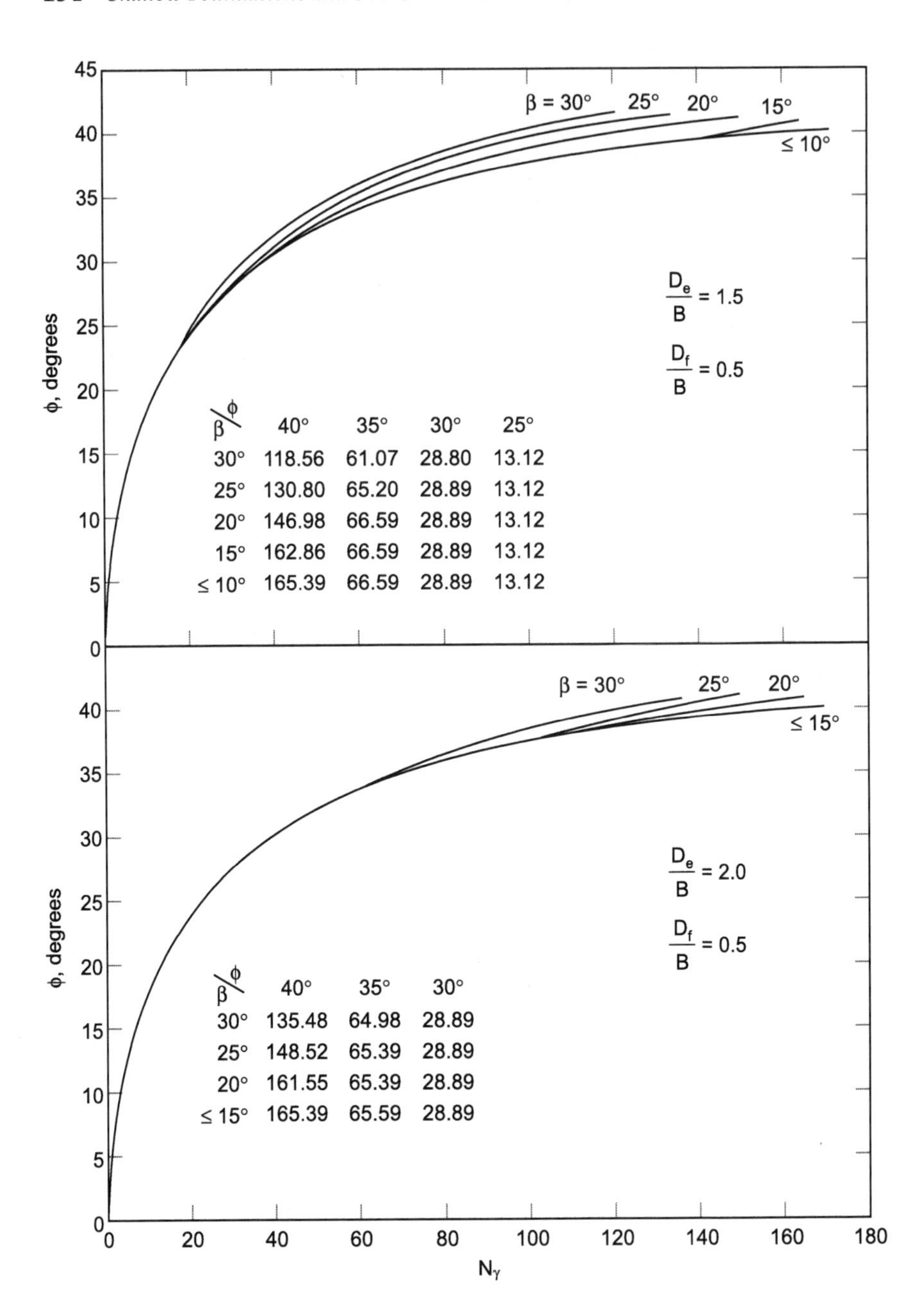

Fig. A6.7. N_r vs ϕ ($D_f/B = 0.5$ and $D_e/B = 1.5, 2.0$).

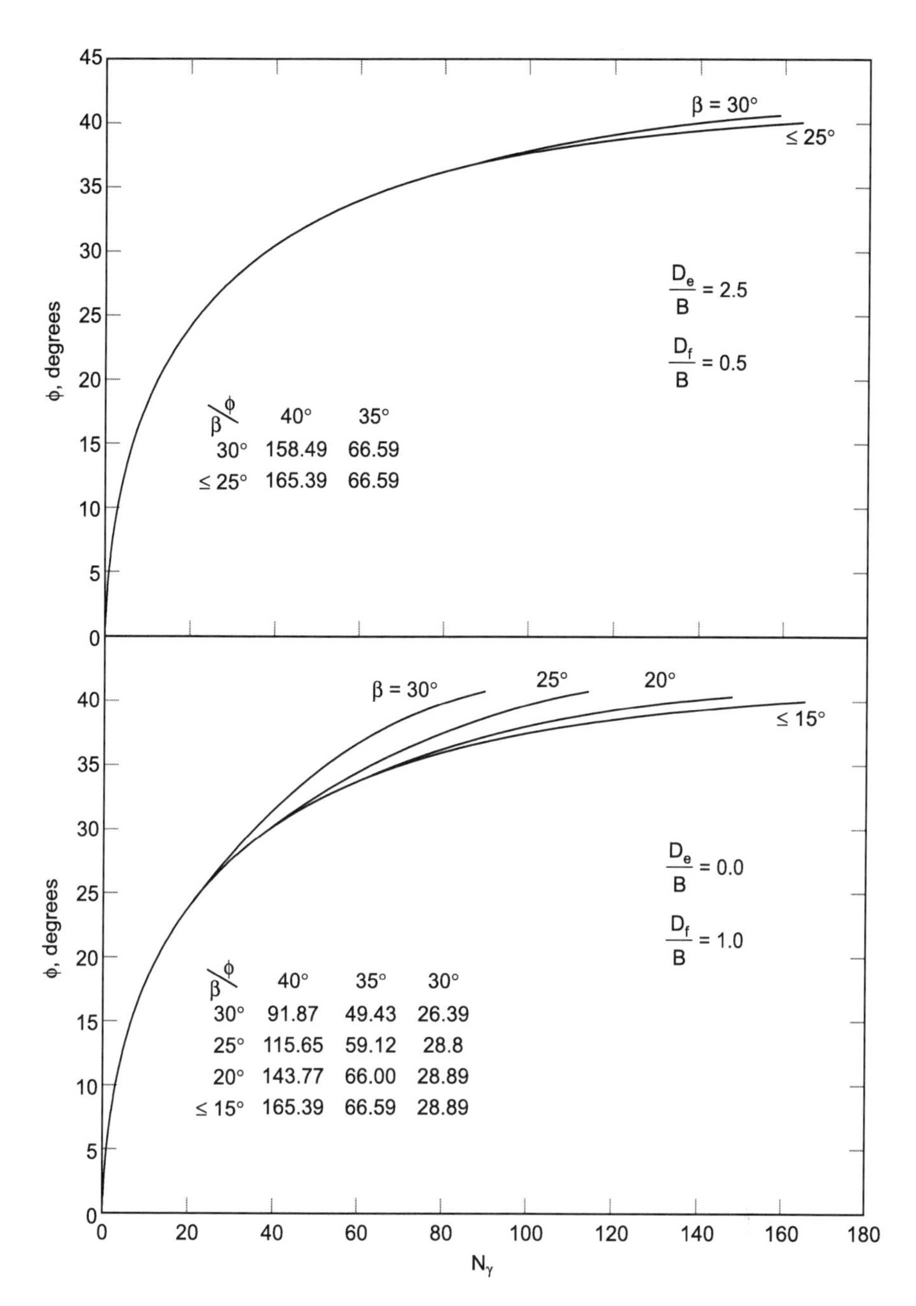

Fig. A6.8. N_r vs ϕ ($D_f/B = 0.5$, $D_e/B = 2.5$ and $D_f/B = 1.0$, $D_e/B = 0.0$).

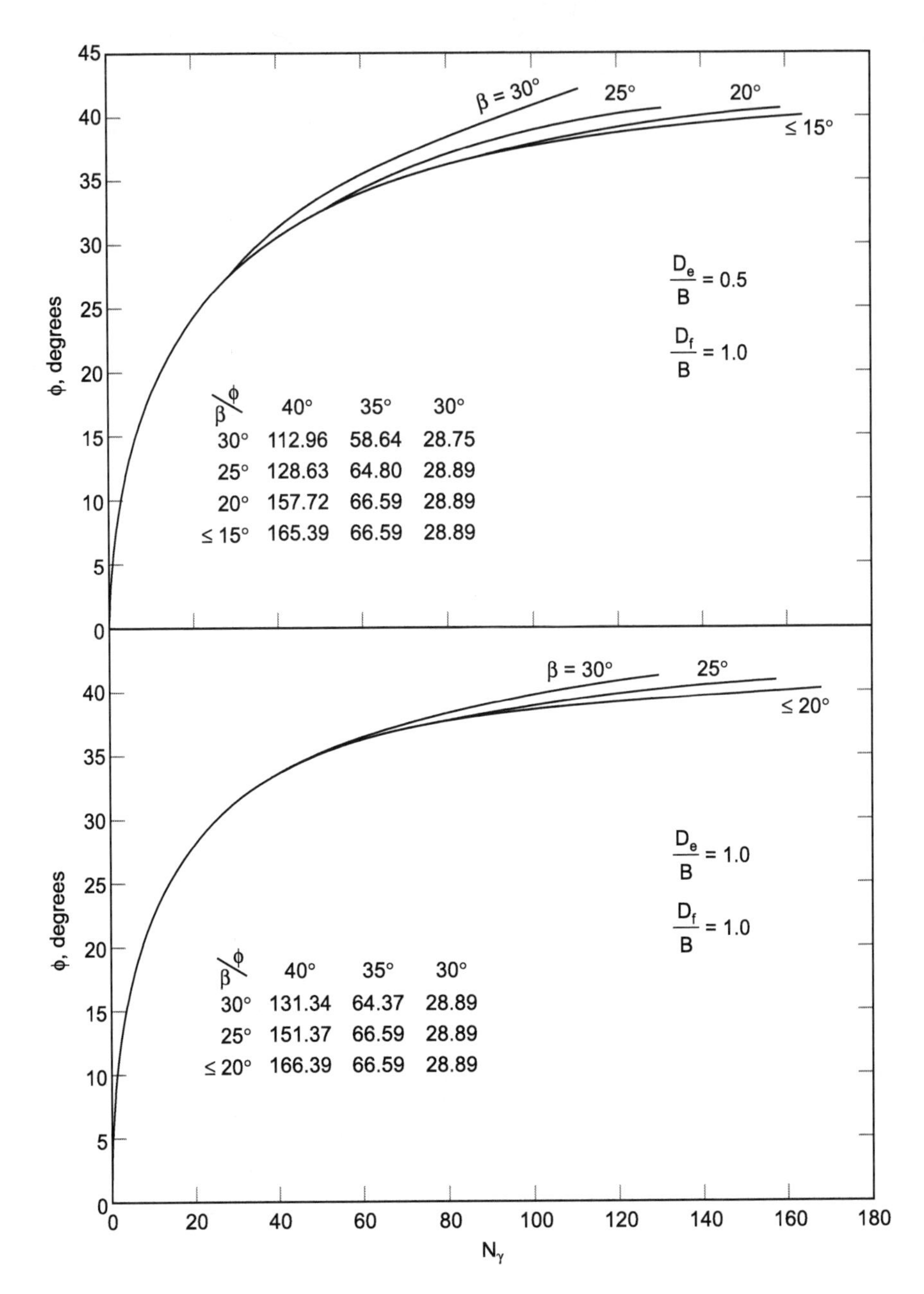

Fig. A6.9. N_r vs ϕ ($D_f/B = 1.0$ and $D_e/B = 0.5, 1.0$).

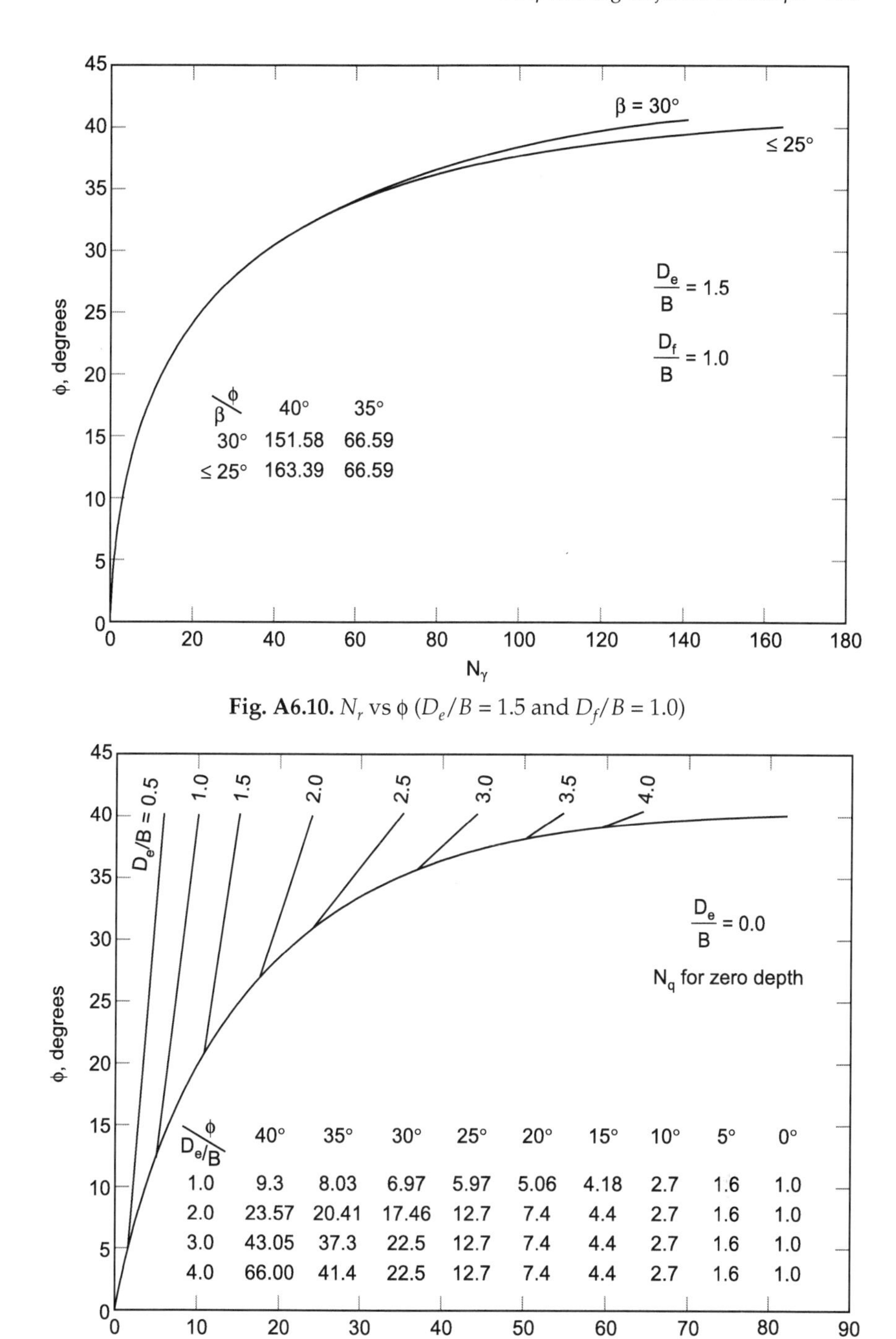

Fig. A6.10. N_r vs φ ($D_e/B = 1.5$ and $D_f/B = 1.0$)

Fig. A6.11. N_q vs φ for various values of D_e/B ($D_f/B = 0.0$).

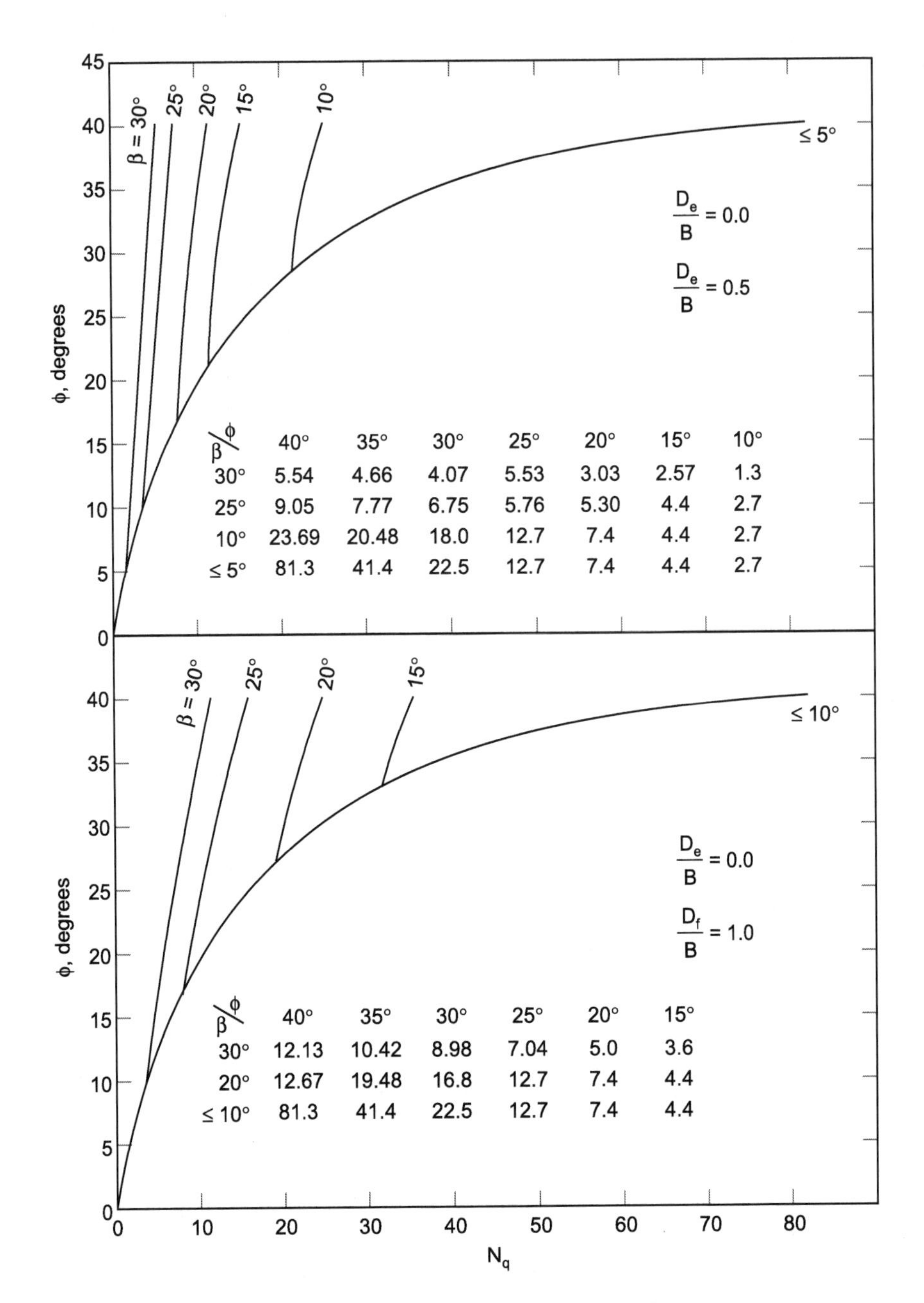

Fig. A6.12. N_q vs ϕ ($D_f/B = 0.5$, 1.0 and $D_e/B = 0.0$).

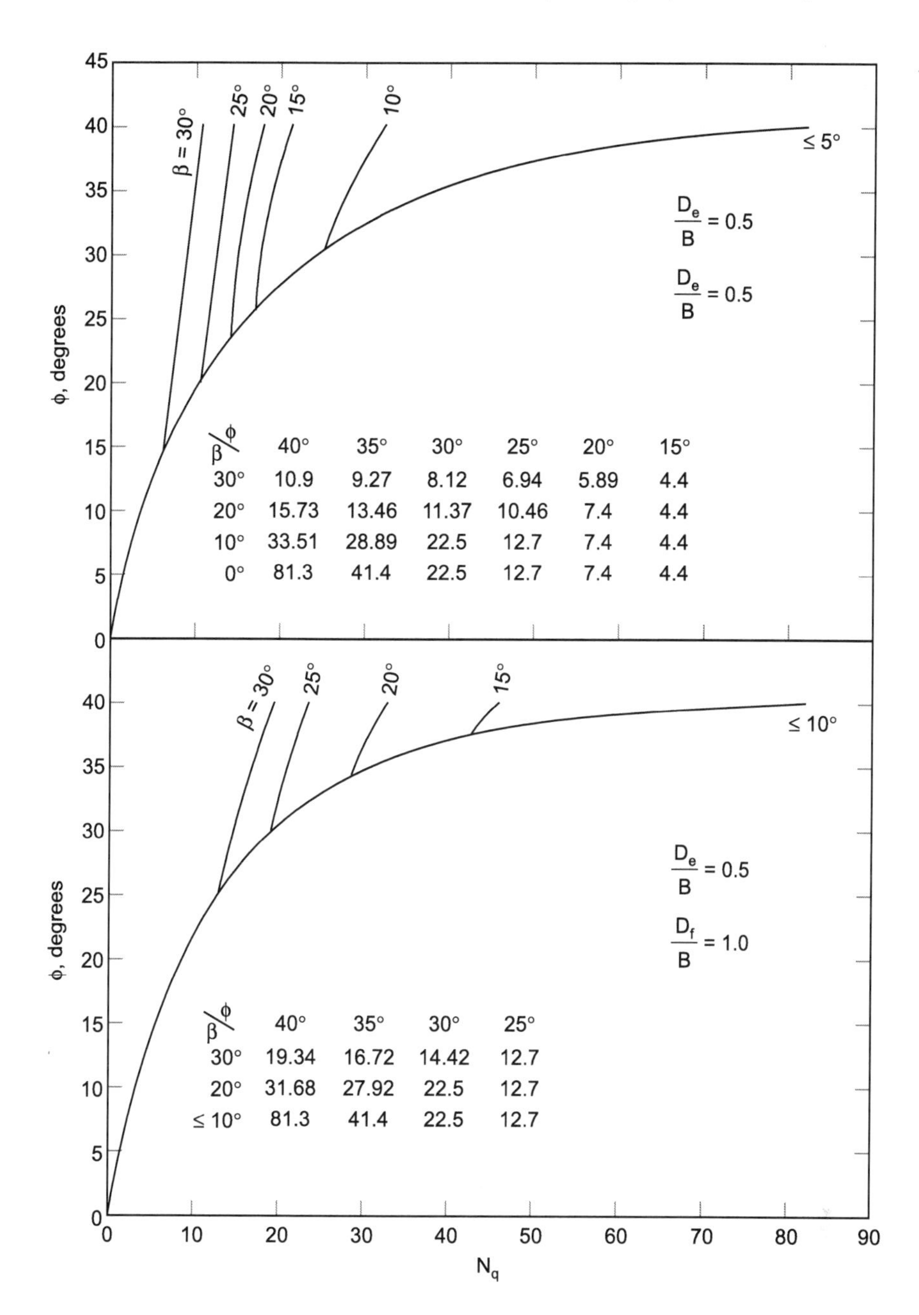

Fig. A6.13. N_q vs ϕ ($D_f/B = 0.5, 1.0$ and $D_e/B = 0.5$).

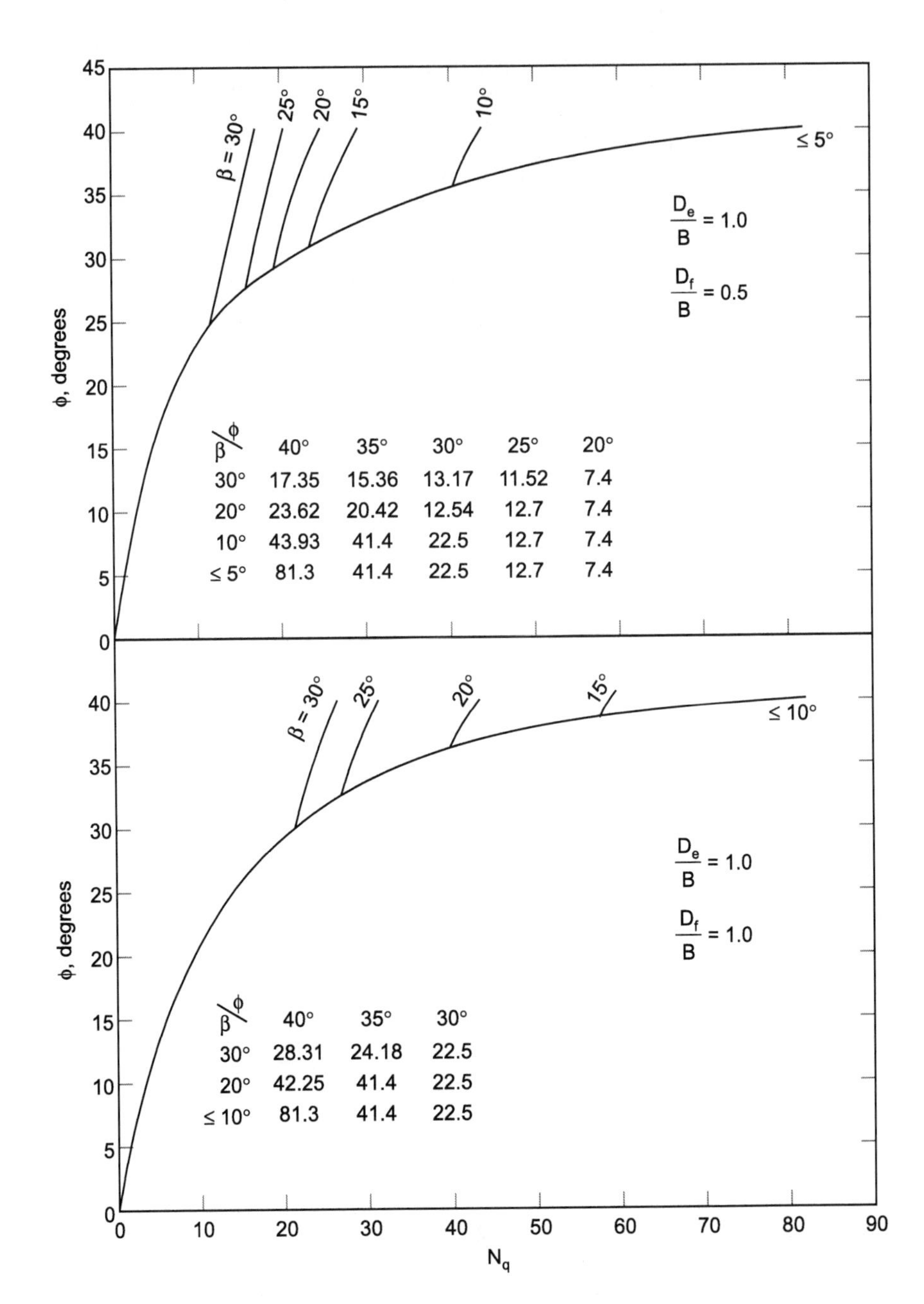

Fig. A6.14. N_q vs ϕ (D_f/B = 0.5, 1.0 and D_e/B = 1.0).

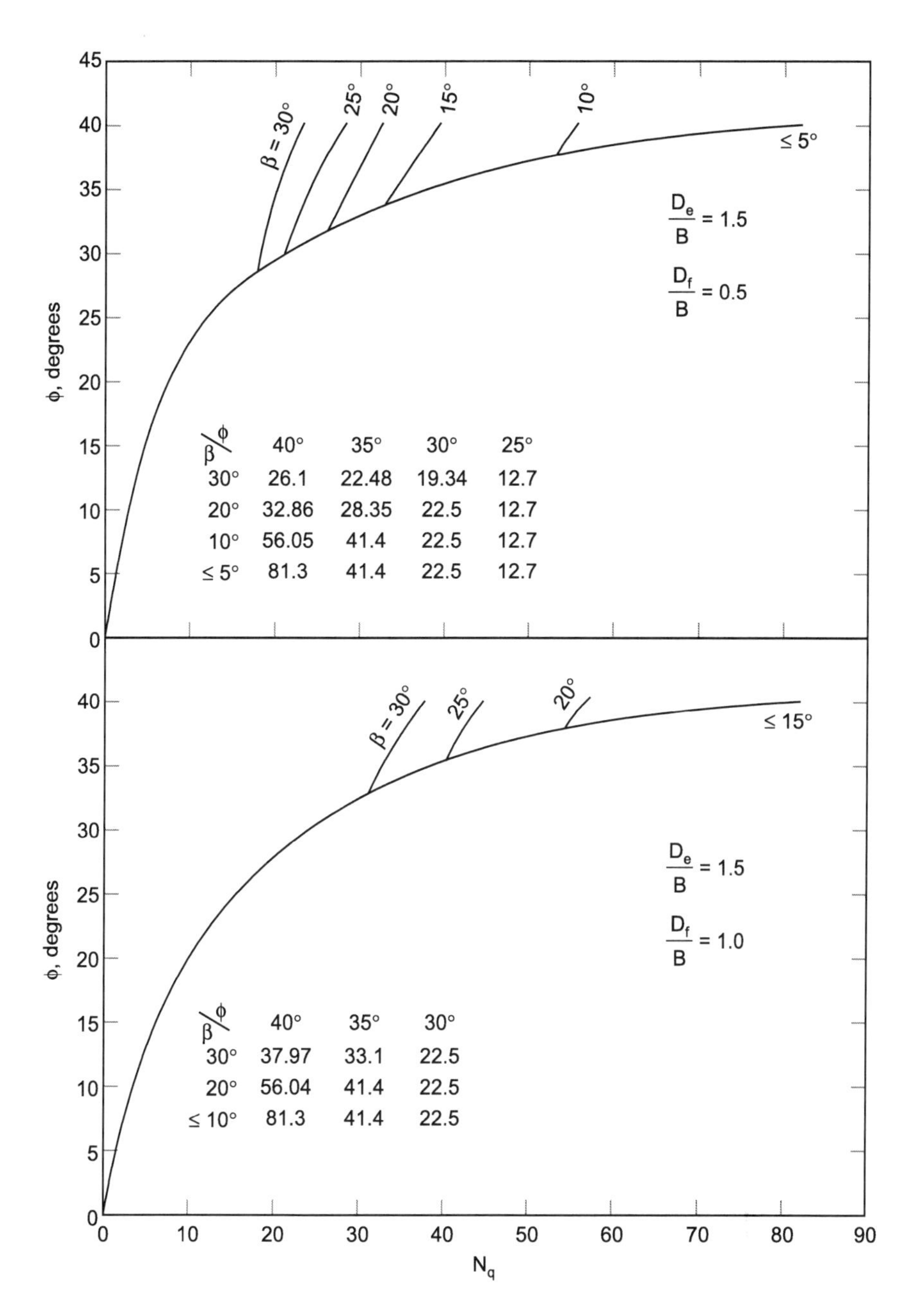

Fig. A6.15. N_q vs ϕ ($D_f/B = 0.5, 1.0$ and $D_e/B = 1.5$).

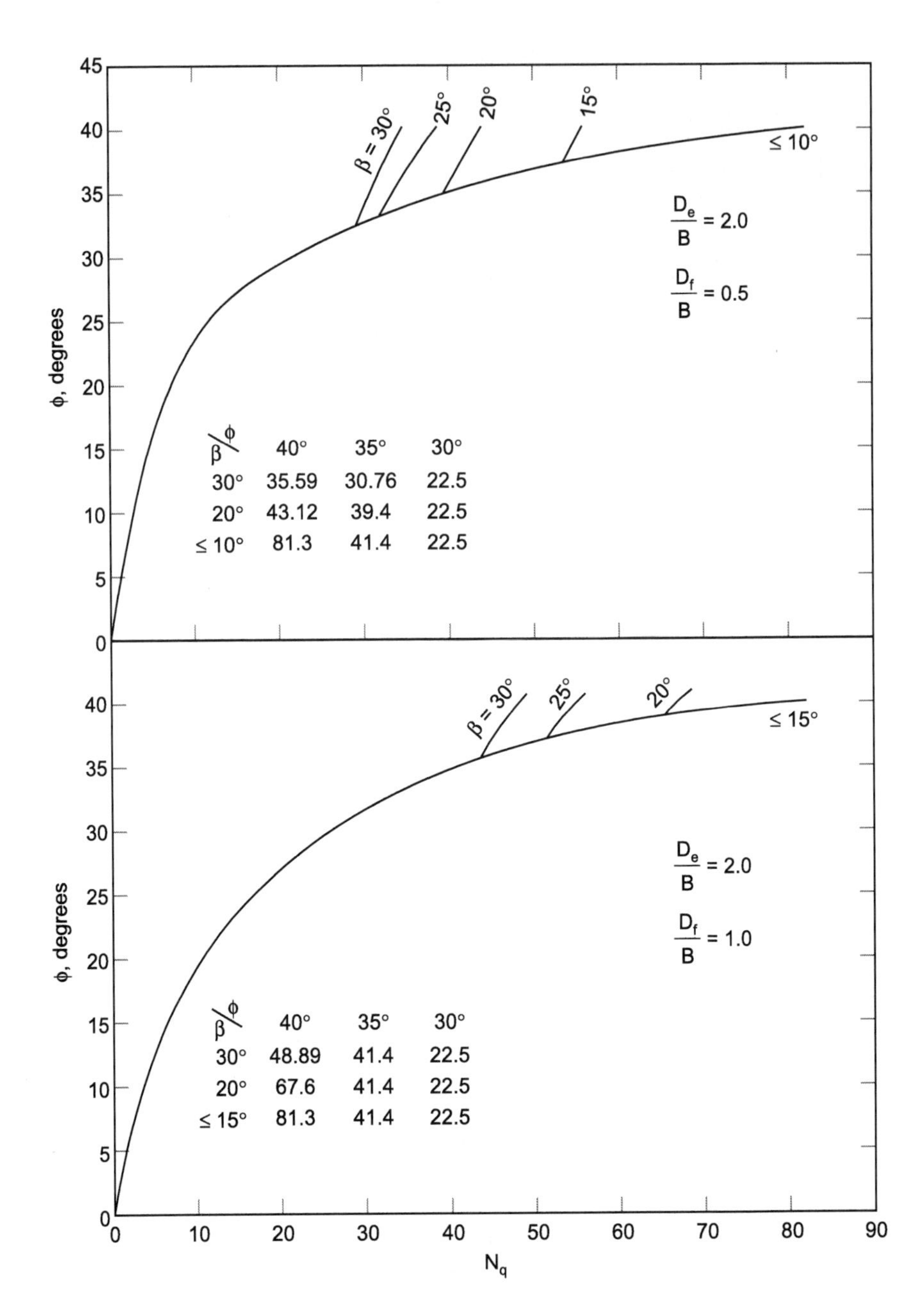

Fig. A6.16. N_q vs φ (D_f/B = 0.5, 1.0 and D_e/B = 2.0).

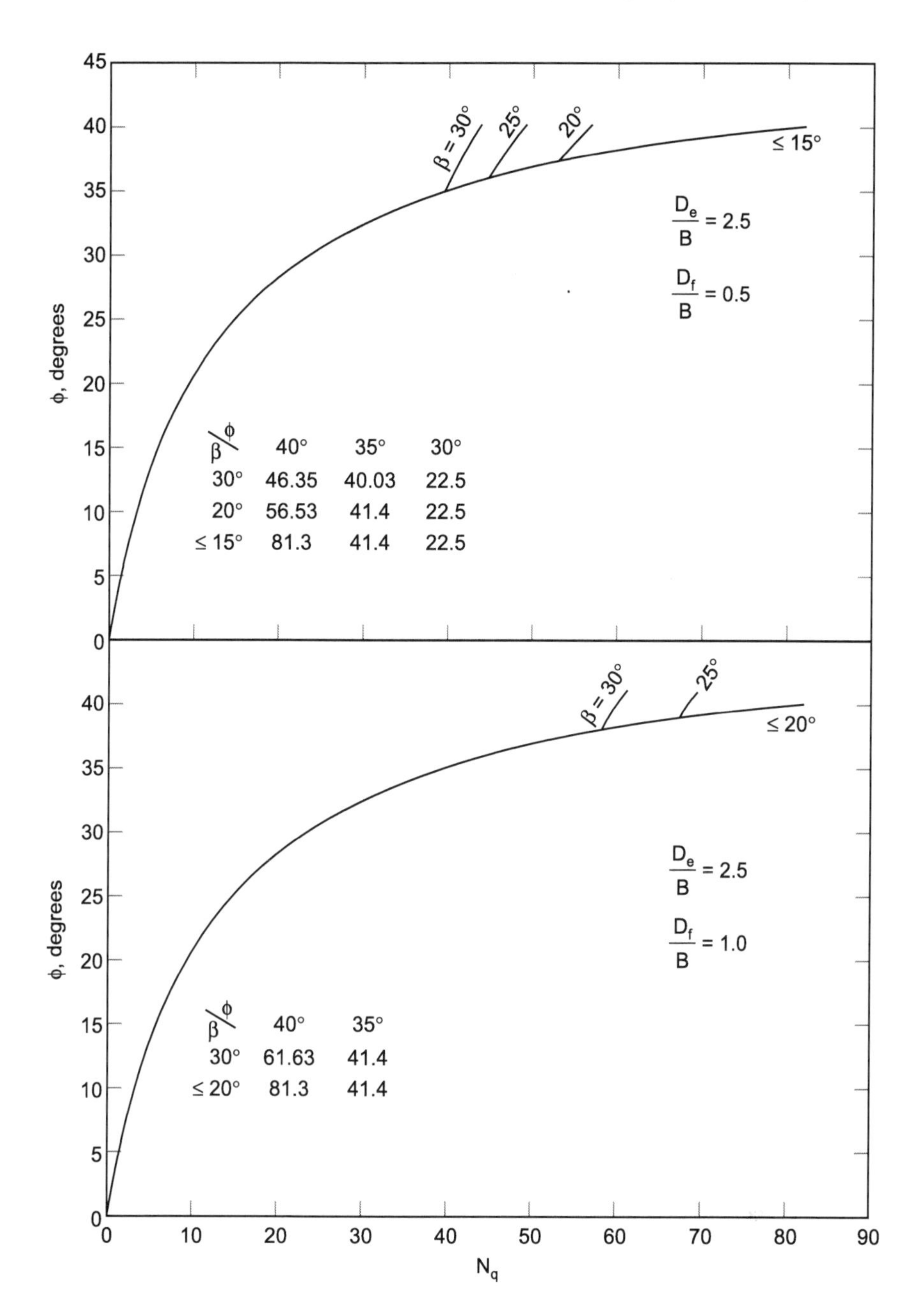

Fig. A6.17. N_q vs ϕ (D_f/B = 0.5, 1.0 and D_e/B = 2.5).

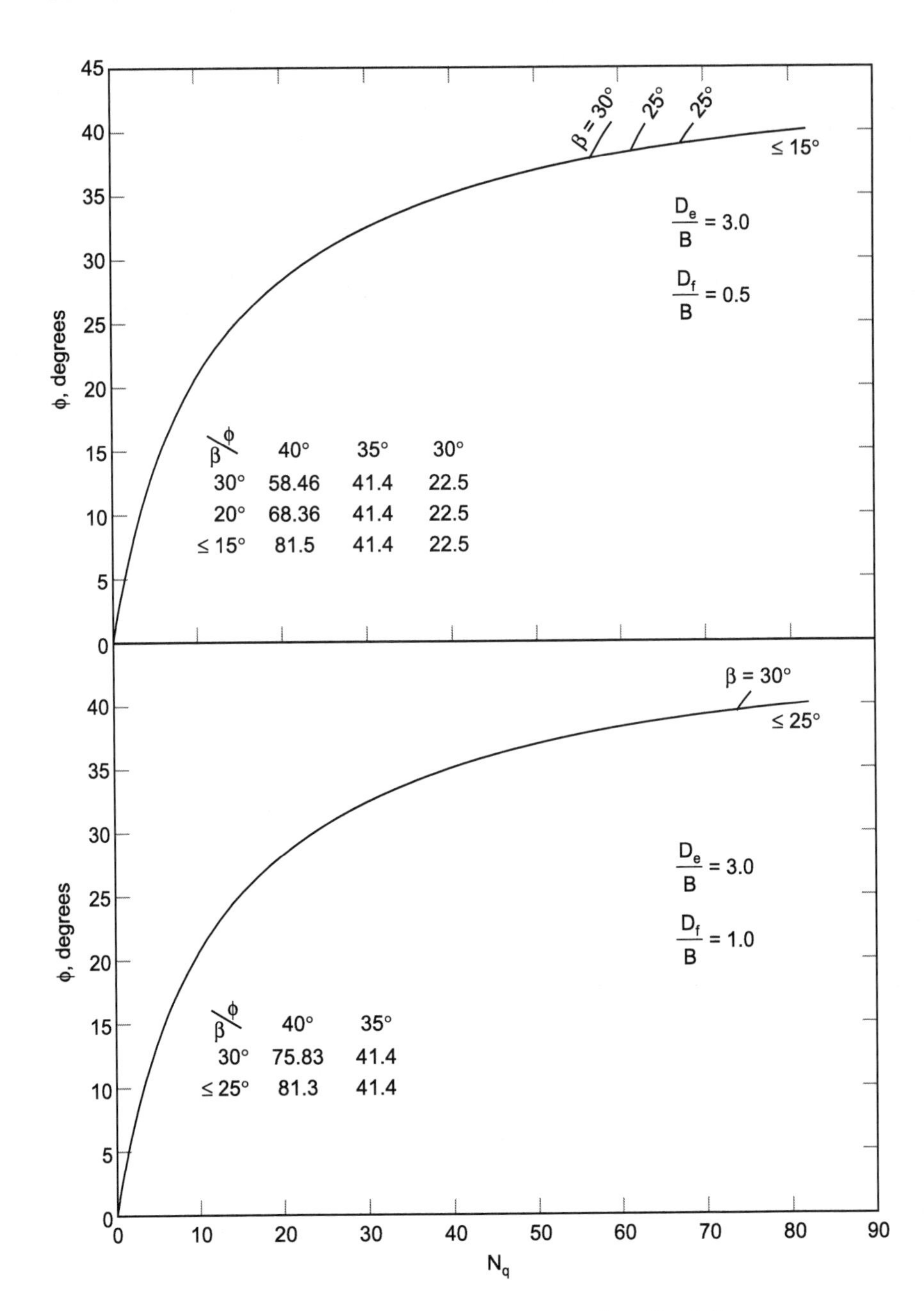

Fig. A6.18. N_q vs ϕ (D_f/B = 0.5, 1.0 and D_e/B = 3.0).

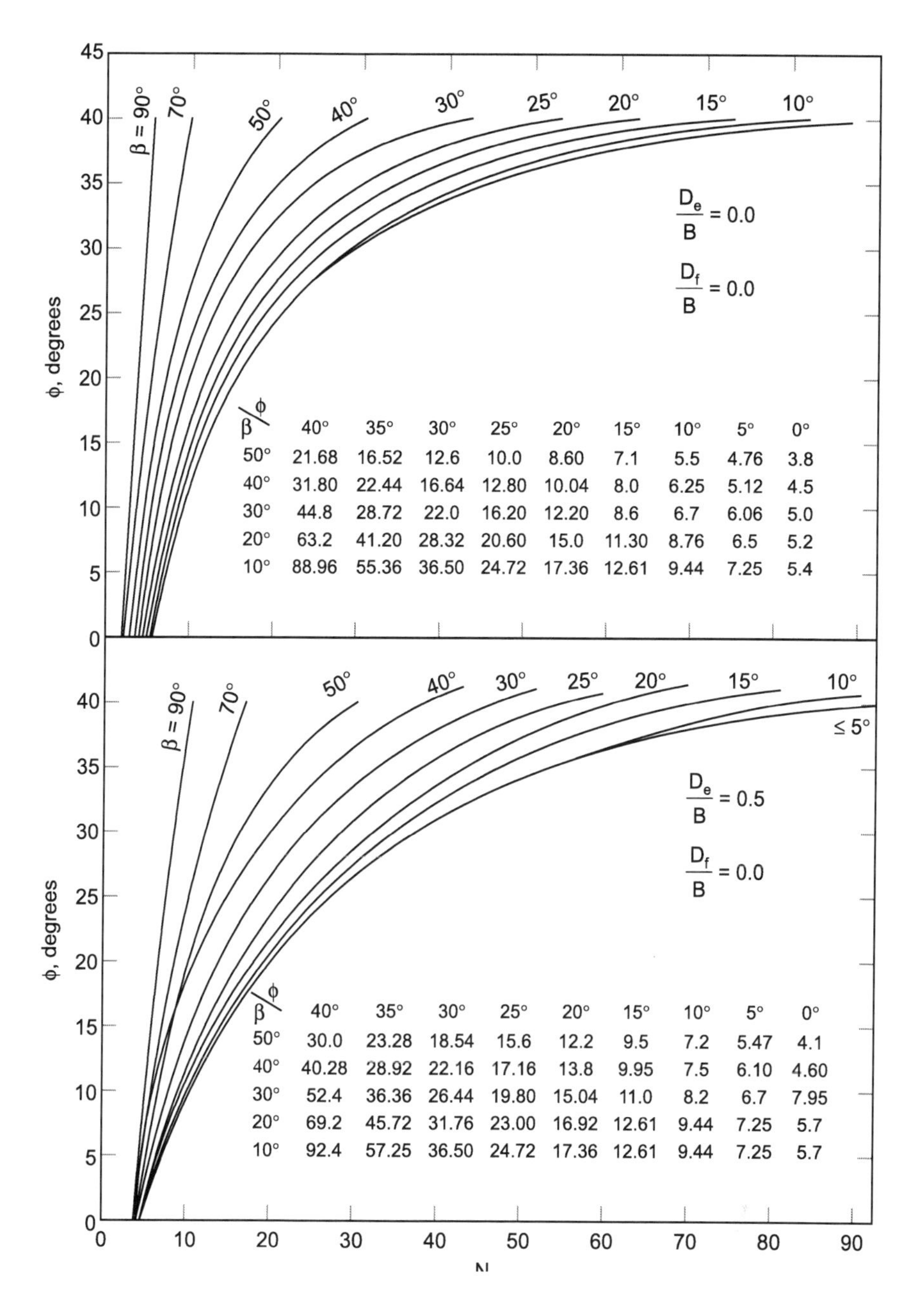

β \ φ	40°	35°	30°	25°	20°	15°	10°	5°	0°
50°	21.68	16.52	12.6	10.0	8.60	7.1	5.5	4.76	3.8
40°	31.80	22.44	16.64	12.80	10.04	8.0	6.25	5.12	4.5
30°	44.8	28.72	22.0	16.20	12.20	8.6	6.7	6.06	5.0
20°	63.2	41.20	28.32	20.60	15.0	11.30	8.76	6.5	5.2
10°	88.96	55.36	36.50	24.72	17.36	12.61	9.44	7.25	5.4

β \ φ	40°	35°	30°	25°	20°	15°	10°	5°	0°
50°	30.0	23.28	18.54	15.6	12.2	9.5	7.2	5.47	4.1
40°	40.28	28.92	22.16	17.16	13.8	9.95	7.5	6.10	4.60
30°	52.4	36.36	26.44	19.80	15.04	11.0	8.2	6.7	7.95
20°	69.2	45.72	31.76	23.00	16.92	12.61	9.44	7.25	5.7
10°	92.4	57.25	36.50	24.72	17.36	12.61	9.44	7.25	5.7

Fig. A6.19. N_C vs ϕ ($D_f/B = 0.0$ and $D_e/B = 0.0, 0.5$).

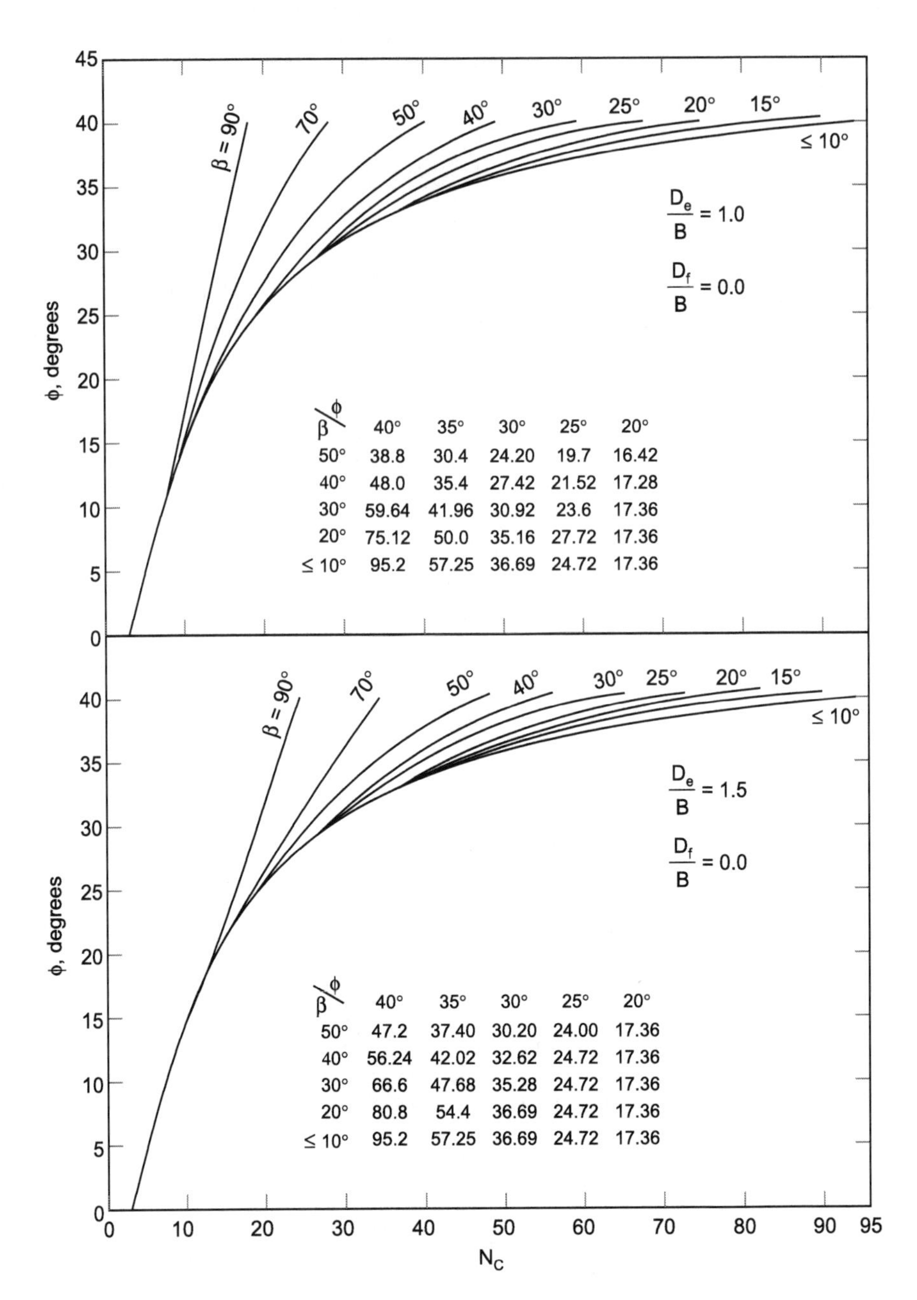

Fig. A6.20. N_C vs φ (D_f/B = 0.0 and D_e/B = 1.0, 1.5).

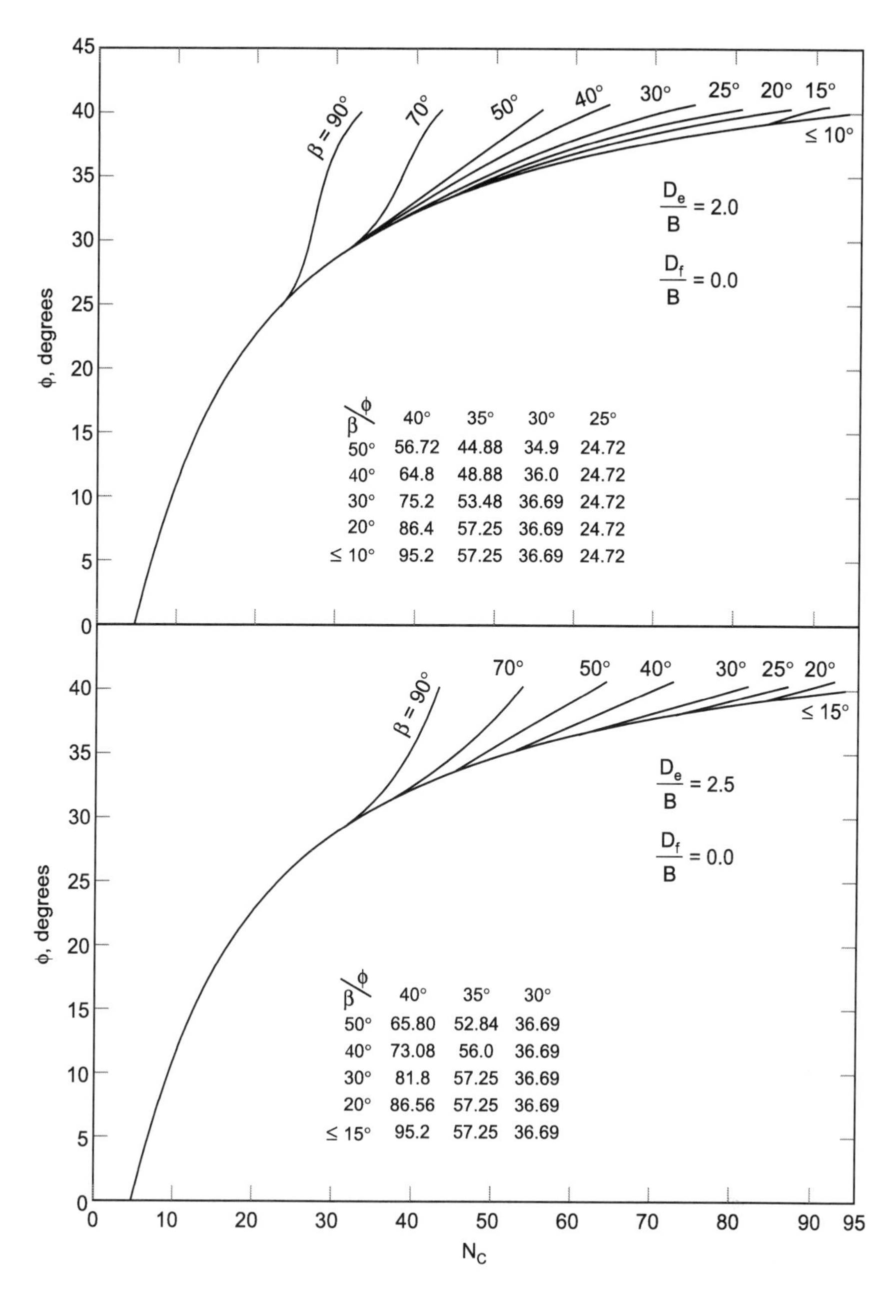

Fig. A6.21. N_C vs ϕ ($D_f/B = 0.0$ and $D_e/B = 2.0, 2.5$).

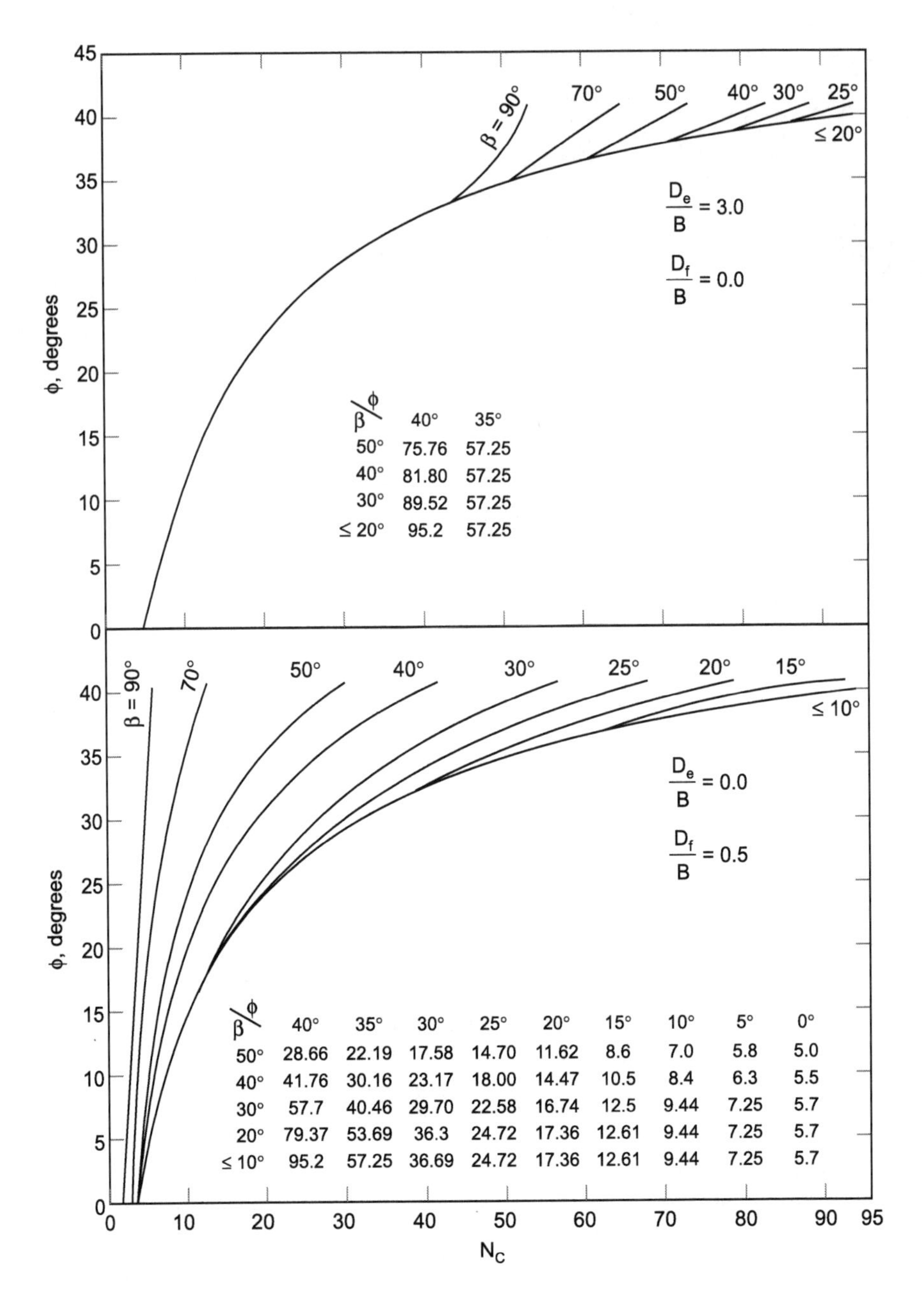

β \ φ	40°	35°
50°	75.76	57.25
40°	81.80	57.25
30°	89.52	57.25
≤ 20°	95.2	57.25

β \ φ	40°	35°	30°	25°	20°	15°	10°	5°	0°
50°	28.66	22.19	17.58	14.70	11.62	8.6	7.0	5.8	5.0
40°	41.76	30.16	23.17	18.00	14.47	10.5	8.4	6.3	5.5
30°	57.7	40.46	29.70	22.58	16.74	12.5	9.44	7.25	5.7
20°	79.37	53.69	36.3	24.72	17.36	12.61	9.44	7.25	5.7
≤ 10°	95.2	57.25	36.69	24.72	17.36	12.61	9.44	7.25	5.7

Fig. A6.22. N_C vs φ ($D_f/B = 0.0$, $D_e/B = 3.0$ and $D_f/B = 0.5$, $D_e/B = 0.0$).

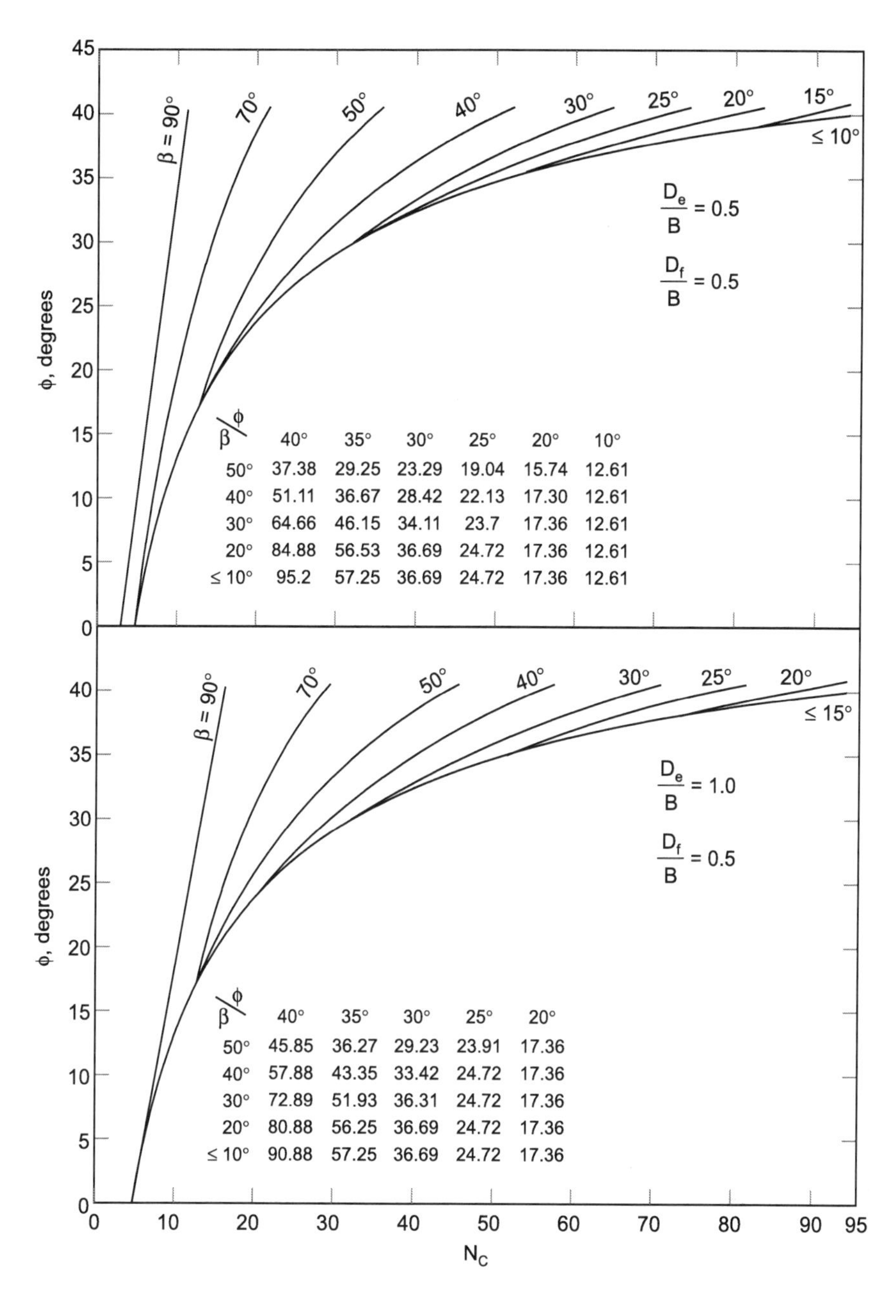

Fig. A6.23. N_C vs ϕ ($D_f/B = 0.5$ and $D_e/B = 0.5, 1.0$).

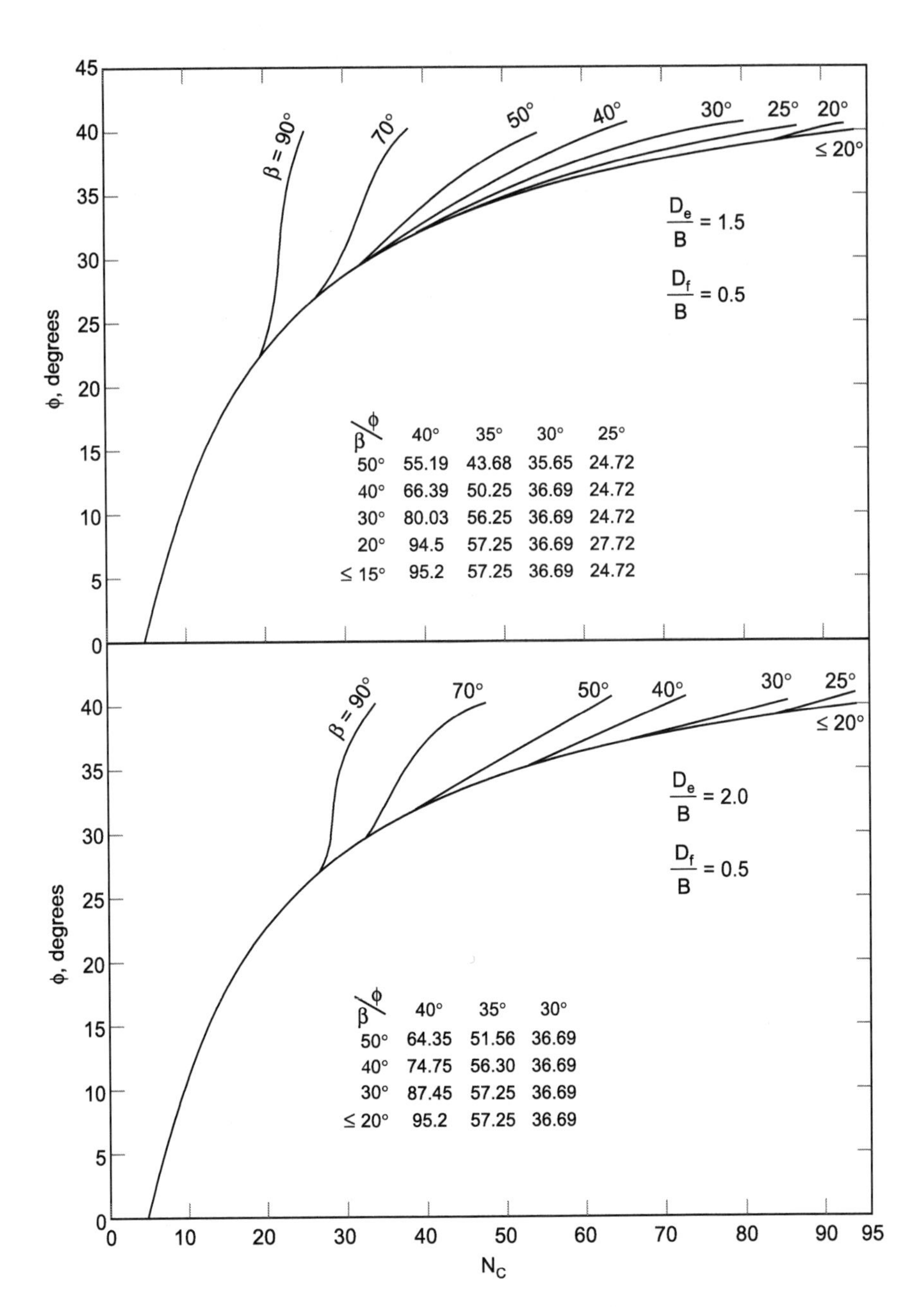

β \ φ	40°	35°	30°	25°
50°	55.19	43.68	35.65	24.72
40°	66.39	50.25	36.69	24.72
30°	80.03	56.25	36.69	24.72
20°	94.5	57.25	36.69	27.72
≤ 15°	95.2	57.25	36.69	24.72

β \ φ	40°	35°	30°
50°	64.35	51.56	36.69
40°	74.75	56.30	36.69
30°	87.45	57.25	36.69
≤ 20°	95.2	57.25	36.69

Fig. A6.24. N_C vs ϕ ($D_f/B = 0.5$ and $D_e/B = 1.0, 2.0$).

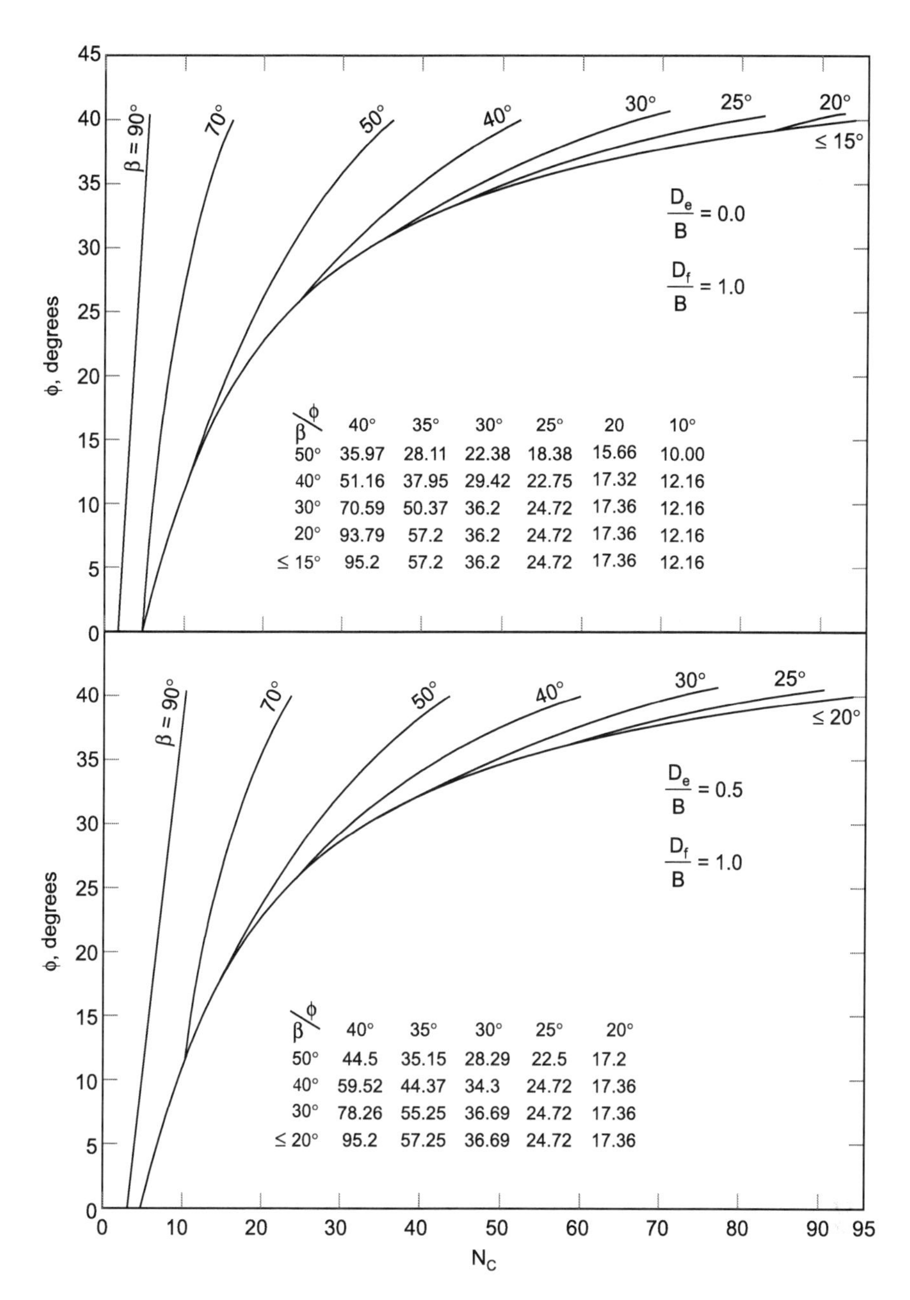

β \ φ	40°	35°	30°	25°	20	10°
50°	35.97	28.11	22.38	18.38	15.66	10.00
40°	51.16	37.95	29.42	22.75	17.32	12.16
30°	70.59	50.37	36.2	24.72	17.36	12.16
20°	93.79	57.2	36.2	24.72	17.36	12.16
≤ 15°	95.2	57.2	36.2	24.72	17.36	12.16

β \ φ	40°	35°	30°	25°	20°
50°	44.5	35.15	28.29	22.5	17.2
40°	59.52	44.37	34.3	24.72	17.36
30°	78.26	55.25	36.69	24.72	17.36
≤ 20°	95.2	57.25	36.69	24.72	17.36

Fig. A6.25. N_C vs ϕ ($D_f/B = 1.0$ and $D_e/B = 0.0, 0.5$).

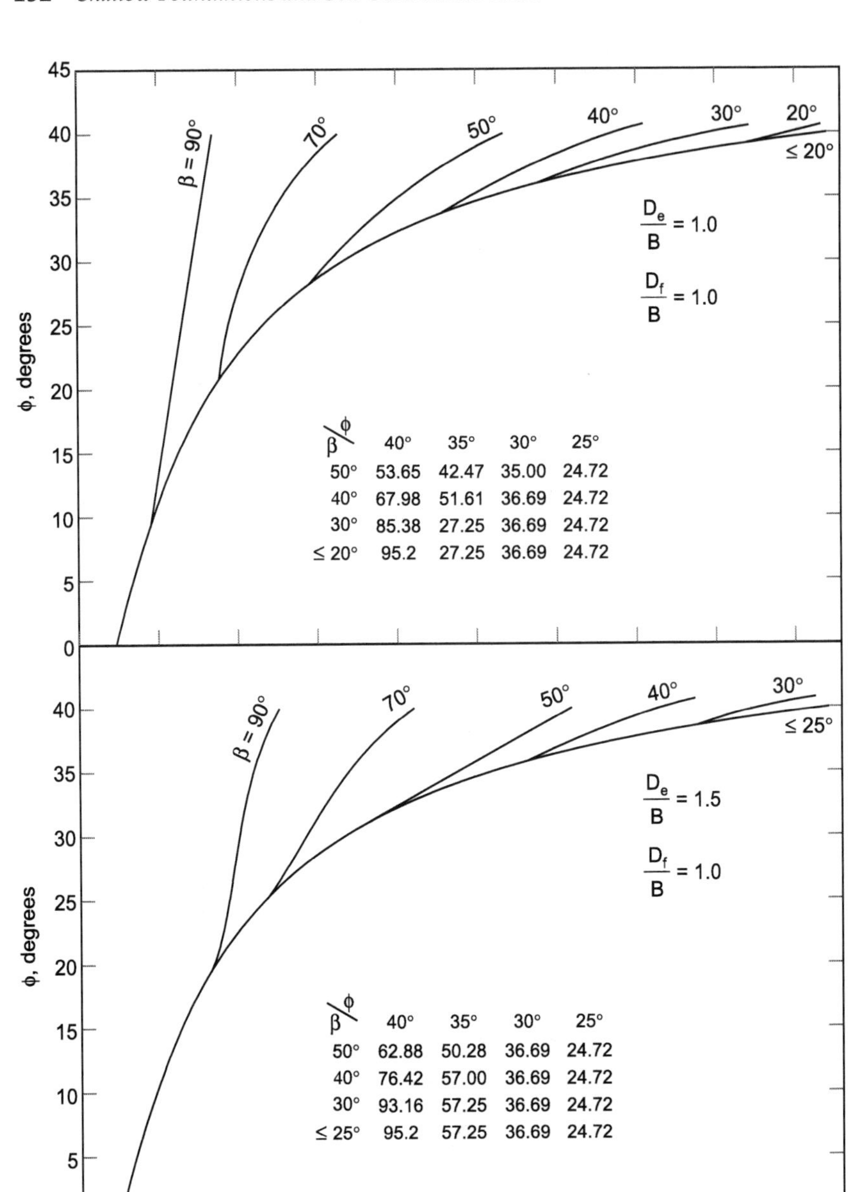

Fig. A6.26. N_C vs ϕ ($D_f/B = 1.0$ and $D_e/B = 1.0, 1.5$).

CHAPTER

7

Square and Rectangular Footings

7.1 General

It is common to use square and rectangular footings for columns depending on their shapes. Bearing capacity and settlement are their basic design criteria. So far both bearing capacity and settlement are obtained in two independent steps.

For obtaining bearing capacity of square and rectangular footings, the expression of bearing capacity obtained for a strip footing has been modified using suitable shape factors (Terzaghi, 1943; Skempton, 1951; Meyerhof, 1951, 1963, 1965; Hansen, 1970; De Beer, 1970). Michalowski (2001) carried out limit analysis for obtaining upper bound estimates of bearing capacity of square and rectangular footings. He gave the results in terms of bearing capacity coefficients, and shape factors applicable as modifiers in the bearing capacity solution for strip footings. Michalowski and Dawson (2002) analyzed the problem of square footing using (i) through code FLAC and (ii) limit analysis independently. Using elasto-plastic model and finite element analysis, Zhu and Michalowski (2005) gave new suggestions for the shape factors. They found that the earlier factors modifying the contribution of cohesion and overburden are conservative. However, the earlier shape factor that affects the contribution of soil weight to the bearing capacity indicate contradictory trends. Gourvenec et al. (2006) analyzed the square and rectangular footings resting on undrained clay using finite element technique and found the results in close proximity of equation developed by Skempton (1951).

Using 3-D random finite element method (RFEM), Griffiths and Fenton (2005) developed a general probabilistic design framework for assessing settlements of rectangular footings.

It was found from earlier studies that the base roughness of the footing has no significant effect on its pressure-settlement characteristics. Therefore, the case of smooth footing has only been studied in square and rectangular footings. Firstly, the case of flexible smooth square footing is considered the contact pressure distribution for centrally-loaded square footing as being uniform but three dimensional. All stress equations are not available for such a case so as to use constitutive laws in the analysis. All stress equations for point load are available. Therefore, stresses in soil mass due to centrally loaded square footing have been computed by dividing the whole base area into n equal squares as shown in Fig. 7.1. The squares are very small; therefore, the total load in each square has been taken as point load (Sharan, 1977). The analysis has been extended for rectangular footings and considered resting either on saturated clay or sand.

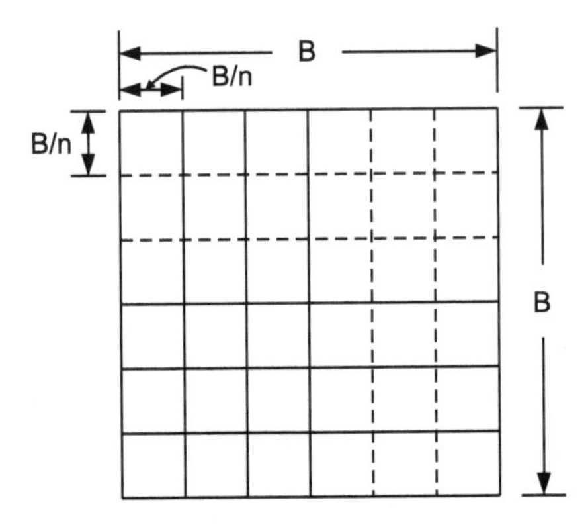

Fig. 7.1. Division of square footing in small sqaures.

7.2 Smooth Flexible Square Footing on Clay

The following procedure has been adopted for the analysis of this case:

Step 1: Evaluation of contact pressure

The contact pressure distribution has been assumed uniform because the footing is considered flexible.

Step 2: Division of soil strata

The soil mass supporting the footing has been divided into N layers as shown in Fig. 7.2. The number of vertical section has been considered passing through the different points on the base of the footing (a_1, a_2 ...; b_1, b_2 ...; etc.) (Fig. 7.2).

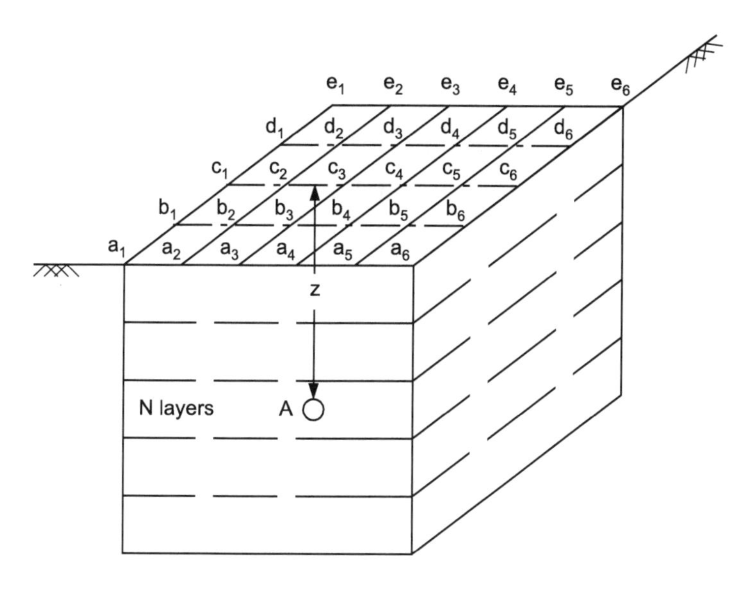

Fig. 7.2. Soil strata divided into n thin layers.

Step 3: Division of base of the footing

Base area of the footing has been divided into n small areas as shown in Figure 7.1. Area of each small square will be A/n^2. Let the base pressure intensity be q $(=V/A)$, V being the vertical load on the footing, and A is the base area of the footing equal to $B \times B$.B which is the width of the footing.

Step 4: Evaluation of stresses

Each small base area of the footing will be subjected to a point load $P(= q.A/n^2)$. The stresses due to point loads at the centre of each layer along a vertical section have been computed using Eqs (7.1) to (7.6) as given below.

$$\sigma_x = \frac{3P}{2\pi} \times \left[\frac{x^2 z}{R^5} - \left(\frac{m-2}{3m} \right) \left\{ -\frac{1}{R(R+z)} + \frac{(2R+z)x^2}{(R+z)^2 R^3} + \frac{z}{R^3} \right\} \right] \tag{7.1}$$

$$\sigma_y = \frac{3P}{2\pi} \times \left[\frac{y^2 z}{R^5} - \left(\frac{m-2}{3m} \right) \left\{ -\frac{1}{R(R+z)} + \frac{(2R+z)y^2}{(R+z)^2 R^3} + \frac{z}{R^3} \right\} \right] \tag{7.2}$$

$$\sigma_z = \frac{3P}{2\pi} \times \frac{z^3}{R^5} \tag{7.3}$$

$$\tau_{xz} = \frac{3P}{2\pi} \times \frac{xz^2}{R^5} \tag{7.4}$$

$$\tau_{yz} = \frac{3P}{2\pi} \times \frac{yz^2}{R^5} \tag{7.5}$$

$$\tau_{xy} = \frac{3P}{2\pi} \times \left[\frac{xyz}{R^5} - \left(\frac{m-2}{3m} \right) \times \left\{ \frac{(2R+z)xy}{(R+z)^2 R^3} \right\} \right] \tag{7.6}$$

Here x, y and z are the orthogonal coordinates,

$$R = \sqrt{r^2 + z^2} = \sqrt{x^2 + y^2 + z^2} \tag{7.7}$$

$$m = \frac{1}{\mu} = \text{Poisson's coefficient, and } \mu = \text{Poisson's ratio.}$$

Step 5: Evaluation of principal stresses

The principal stresses and their directions with respect to vertical Z-axis have been computed using equations of the theory of elasticity (Durelli, 1958; Sokolniko, 1956; Selby, 1972).

The stresses in variants are:

$$I_1 = \sigma_x + \sigma_y + \sigma_z \tag{7.8}$$

$$I_2 = \sigma_x . \sigma_y + \sigma_y . \sigma_z + \sigma_z . \sigma_x - \tau^2_{xy} - \tau^2_{yz} - \tau^2_{zx} \tag{7.9}$$

$$I_3 = \sigma_x . \sigma_y . \sigma_z - \sigma_x . \tau^2_{yz} - \sigma_y . \tau^2_{zx} - \sigma_z . \tau^2_{xy} + 2\tau_{xy} . \tau_{yz} . \tau_{zx} \tag{7.10}$$

Also

$$I_1 = -\sigma_1 + \sigma_2 + \sigma_3 \tag{7.11}$$

$$I_1 = \sigma_1 . \sigma_2 + \sigma_2 . \sigma_3 + \sigma_3 . \sigma_1 \tag{7.12}$$

$$I_3 = \sigma_1 . \sigma_2 . \sigma_3 \tag{7.13}$$

Solving for σ_1, σ_2 and σ_3 we have

$$\sigma_1^3 - I_1 \sigma_1^2 + I_2 \sigma_1 - I_3 = 0 \tag{7.14}$$

$$\sigma_2^3 - I_1 \sigma_2^2 + I_2 \sigma_2 - I_3 = 0 \tag{7.15}$$

$$\sigma_3^3 - I_1 \sigma_3^2 + I_2 \sigma_3 - I_3 = 0 \tag{7.16}$$

Thus σ_1, σ_2 and σ_3 are the roots of equation

$$\sigma^3 + p\sigma^2 + q\sigma + r = 0 \tag{7.17}$$

where $p = -I_1$, $q = I_2$, and $r = -I_3$

Values of I_1, I_2 and I_3 are taken from Eqs (7.8), (7.9) and (7.10) respectively. Solving Eq. (7.17), three values of σ_1, σ_2 and σ_3 are obtained.

Solving for direction cosines with respect to z axis, we have (Sokolnikoff, 1956):

$$A_1 = (\sigma_y - \sigma_1)(\sigma_z - \sigma_1) - \tau_{yz} . \tau_{zy} \tag{7.18}$$

$$B_1 = -\tau_{xy}(\sigma_z - \sigma_1) + \tau_{yz}.\tau_{xz} \tag{7.19}$$

$$C_1 = \tau_{xy}.\tau_{yz} - (\sigma_y - \sigma_1)\tau_{xz} \tag{7.20}$$

$$A_2 = (\sigma_y - \sigma_2)(\sigma_z - \sigma_2) - \tau_{zy}.\tau_{xz} \tag{7.21}$$

$$B_2 = -\tau_{xy}(\sigma_z - \sigma_2) + \tau_{yz}.\tau_{xz} \tag{7.22}$$

$$C_2 = \tau_{xy}.\tau_{yz} - (\sigma_y - \sigma_2)\tau_{xz} \tag{7.23}$$

$$A_3 = (\sigma_y - \sigma_3)\,(\sigma_z - \sigma_3) - \tau_{yz}.\tau_{zy} \tag{7.24}$$

$$B_3 = -\tau_{xy}(\sigma_z - \sigma_3) + \tau_{zy}.\tau_{xz} \tag{7.25}$$

$$C_3 = \tau_{xy}.\tau_{yz} - (\sigma_y - \sigma_3)\tau_{xz} \tag{7.26}$$

And,

$$\cos\theta_1 = \frac{C_1}{\sqrt{A_1^2 + B_1^2 + C_1^2}} \tag{7.27}$$

$$\cos\theta_1 = \frac{C_2}{\sqrt{A_2^2 + B_2^2 + C_2^2}} \tag{7.28}$$

$$\cos\theta_1 = \frac{C_3}{\sqrt{A_3^2 + B_3^2 + C_3^2}} \tag{7.29}$$

Step 6: Evaluation of vertical strain

$$\varepsilon_z = \varepsilon_1\cos^2\theta_1 + \varepsilon_2\cos^2\theta_2 + \varepsilon_3\cos^2\theta_3 \tag{7.30}$$

The principal strains, ε_1, ε_2 and ε_3 were evaluated using the following equations:

$$\varepsilon_1 = \frac{a(\sigma_1 - \sigma_{3a})}{1 - b(\sigma_1 - \sigma_{3a})} \tag{7.31}$$

$$\varepsilon_2 = \frac{\sigma_2 - \mu(\sigma_3 + \sigma_1)}{\sigma_1 - \mu(\sigma_2 + \sigma_3)}.\varepsilon_1 \tag{7.32}$$

$$\varepsilon_1 = \frac{\sigma_3 - \mu(\sigma_1 + \sigma_2)}{\sigma_1 - \mu(\sigma_2 + \sigma_3)}.\varepsilon_1 \tag{7.33}$$

where $$\varepsilon_{3a} = \frac{\sigma_2 + \sigma_3}{2}$$

a and b are Kondner's hyperbola constants; μ is the Poisson's ratio.

Vertical strain ε_z at the point under consideration is then given by

$$\varepsilon_z = \varepsilon_1\cos^2\theta_1 + \varepsilon_2\cos^2\theta_2 + \varepsilon_3\cos^2\theta_3 \tag{7.34}$$

Step 7: Evaluation of total settlement

The total settlement along axis vertical section is obtained by numerical integration.

$$S = \int_{0}^{N} \varepsilon_z . dz \tag{7.35}$$

N represents the number of layers and dz the thickness of each layer.

Step 8: Evaluation of average settlement:

The total settlement was computed along the vertical section passing through the different points at the base of the footing. The average settlement is then computed by dividing the area of settlement diagram by the width of footing.

The evaluation of constitutive equations, settlement and pressure settlement curves is done as in case (a) of smooth flexible strip footing.

7.3 Smooth Rigid Square Footing on Clay

The analysis of rigid square footing on clay has also been done by dividing the whole area of foundation in n equal parts as shown in Fig. 7.1. The loads in each small square have been taken as point loads.

The contact pressure distribution has been assumed as shown in Fig. 7.3 in three-dimensional space and has been defined by two coefficients α_1 and α_2. The coefficients have been evaluated by the following assumptions:

(i) The total vertical load equals the volume of the contact pressure diagram (Fig. 7.3).

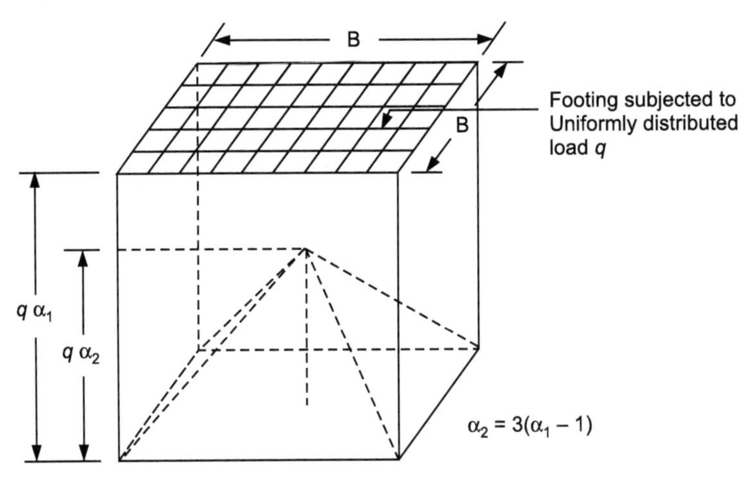

Fig. 7.3. Contact pressure distribution for uniformly loaded rigid square footing.

$$q.A = \alpha_1.q.A - \frac{1}{3}.\alpha_2.q.A$$

$$\alpha_2 = 3(\alpha_1 - 1) \tag{7.36}$$

When $\alpha_1 = 1$, $\alpha_2 = 0$ and when $\alpha_1 = 1.5$, $\alpha_2 = 1.5$

(ii) The values of settlements at different point on the base of the footing have been taken as equal since the footing is rigid. It facilitates the determination of α_1 and thus the contact pressure distribution. The evaluation of stresses at the centre of each layer, the principal stresses and vertical strains have been done as in the case of smooth flexible square footing.

7.4 Smooth Flexible and Rigid Rectangular Footings on Clay

Procedures for obtaining the pressure-settlement characteristics of smooth flexible and rigid rectangular footings are same as illustrated in Sections 7.2 and 7.3 respectively. In this case also, the division of the base of the rectangular footing is done in n small rectangular areas as shown in Fig. 7.4. Each area is assumed to be acted upon the concentrated load whose magnitude is equal to average pressure intensity acting on the considered small portion of the footing multiplied with its area. Values of σ_x, σ_y, σ_z, τ_{xz}, τ_{zy}, and τ_{xy} were then obtained using Eqs (7.1) to (7.6). Base contact pressure in rigid rectangular footing will be as shown in Fig. 7.5.

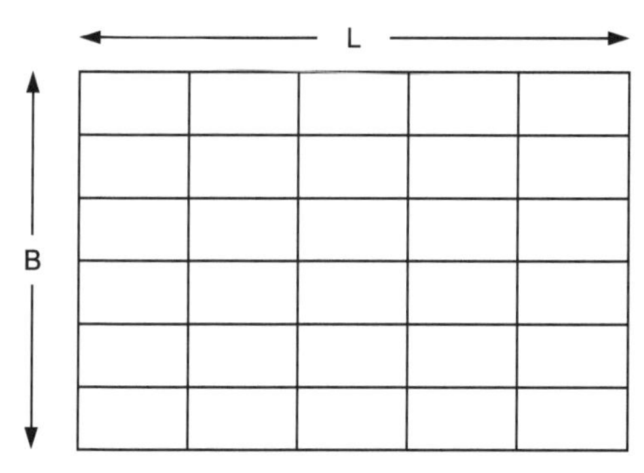

Fig. 7.4. Division of rectangular footing in small rectangles.

Rest of the procedure will be the same as discussed earlier for square footings.

7.5 Square and Rectangular Footings on Sand

The following steps have been adopted for getting the pressure-settlement characteristics of square and rectangular footings resting on sand:

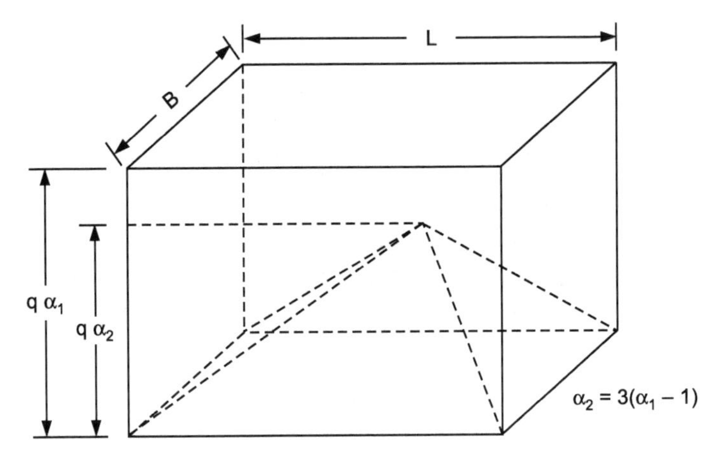

Fig. 7.5. Contact pressure distribution for uniformly loaded rigid rectugular footing.

1. The ultimate bearing capacity (q_u) has been computed using the following equation:

$$q_u = \frac{1}{2}.\gamma.B.N_\gamma.S_\gamma.r'_w \tag{7.37}$$

where γ = Unit weight of soil, kN/m^3

B = Width of the footing, m

N_γ = Non-dimensional bearing capacity factor (Table 7.1)

S_γ = Shape factor, Table 7.2

r'_w = Water table correction factor

$$r'_w = 0.5 + 0.5\frac{d_b}{B} \tag{7.38}$$

when d_b represents the position of water table with respect to the base of footing (Fig. 7.6).

Table 7.1. Bearing capacity factor N_γ

ϕ*(Deg)*	N_γ
0	0.00
5	0.45
10	1.22
15	2.65
20	5.39
25	10.88
30	22.40
35	48.03
40	109.41
45	271.76

Table 7.2. Shape factors

S.No.	*Shape of base of footing*	*Shape factors*
1	Continuous strip	1.0
2	Rectangular	1-0.4B/L
3	Square	0.8
4	Circle (B=diameter)	0.6

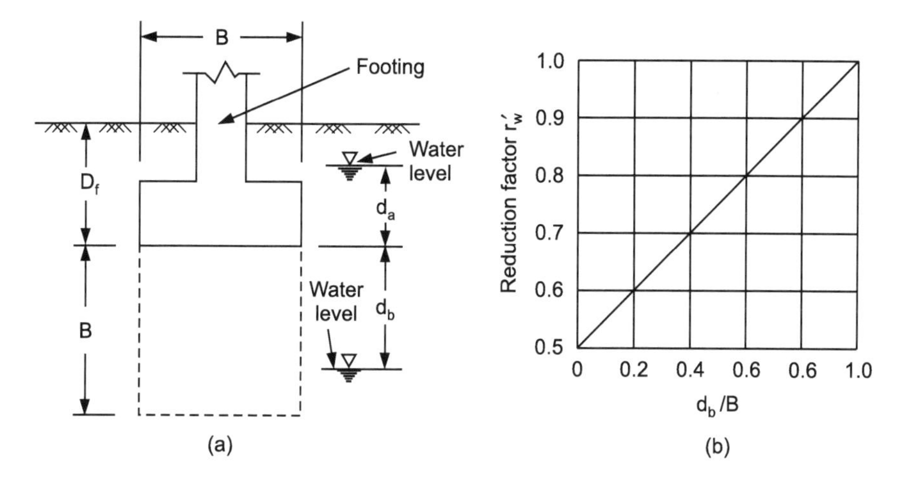

Fig. 7.6. Correction factor for position of water table.

2. Footing base is divided into *n* small and rectangular areas depending on the shape of footing base. Each area is subjected to a concentrated load whose magnitude is equal to average pressure intensity multiplied by the area of the small area. Due to these concentrated loads, the stresses σ_x, σ_y, σ_z, τ_{xz}, τ_{zy} and τ_{xy} were obtained using Eqs (7.1) to (7.6) at the centre of each layer along the vertical axis passing through the different points on the base of the footing. Overburden stresses are added in σ_x, σ_y and σ_z. Stresses σ_1, σ_2 and σ_3 and direction cosines cos θ_1, cos θ_2 and cos θ_3, were obtained solving Eqs (7.8) to (7.17) and Eqs (7.18) to (7.29) respectively.
3. A coefficient *F* for a given load intensity (*q*) is computed from the following relationship:

$$\frac{q_u}{q} = F \tag{7.39}$$

4. The modulus of elasticity (*E*) is calculated from Fig. 7.7 at stress level of *a*/*F*

$$E = \frac{1 - b\left(\frac{\sigma_u}{F}\right)}{a} \quad (7.40)$$

where a and b are the constants of hyperbola whose values depend upon confining pressure.

Average confining pressure in the analysis may be assumed as $(\sigma_2 + \sigma_3)/2$, say, equal to σ_{3a}.

5. The strain in each layer at the point under consideration in the direction of major principle stress is calculated from the following equation:

$$\varepsilon_1 = \frac{\sigma_1 - \sigma_{3a}}{E_s} \quad (7.41)$$

6. Evaluation of strain in the direction of principle stress σ_2 and σ_3. Assuming a suitable value of μ, the strains in the directions of other principal stresses (i.e. ε_2 and ε_3) may be obtained using Eqs (7.32) and (7.33).

   The value of vertical strain ε_z and total settlement (S) along any vertical section are obtained using Eqs (7.34) and (7.35).
7. The average settlement is then computed by dividing the area of settlement diagram by the width of footing.

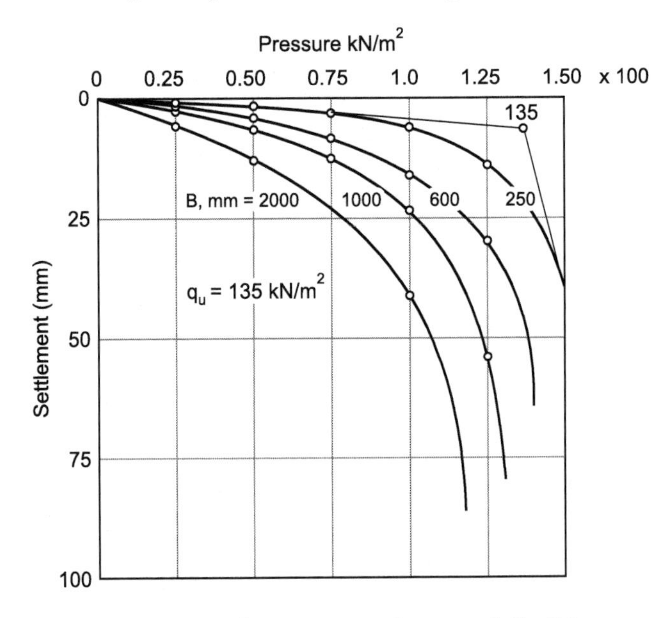

Fig. 7.7. Pressure-settlement curves for smooth flexible square footing resting on Buckshot clay.

Some typical bearing pressure versus settlement curves obtain for Buckshot clay are shown in Figs 7.7 and 7.8 respectively for smooth flexible and rigid footing. It is evident from these figures that effect of flexibility is insignificant.

Illustrative Examples

Example 7.1

A square footing of width 1.0 m is subjected to a uniform pressure intensity of magnitude 100 kN/m^2 and resting on saturated clay. Soil strata has been divided in several layers, each of 0.25 m thickness. Determine the settlement of point lying along the centre line of the footing at the mid point of the fourth layer.

Values of Kondner's hyperbola constants are: $a = 9.86 \times 10^{-5}$ m^2/kN and $b = 0.014$ m^2/kN.

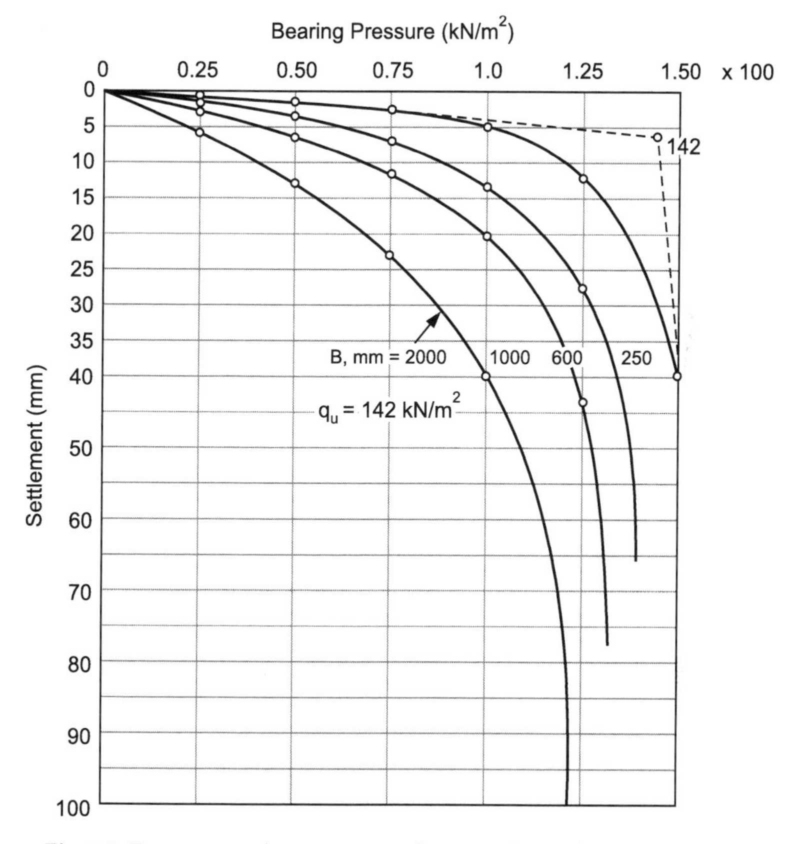

Fig. 7.8. Pressure-settlement curves for smooth rigid square footings resting on Buckshot clay.

Solution

Divide the base of the footing into four small squares, each of 0.5 m width. The concentrated load acting at the centre of this small portion will be $0.5 \times 0.5 \times 100 = 25$ kN.

Therefore, in Eqs (7.1) to (7.6), for the centre of footing,

$$x = 0.25 \text{ m}, y = 0.25 \text{ m}, z = 0.875 \text{ m}, \mu = 0.4, \text{m} = \frac{1}{\mu} = \frac{1}{0.4} = 2.5$$

$$r = \sqrt{x^2 + y^2} = \sqrt{0.25^2 + 0.25^2} = 0.353 \text{ m}$$

$$R = \sqrt{x^2 + y^2 + z^2} = \sqrt{0.25^2 + 0.25^2 + 0.875^2} = 0.944 \text{ m}$$

$$\sigma_x = \frac{3P}{2\pi} \times \left[\frac{x^2 z}{R^5} - \left(\frac{m-2}{3m} \right) \left\{ -\frac{1}{R(R+z)} + \frac{(2R+z)y^2}{(R+z)^2 R^3} + \frac{z}{R^3} \right\} \right]$$

$$\sigma_x = \frac{3 \times 25}{2\pi} \times \left[\frac{0.25^2 \times 0.875}{0.944^5} - \left(\frac{2.5-2}{3 \times 2.5} \right) \left\{ -\frac{1}{0.944(0.944+0.875)} + \frac{(2 \times 0.944 + 0.875) \times 0.25^2}{(0.944+0.875)^2 \times 0.944^3} + \frac{0.875}{0.944^3} \right\} \right]$$

$$\sigma_x = 0.45763 \text{ kN/m}^2$$

$$\sigma_y = \frac{3P}{2\pi} \times \left[\frac{x^2 z}{R^5} - \left(\frac{m-2}{3m} \right) \left\{ -\frac{1}{R(R+z)} + \frac{(2R+z)y^2}{(R+z)^2 R^3} + \frac{z}{R^3} \right\} \right]$$

Because $x = y$, therefore $\sigma_y = \sigma_x = 0.45763$ kN/m^2.

$$\sigma_z = \frac{3P}{2\pi} \times \frac{z^3}{R^5}$$

$$\sigma_z = \frac{3 \times 25}{2\pi} \times \frac{0.875^3}{0.944^5} = 10.67 \text{ kN/m}^2$$

$$\tau_{xz} = \frac{3P}{2\pi} \times \frac{xz^2}{R^5} = \frac{3 \times 25}{2\pi} \times \frac{0.25 \times 0.875^2}{0.944^5} = 3.049 \text{ kN/m}^2$$

$$\tau_{xz} = \tau_{yz} \text{ (because } x = y\text{)}$$

$$\tau_{xy} = \frac{3P}{2\pi} \times \left[\frac{xyz}{R^5} - \left(\frac{m-2}{3m} \right) \times \left\{ \frac{(2R+z)xy}{(R+z)^2 R^3} \right\} \right]$$

$$\tau_{xy} = \frac{3\times 25}{2\pi}\times\left[\frac{0.25\times 0.25\times 0.875}{0.944^5} - \left(\frac{2.5-2}{3\times 2.5}\right) \times\left\{\frac{(2\times 0.944+0.875)\times 0.25\times 0.25}{(0.944+0.875)^2\times 0.944^3}\right\}\right]$$

$$\tau_{xy} = 0.7813 \text{ kN/m}^2$$

Total values of stresses due to four squares:

$$\sigma_x = \sigma_y = 4\times 0.45763 = 1.83 \text{ kN/m}^2$$

$$\sigma_z = 4\times 10.67 = 42.68 \text{ kN/m}^2$$

Due to symmetry there will not be any total shear stresses on the vertical section passing through the centre of footing

i.e. $\tau_{xz} = \tau_{yz} = \tau_{xy} = 0$

Therefore

$$\sigma_1 = 42.68 \text{ kN/m}^2$$

$$\sigma_2 = 1.83 \text{ kN/m}^2$$

$$\sigma_3 = 1.83 \text{ kN/m}^2$$

$$\sigma_1 = \frac{a(\sigma_1-\sigma_{3a})}{1-b(\sigma_1-\sigma_{3a})}$$

$$\varepsilon_1 = \frac{9.86\times 10^{-5}(42.68-1.83)}{1-0.014(42.68-1.83)} = 941.4\times 10^{-5}$$

Lateral strains will not add to the vertical strain, hence

$$\varepsilon_z = \varepsilon_1 = 941.4\times 10^{-5}$$

Example 7.2

Solve Example 7.1, assuming that the footing rests on sand $\phi = 35°$, $\gamma = 16.5$ kN/m^2 with the values of Kondner's hyperbola constants given by the following relation:

$$\frac{1}{a} = 7000(\sigma_{ac})^{0.5}\text{ kN/m}^2;\ \frac{1}{b} = 44\,(\sigma_{ac})\text{ kN/m}^2$$

σ_{ac} is average confining pressure in kN/m^2.

Solution

(i) Ultimate bearing capacity of footing

$$q_u = \frac{1}{2}.\gamma.B.N_\gamma\times 0.8$$

$$= \frac{1}{2} \times 16.5 \times 1.0 \times 48.03 \times 0.8 \qquad [\text{For } \phi = 35°, N_\gamma = 48.03]$$

$$= 317 \text{ kN/m}^2$$

$$F = \frac{317}{100} = 3.17$$

$$K_0 = 1 - \sin\varphi = 1 - \sin 35 = 0.426$$

(ii) Stress due to overburden

$$\sigma_1 = 16.5 \times 0.875 = 14.4 \text{ kN/m}^2$$

$$\sigma_2 = \sigma_3 = 0.426 \times 16.5 \times 0.875 = 6.15 \text{ kN/m}^2$$

Stresses in the ground due to pressure intensity of 100 kN/m^2 will be same as given in Example 7.1.

Therefore total principal stresses:

$$\sigma_1 = 42.68 + 14.4 = 57.08 \text{ kN/m}^2$$

$$\sigma_2 = 1.83 + 6.15 = 7.98 \text{ kN/m}^2$$

$$\sigma_3 = 1.83 + 6.15 = 7.98 \text{ kN/m}^2$$

Average confining pressure is

$$\sigma_{ac} = \frac{7.98 + 7.98}{2} = 7.98 \text{ kN/m}^2$$

$$\frac{1}{a} = 7000(\sigma_{ac})^{0.5} = 7000(7.98)^{0.5} = 19774 \text{ kN/m}^2$$

$$\frac{1}{b} = 44(\sigma_{ac}) = 44 \times 7.98 = 351.2 \text{ kN/m}^2$$

$$a = 5.0 \times 10^{-5} \text{ m}^2/\text{kN}$$

$$b = 2.84 \times 10^{-3} \text{ m}^2/\text{kN}$$

$$\varepsilon_1 = \frac{a(\sigma_1 - \sigma_3)}{1 - b(\sigma_1 - \sigma_3)}$$

$$\varepsilon_1 = \frac{5.0 \times 10^{-5}(57.08 - 7.98)}{1 - 2.84 \times 10^{-3}(57.08 - 7.98)}$$

$$\varepsilon_1 = 285 \times 10^{-5} = \varepsilon_z$$

Practice Problems

1. Describe stepwise the procedure of getting pressure settlement characteristics of a smooth flexible square footing resting on clay.
2. Describe stepwise the procedure of getting pressure settlement characteristics of a rough rigid rectangular footing resting on clay.
3. Give the steps involved in getting pressure settlement characteristics of square/rectangular footing resting on sand.
4. A square footing of 2.0 m width is resting on clay ($a = 10^{-6}$ m^2/kN and $b = 0.09$ m^2/kN). Determine the vertical strain on a point below 1.0 m depth below the base of footing and lying on vertical section passing through (i) the centre of footing, and (ii) one corner of the footing.
5. A square footing of 1.5 m width is resting on clay ($1/a = 50.3 \times 10^3$ kN/m^2 and $1/b = 724$ kN/m^2). Determine the vertical strain on a point below 1.5 m depth below the base of footing along with the vertical section passing through the centre of footing.

REFERENCES

De Beer, E.E. (1970). Experimental determination of the shape factors and the bearing capacity factors of sand. *Geotechnique*, **20(4):** 387-411.

Durelli, A.J. and Phillips, T.S.A. CH. (1958). Introduction to the Theoretical and Experimental Analysis of Stress and Strain. McGraw Hill Book Company, Inc., New York.

Griffiths, D.V. and Fenton, G.A. (2005). Probabilistic Settlement Analysis of Rectangular Footing. 16th ICSMGE. Osaka, Japan, **2:** 1041-1044.

Gourvenec, S., Randolph, M. and Kingsnorth, O. (2006). Undrained bearing capacity of square and rectangular footings. *Int. J. Geomech.*, ASCE, **6(3):** 147-157.

Hansen, J.B. (1970). A Revised and Extended Formula for Bearing Capacity. Danish Geotechnical Institute, **28:** 5-11.

Meyerhof, G.G. (1951). The ultimate bearing capacity of foundations. *Géotechnique*, **2(4):** 301-332.

Meyerhof, G.G. (1963). Some recent research on bearing capacity of foundations. *Canadian Geotechnical Journal*, **1(1):** 16-26.

Meyerhof, G.G. (1965). Shallow Foundations. *J. SMFE*, ASCE, **91(2):** 21-31.

Michalowski, R.L. and Dawson, E.M. (2002). Ultimate loads on square footings. *Proc., 8th Int. Symp. on Numerical Models in Geomechanics*, Rome. Balkema, Rottderdam. The Netherlands, 415-418.

Michalowski, R.L. (2001). Upper-bound load estimates on square and rectangular footings. *Geotechnique*, **51(9):** 787-798.

Selby, Samuel, M. (1972). Standard Mathematical Tables. The Chemical Rubber Co. Cleveland, Ohio.

Sharan, U.N. (1977). Pressure Settlement Characteristics of Surface Footings Using Constitutive Laws. Ph.D. Thesis, University of Roorkee, Roorkee.

Skempton, A.W. (1951). The Bearing Capacity of Clays. Building Research Congress, London, **I:** 180-189.

Sokolnikoff (1956). Mathematical Theory of Elasticity. McGraw Hill Book Company, New York.

Terzaghi, K. (1943). Theoretical Soil Mechanics. John Wiley and Sons Inc., New York.

Zhu, M. and Michalowski, R. (2005). Shape factors for limit loads on square and rectangular footings. *J. Geotech. Geoenviron. Eng.*, ASCE, **131(2):** 223-231.

CHAPTER 8

Interaction between Adjacent Footings

8.1 General

Interference between geotechnical structures and the supporting soil media is of fundamental importance to both geotechnical and structural engineering. Interference between geotechnical structures is considered to exist if there is overlapping of their potential failure surfaces.

Construction of building foundations at close spacings is a common feature in built-up areas (Fig. 8.1a). The occurrence of adjacent footings carrying loads of different magnitudes is not a rare feature in building design. A covered colonnade adjacent to a multistorey building produces such circumstances, as does a covered loading area adjacent to an industrial building (Fig. 8.1b). Such type of problems are also encountered at many places in civil engineering constructions, such as grillage foundations, runway strips and railway crossties etc. (Figs 8.1c and 8.1d). Information regarding bearing capacity, settlement, tilt, contact stress distribution and extent of failure surfaces is required for an adequate design of the foundation. The mutual interference of foundations in a group has a significant influence on these design factors.

8.2 Brief Review

Many theories are available to find ultimate bearing capacity of two and three interfering strip footings loaded simultaneously with equal loads (Stuart, 1962; Mandel, 1963; West and Stuart, 1965; Amir, 1967; Siva Reddy and Mogaliah, 1976; Khadilkar and Varma, 1977; Patankar and Khadilkar, 1981; Graham, et al. 1984). By using an upper bound limit analysis in conjunction with finite element and linear programming, the ultimate bearing capacity of two interfering rough strip footings, resting on a cohesion-less medium, was computed by Kumar and Kouzer (2007).

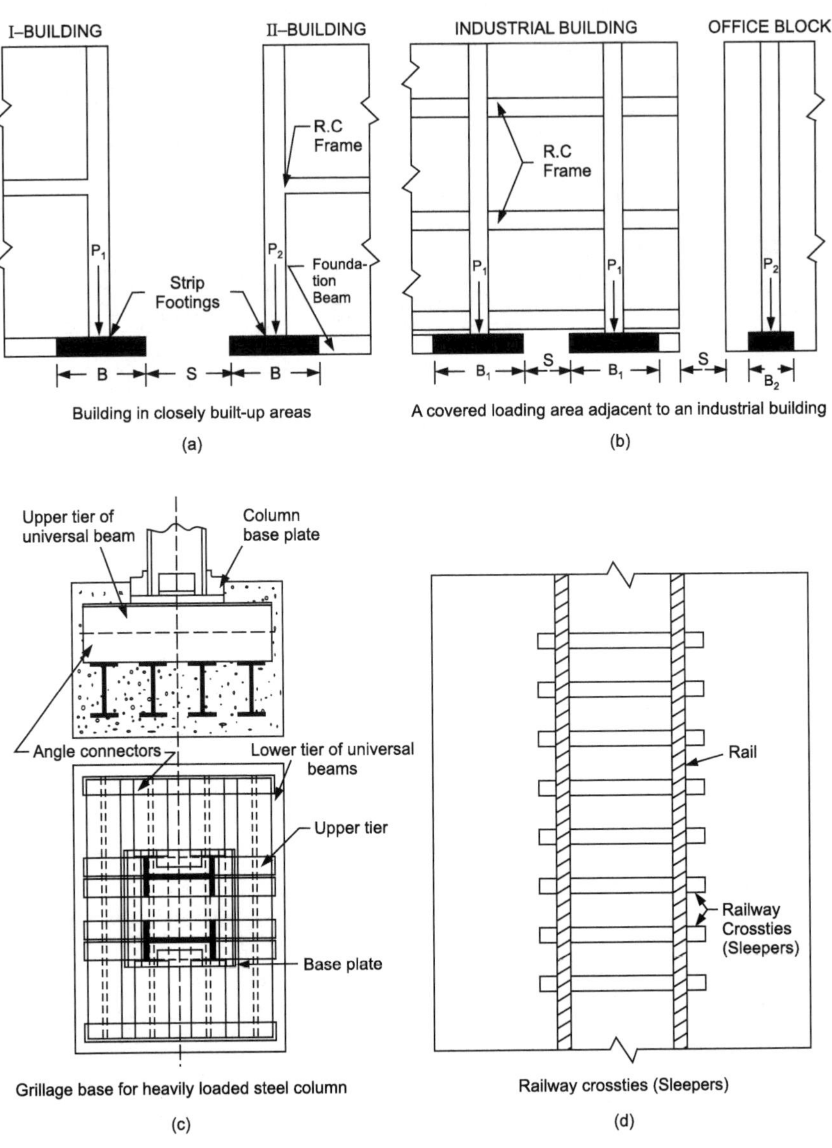

Fig. 8.1. Examples of mutual interference between stripe foundation.

Kumar and Ghosh (2007) obtained the bearing capacity of two interfering footings using the method of stress characteristics. Problem of interfering multiple strip footings have been solved using an upper-bound theorem and lower-bound limit analysis (Kouzer and Kumar, 2007).

Dash (1981, 1982) investigated into the problem of determining ultimate bearing capacity of two strip footings when one of the footings carried certain load located nearby. Kumar and Kouzer (2010) obtained the ultimate bearing capacity of footing considering the interference of an existing footing on sand using an upper-bound finite limit analysis.

According to these studies, the effect of interference of footings is, in general, to cause an increase in the bearing capacity with reduction of spacing.

Many investigators have experimentally studied the problem of two and three interfering footings loaded simultaneously (Stuart, 1962; West and Stuart, 1965; Dembicki et al., 1971; Myslivec and Kysela, 1973; Singh et al., 1973; Saran and Agarwal, 1974; Deshmukh, 1979; Das and Labri Cherif, 1983; Selvadurai and Rabbaa, 1983; Graham et al., 1984). The effect of footing carrying certain load on the behaviour of an adjacent footing has also been studied experimentally by a few investigators (Murthy, 1970; Das and Labri-Cherif, 1983; Dash, 1990).

No method has been reported to find settlement and tilt of interfering footings except finite element method and finite difference technique.

On the basis of the review of the available literature, it can be concluded that the influence of the interference between foundations on the tilt in particular appears to have received little or no attention. So far no attention has been given to the problems of interfering footings of different widths. Footings carry loads of different magnitudes and interference of neighboring foundation under unequal loads (Amir,1992). To date no rational method exists to evaluate the pressure-settlement and pressure-tilt characteristics of interfering footings using constitutive laws of soils. Keeping this in view, the solution of this problem is the aim of this work.

Actually interring footings settle non-uniformly thus get tilted. Tilting occurs in the inward direction of footings (Fig. 8.2) due to more concentration of stress on this side.

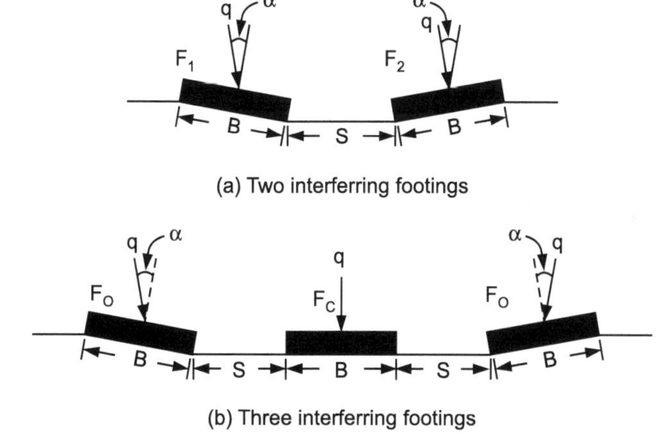

Fig. 8.2. Showing the tilting of footings due to interference.

In this chapter, procedure has been given to obtain the complete pressure-settlement and pressure-tilt characteristics of interring footings. This approach enables one to obtain the ultimate bearing capacity, settlement and tilt of the interring footings. Thus their complete proportioning is possible (Amir, 1992).

There may be numerous situations in which interaction between adjacent footings occurs. Some of the common situations encountered in practice have been identified in the following eight cases:

Case I

Two strip footings of same widths subjected to same pressure intensity, q.

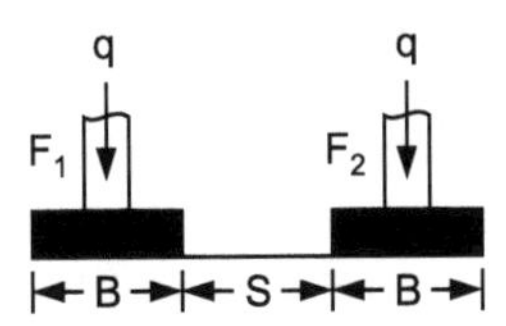

Case II

Two strip footings of the same width, one representing the existing foundation having one-third of failure pressure intensity of isolated footing and the new one subjected to pressure intensity q varying upto failure.

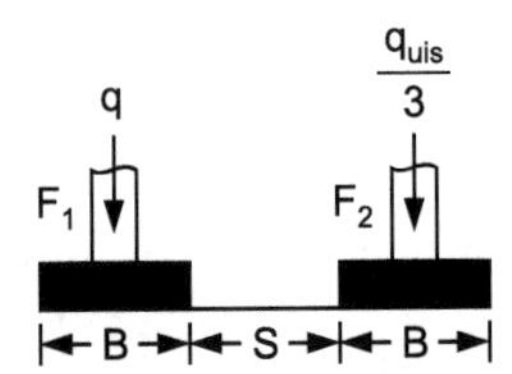

Case III

Two strip footings of the same widths, one representing the existing foundation having one-half of failure pressure intensity of isolated footing and the new one subjected to pressure intensity q varying upto failure.

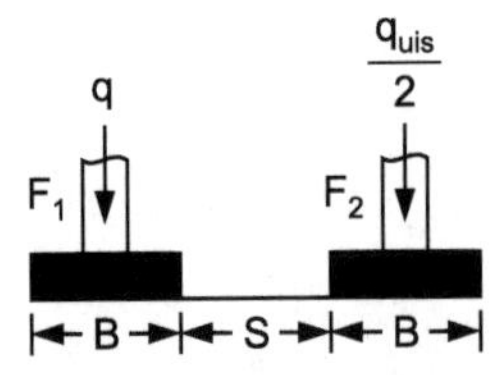

Case IV

Two strip footing of different widths subjected to same pressure intensity, q.

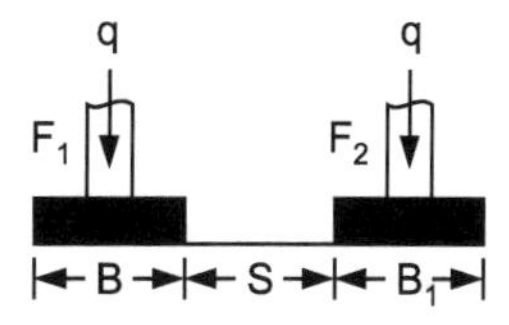

Case V

Two strip footings of different widths, one representing the existing foundation having one-half of failure pressure intensity of isolated footing and the new one subjected to pressure intensity q varying upto failure.

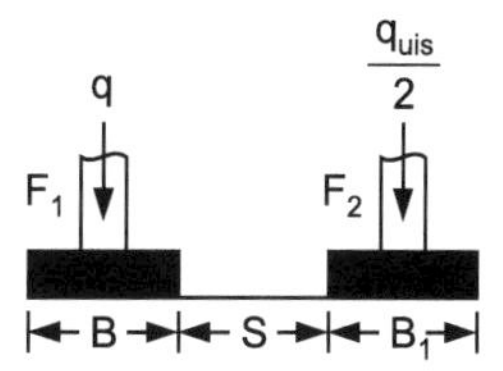

Case VI

Three strip footings of same widths subjected to same pressure intensity, q.

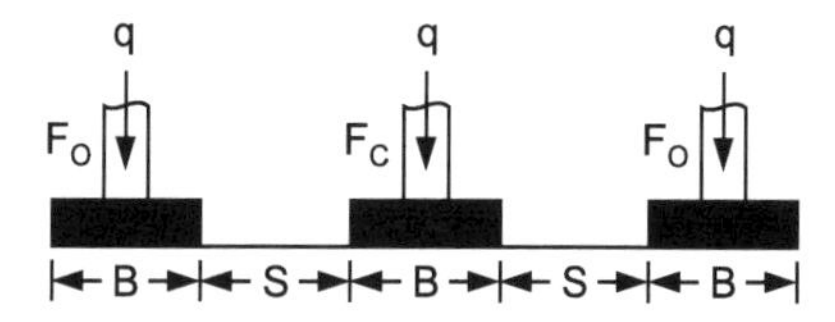

Case VII

Three strip footings central one representing the existing foundation subjected to half of the failure pressure intensity of isolated loaded and the adjacent two subjected to pressure intensity q varying upto failure.

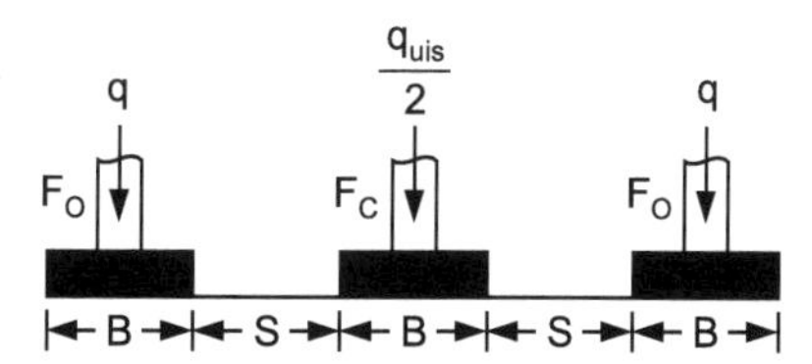

Case VIII

Three strip footings two representing the existing foundations each subjected to half of failure pressure intensity of isolated footing and the central one with pressure intensity, q varying upto failure.

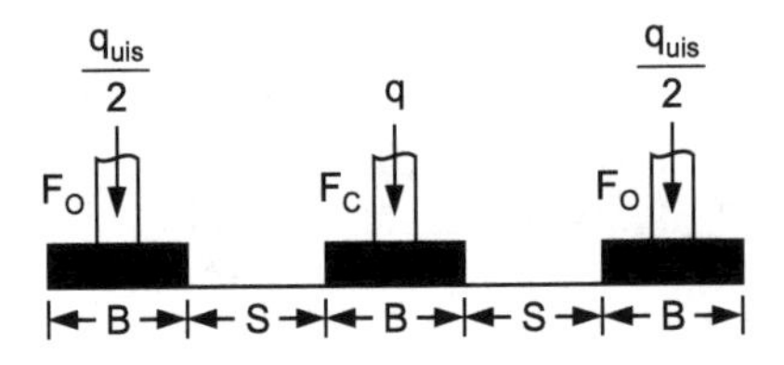

The procedure of solving any of the above case is identical with other. Keeping this fact in view, a general procedure is given is the following sections for footing resting on (i) clays and (ii) sand. The material described herein is taken from Amir (1992).

8.3 Footing on Clays

8.3.1 Assumptions

Following assumptions have been made in the analysis:

- The soil mass is semi-infinite and an isotropic medium.
- The footings base is rough.
- The roughness of footings is assumed to generate uniform tangential stress at the contact surface, which follows the relationship $t_a = c\left(\frac{q}{q_u}\right)$ for cohesive soil in which t_a = the tangential stress at the contact surface. The value of pressure at failure (q_u) may be taken as $c \times N_c$, with N_c = Terzaghi's bearing capacity factor table, c = unit cohesion, and q = applied pressure intensity.

Assumptions of the tangential stress under a set of two and three rough strip footings are shown in Figs 8.3 and 8.4 respectively. The shear stress due to tangential stress is zero at the centre of the isolated footing, acting inwardly (Figs 8.3b and 8.4b). If the clear spacing between the footings, $S \geq 4B$ the effect of stresses caused by one footing on to other is insignificant. Therefore, the direction of the tangential stress is taken to be same as of isolated footing (Figs 8.3b and 8.4b). In case of the spacing of the footings, $S < 4B$, the overlapping of stresses starts. The direction of the tangential stress is assumed as shown in Figs 8.3c and 8.4c and the value of a is linearly interpolated (Figs 8.3d and 8.4d).

If the spacing of the footings is further reduced such that $S = 0$, the pair of footings will act as a single footing and the direction of tangential stress is same as that of isolated footing (Figs 8.3e and 8.4e).

- The contact pressure distribution is uniform. The footing, initially is considered flexible.
- The whole soil mass supporting the footings is divided into a large number of thin horizontal strips up to a depth beyond which the stresses are less than 0.08 q, q being the applied stress on the footing.

- Stresses in each layer of soil mass have been computed using the theory of elasticity since stress equations for various types of loads are available (Carother, 1920; Kolosov, 1935).
- Strains have been computed from the known stress condition using constitutive laws.
- There is no slippage at the interface of layers in the soil mass.

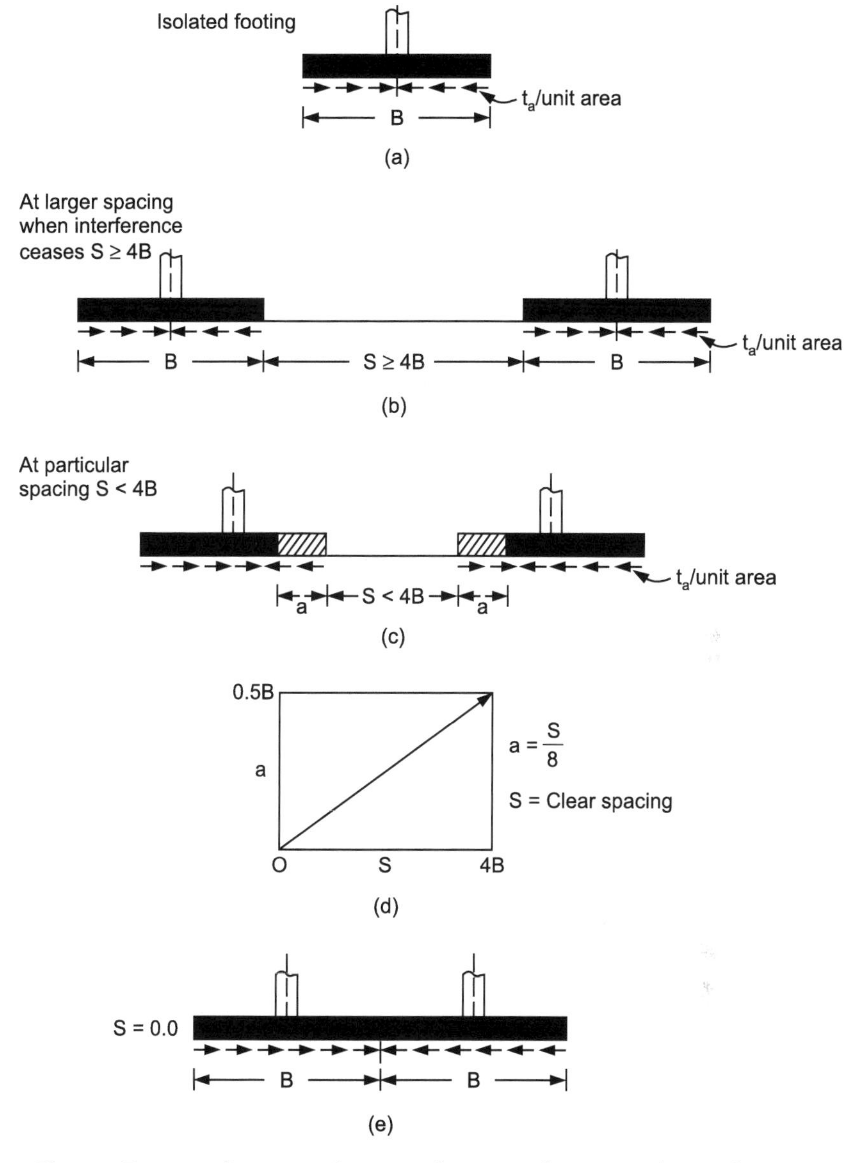

Fig. 8.3. Tangential stress under a set of two interfering rough strip footings.

8.3.2 Vertical Settlement and Tilt

The procedure adopted for analysis of pressure-settlement and pressure-tilt characteristics under a set of two and three interfering footings is described in the following steps:

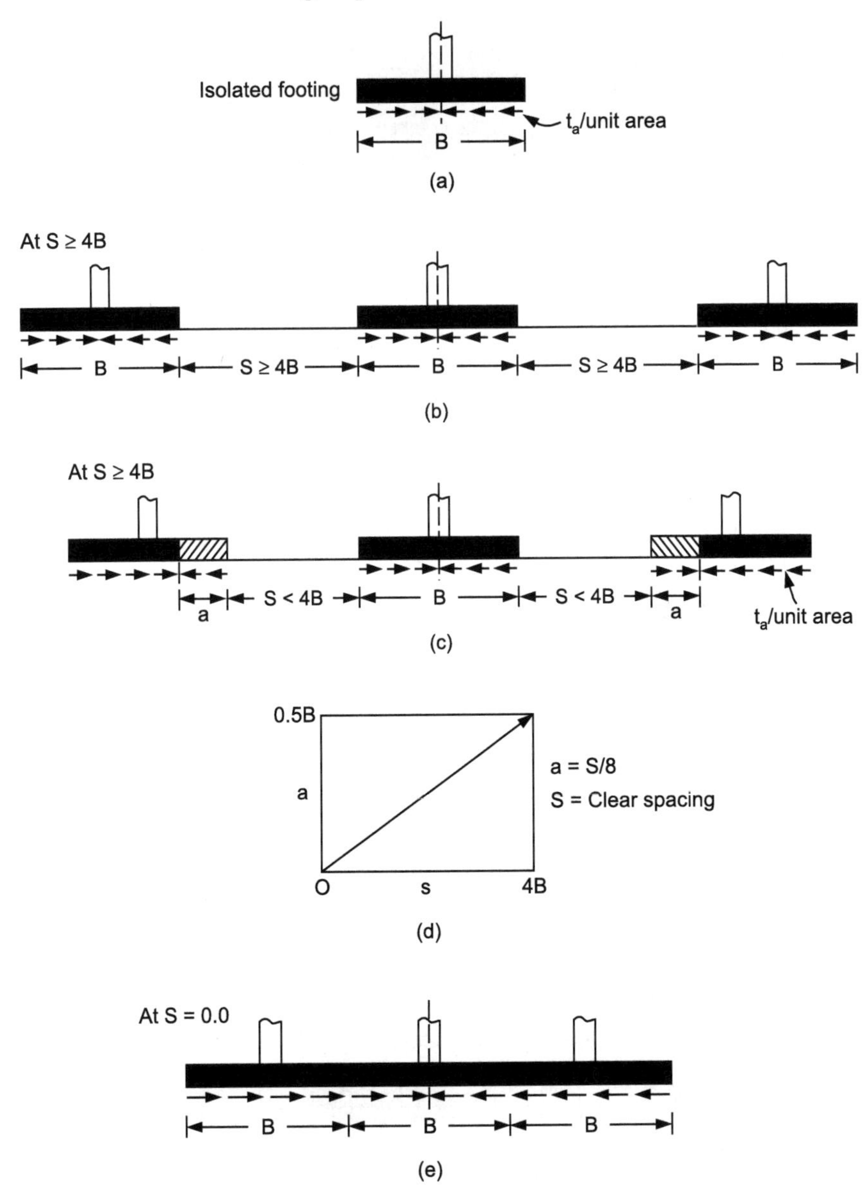

Fig. 8.4. Tangential stress under a set of three interfering rough strip footings.

Step 1

For a given intensity of pressure (q) and spacing of footings (S), the contact pressure distribution and tangential stresses at the interface of footing-bases and supporting soil media are taken as shown in Figs 8.5 and 8.6, which induce stresses in the soil. The variation of the tangential stress (t_a) is assumed according to Figs 8.3d and 8.4d.

Step 2

The soil mass supporting the footing is divided into a large number of thin layers (say n layers) up to a depth at which the pressure intensity is less than 0.05 q (Figs 8.5 and 8.6).

Step 3

Evaluation of stresses σ_z, σ_x and τ_{xz} in each layer of the soil mass (Figs 8.5 and 8.6) at vertical sections due to q and t_a is obtained separately and then added. Superimposing of stresses due to two or three footing (as the case may be) is done to get the total stresses. Equations used to obtain stresses are already mentioned in Chapter 3.

Steps 4 to 7 are same as described in Section 4.2.2 of Chapter 4.

Step 8

Evaluation of settlement and tilt of rigid interfering footing is shown in Figs 8.7b and 8.8b to the values of $(S_{\min})_{in}$ and $(S_{\max})_{in}$ which are obtained by equating:

(i) The area of settlement diagram of Fig. 8.7b to the area of settlement diagram of Fig. 8.7a and
(ii) The distance of center of settlement diagram of Fig. 8.7a from edge of footing point A to the distance of centre of settlement diagram of Fig. 8.7b from the edge of footing point A'. The area of settlement

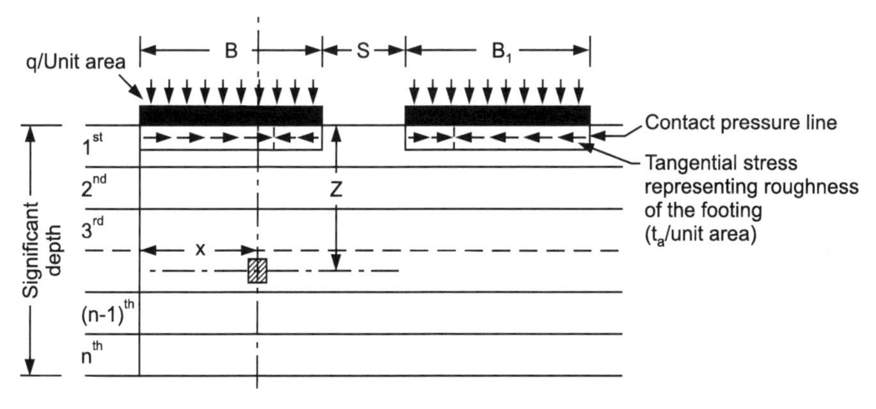

Fig. 8.5. Two interfering rough strip footings and soil media divided into n layers.

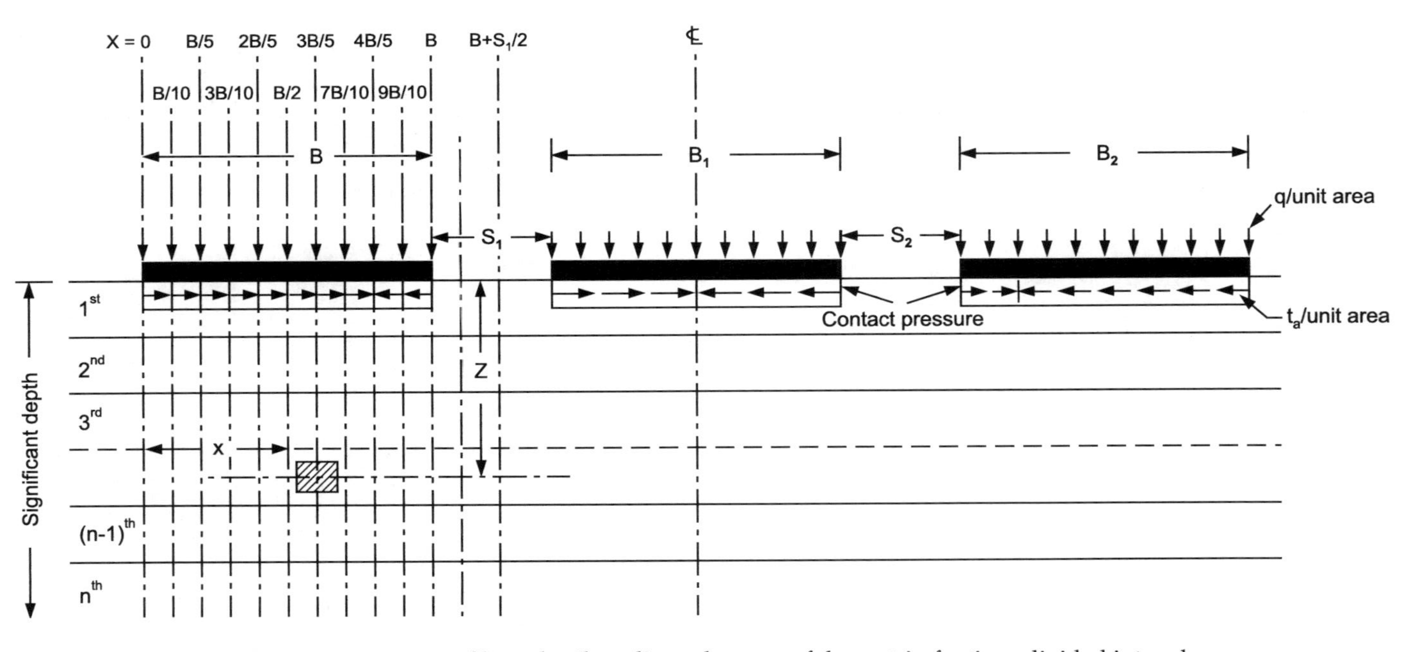

Fig. 8.6. Contact pressure profile and soil media under a set of three strip footings divided into *n* layers.

diagram and the distance of centre of settlement diagram from the edge of the footing are given as:

$$A = \frac{(S_{\text{min}})_{\text{in}} + (S_{\text{max}})_{\text{in}}}{2} \times B \tag{8.1}$$

$$\text{c.g.} = \frac{(S_{\text{min}})_{\text{in}} + 2(S_{\text{max}})_{\text{in}}}{(S_{\text{min}})_{\text{in}} + (S_{\text{max}})_{\text{in}}} \times \frac{B}{3} \tag{8.2}$$

where $(S_{max})_{in}$ = maximum settlement of one of the interfering footings

$(S_{min})_{in}$ = minimum settlement of one of the interfering footings

B = width of the footing

c.g. = center of gravity of the settlement diagram of Fig. 8.7a or Fig. 8.8a from point A'.

A = area of settlement diagram of Fig. 8.7a or Fig. 8.8a.

Knowing the value of $(S_{min})_{in}$ and $(S_{max})_{in}$, average settlement (S_{av}) and tilt (t) of the rigid footing may be computed as follows:

$$S_{av} = \frac{(S_{\text{min}})_{\text{in}} + (S_{\text{max}})_{\text{in}}}{2} \tag{8.3}$$

$$\text{Tan } t = \frac{(S_{\text{max}})_{\text{in}} - (S_{\text{min}})_{\text{in}}}{B} \tag{8.4}$$

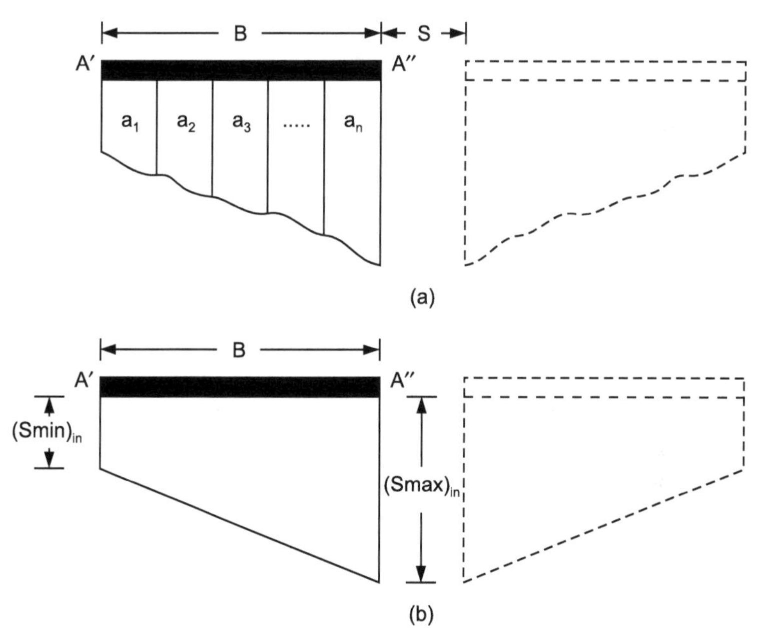

Fig. 8.7. Settlements diagram under a set of two interfering strip footings.

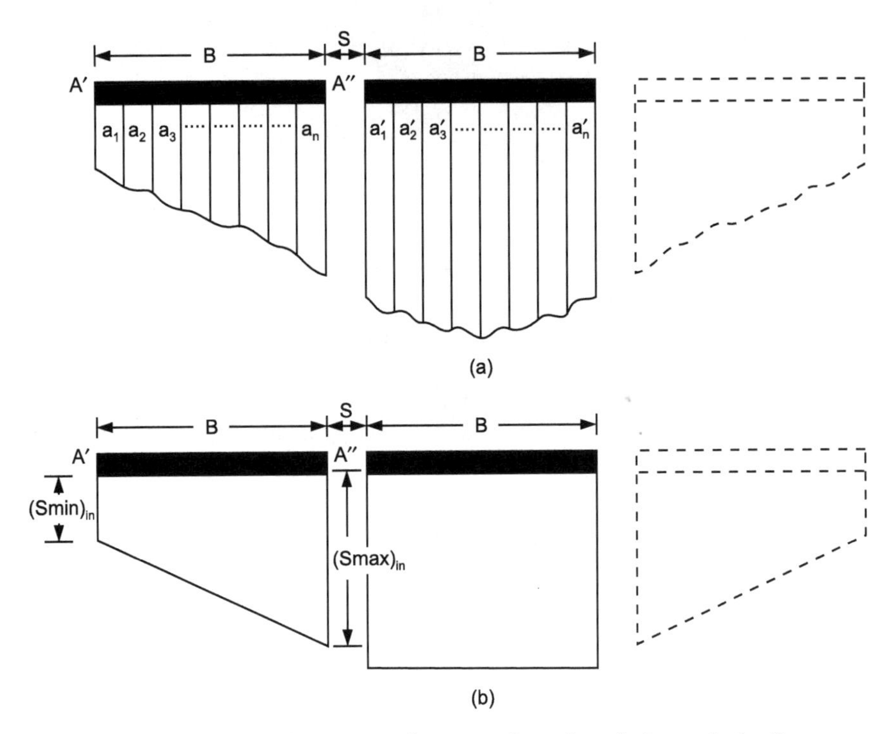

Fig. 8.8. Settlements diagram under a set of two interfering strip footings.

It may be mentioned that the maximum settlement of interfering rigid footing will be given by

$$(S_{max})_{in} = S_{av} + \frac{B}{2}\tan t \tag{8.5}$$

Step 10

The average settlement (S_{av}) and tilt (t) for various pressure intensities on footings were computed by repeating Steps 1 to 8. The pressure versus settlement and pressure versus tilt were obtained.

Step 11

The clear spacing of the footings is changed and Steps 1 to 9 were repeated.

8.4 Footing on Sand

8.4.1 Introduction

As mentioned in the earlier chapters, the procedure adopted for the analysis of interfering footings on clay cannot be applied in sand directly as its modulus of elasticity varies with depth. For such materials, stress equations are not available. Therefore, an empirical approach has been developed to obtain pressure-settlement and pressure-tilt characteristics of interfering footings resting on sand, which has been found to yield satisfactory results.

The constitutive relations of sand represented by hyperbola were established from consolidated drained triaxial compression tests.

8.4.2 Assumption

1. Contact pressure distribution below the footings has been assumed as in the case of clay.
2. The effect of weight of soil mass is considered in determination of stresses. Vertical stress due to weight of the soil is taken equal to γz where γ is the unit weight of the soil and z is the depth of soil layer. The horizontal stress due to weight of the soil has been taken to be equal to $K_0\gamma z$.
   Where K_0 is the coefficient of earth pressure at rest, it may be taken as
   $$K_0 = 1 - \sin\phi \tag{8.6}$$
   where ϕ is angle of internal friction.
3. The roughness of footings has been assumed to generate tangential stress at the contact surface, which follows the relationship
   $$t_a = q.\tan\left(\frac{2}{3}\phi\right)$$
   where t_a = tangential stress at the contact surface, and
   ϕ = angle of internal friction.

   The ultimate bearing pressure (q_{uin}) of interfering footings on sand is computed using
   $$q_{uin} = \frac{1}{2}\gamma B N_\gamma \xi_\gamma \tag{8.7a}$$
   N_γ is bearing capacity factor (Terzaghi's values, Table A4.1) and ξ_γ is the efficiency factor which is defined as the ratio of the interfering to isolated values of bearing capacity coefficients, the value of ξ_γ may be obtained from charts given in Fig. 8.9.

   A factor of safety (F) has been introduced such that at all stress levels the following relationship is satisfied:
   $$\frac{q_{uin}}{q} = \frac{\sigma_u}{\sigma_1 - \sigma_3} = F \tag{8.7b}$$
   where q = intensity of surface load
   σ_u = ultimate stress from hyperbolic relation and is equal to $(1/b)$; b being Kondner's hyperbola constants.
   σ_1 and σ_3 = major and minor principal stress in the soil mass.

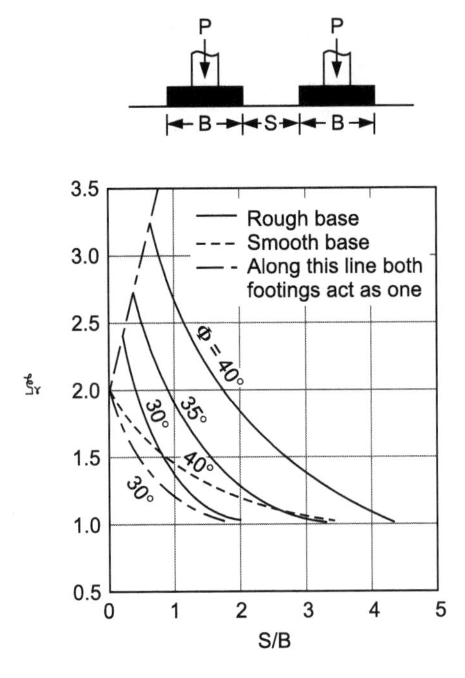

Fig. 8.9. Theoretical values of the bearing capacity efficiency factors ξ_γ and ξ_q versus *S/B* (after stuart, 1962).

8.4.3 Vertical Settlement and Tilt

Procedure adopted for obtaining pressure-settlement and pressure-tilt characteristics under a set of two and three rough flexible strips interfering footings are given in the following steps:

Step 1

Contact pressure distribution, stress in each layer, principal stresses and their directions have been obtained by using Steps 1 to 3 as described in the case of footings on clays in Section 8.2.2.

5. The value of Poisson's ratio μ has been obtained using Eq. (8.8).

Step 2

The factor *F* for the given surface load intensity (q) has been obtained from the following relationship:

$$\frac{q_{uin}}{q} = F \tag{8.8}$$

where q_{uin} = ultimate bearing pressure interfering footings as per step (3) of Section 8.4.2

q = applied load intensity

Step 3

The modulus of elasticity (E_s) is calculated by the following equation:

$$E_s = \frac{1-b(\sigma_u/F)}{a} \tag{8.9}$$

where a and b are the constants of hyperbola, whose values depend upon confining pressure.

Step 4

The strain in each layer in the direction of major principal stress has been obtained from the following equation:

$$\varepsilon_1 = \frac{\sigma_1 - \sigma_3}{E_s} \tag{8.10}$$

The strain in the direction of minor principal stress is calculated from the following relationship

$$\varepsilon_3 = -\mu_2.\varepsilon_1 \tag{8.11}$$

where,

$$\mu_2 = \frac{-\sigma_3 + \mu_1\sigma_1}{\sigma_1 + \mu_1\sigma_3} \tag{8.12}$$

$$\mu_1 = \frac{\mu_1}{1-\mu} \tag{8.13}$$

$$\mu = \frac{K_0}{1+K_0} \tag{8.14}$$

Step 5

Vertical strain, settlement and tilt have been obtained using the Steps 7 to 11 as described for clay in Section 4.2.3.

8.5 Results and Interpretation

Amir (1992) obtained the pressure-settlement characteristics of surface footing resting on three types of soil namely (i) Dhanori clay (LL = 55%, PL = 25%, ϕ = 25.5°, c_u = 55 kN/m^2 and γ = 17.6 kN/m^3), (ii) Buckshot clay and (iii) Amanatgarh sand (SP, DR = 70% and γ = 16 kN/m^3). Using triaxial tests, Kondner (1962) hyperbola constants a and b were obtained as given below:

	1/a (kN/m^2)	*1/b (kN/m^2)*
Dhanori clay	18180	160
Buckshot clay	10510	187
Amanatgarh sand	3810 $(\sigma_3)^{0.57}$	80+3.9 (σ)

He obtained the solutions of all the eight cases (Amir, 1992). However, here only three cases have been presented—Case III, Case V and VII are discussed.

Case III

Two footings of the same width, one of the footings (F_2) represent the existing foundation loaded half the failure load of isolated footing and adjacent footing (F_1) loaded upto failure.

Assume that soil is cohesive frictional (c-φ) and both the footings are flexible, i.e. contact pressure will be of uniform intensities q and $q_{uis}/2$ respectively. Values of q_{uis} may be obtained, using Terzaghi's bearing capacity equation, i.e.

$$q_{uis} = \left(cN_c + \frac{1}{2}\gamma BN_\gamma \right) \tag{8.15}$$

where, N_c and N_γ are Terzaghi's bearing capacity factors (Table); c, γ and B are respectively the unit cohesion, unit weight of soil and width of the footings.

As mentioned earlier, $a = S/8$, S being the clear spacing between the footings. Tangential stresses are:

$$t_{al} = \frac{q}{q_{uis}} c_a + q \tan \delta_m; \ \delta_m = \tan^{-1}\left(\frac{q}{q_{uis}} \tan \delta \right) \tag{8.16}$$

$$t_{al} = \frac{q_{uis}/2}{q_{uis}} c_a + \frac{q_{uis}}{2} \tan(\delta_m'); \delta_m = \tan^{-1}\left(\frac{\tan \delta}{2} \right) \tag{8.17}$$

where, c_a and δ are base adhesion and base friction coefficient; their values may be taken as $2/3\,c$ and $2/3\,\varphi$. Directions of tangential shear stresses will be as shown in Fig. 8.3c. Finally, it may be said that footing F_2 is subjected to uniform pressure intensity ($q_{uis}/2$) and tangential stresses t_{a2}, while footing F_1 is loaded in various stages with uniform pressure intensities (q) of $q_1, q_2, \ldots, q_n$ along with tangential stresses t_{a1}. Procedure for solving the problem is as given below:

1. Divide the soil stratum in thin layers upto significant depth.
2. Select vertical sections aa, bb ... and $a'a'$, $b'b'$ below the footing F_1 and F_2 respectively.
3. For illustration, procedure of determining settlement of a point on the base of the footing where Section dd touches Fig. 8.9 is discussed. Consider a point at the centre of the first layer on this vertical section. Let the depth of this point below ground surface be Z_1. At this point, using the theory of elasticity, determine the values of stresses σ_{z1}, σ_{x1} and τ_{xz1} due to applied stress q_1, t_{a1}, $\frac{q_{uis}}{2}$ and t_{a2}.

Add $\gamma_e z_1$ in σ_{z1} $K_0\gamma_e z_1$ in σ_{x1} where γ_e is the effective unit of soil and K_0 is the coefficient of earth pressure at rest; say these stresses are σ'_{z1} and σ'_{x1}.

Using stresses σ'_{z1}, σ'_{x1} and τ_{xz1} obtain principal stress σ_{11} and σ_{31} along with θ_{11} and θ_{31}.

Rest of the procedure is same for determining the vertical strain at this point, say ε_{z1} as discussed in the earlier chapters i.e. using constitutive law of soil. On multiplying ε_{z3} with thickness of horizontal strip, settlement of this strip along with the vertical section may be obtained, say it is S_{d1}.

The above procedure is repeated for the points considered at the centre of other layers along this vertical section and settlement S_{d1}, S_{d3} ... S_{dn} were obtained as all these settlement values will give the settlement of the base point passing though the section *dd*. Proceeding like this, settlement of other points on the base of the footing were obtained and it resulted settlement patterns as shown in Fig. 8.7. From these settlement diagrams, average settlements and tilts of both the footing are obtained. Thus pressure versus settlement and pressure versus tilt curves were plotted.

In Figs 8.10 and 8.11, pressure versus settlement and pressure versus tilt curves are shown for a footing of width 300 mm resting on the surface of Dhanori clay for different spacing/width ratios. It is evident from Fig. 8.10 that for a given pressure intensity acting on the footing, settlement increases with decrease in the S/B ratio. Further, existing footing, i.e. F_2 settles by less amount in comparison to footing F_1. Figure 8.11a shows that the tilt of footing F_1 increases with decrease in the S/B ratio, and for a particular S/B ratio, tilt remains almost same for a wide range of pressure intensities (40 kN/m^2 – 200 kN/m^2). Beyond the pressure intensity of 200 kN/m^2 the tilt increases at a very fast rate. Footing F_2 (Fig. 8.11b) tilts almost at uniform rate with the increase in pressure intensity. In Figs 8.12 and 8.13 pressure versus settlement and pressure versus tilt curves are shown where footings of widths equal to 100 mm each are resting on sand. The trend of variation of settlement and tilt shown in these figures are almost same as it was in the footings resting on clay.

Case V

Two strip footings of different widths with a wider footing (F_2) representing the existing foundation loaded half of the failure load of isolated footing and adjacent footing (F_1) loaded upto failure.

This case is similar to Case III except that the existing footing (F_2) is considered wider. In Figs 8.14 and 8.15, pressure-settlement and pressure versus tilt curves are shown when taking the footings of widths 300 mm and 600 mm. In this case also for a given pressure intensity, both the settlement and tilt increase with decrease in spacing. Further, wider

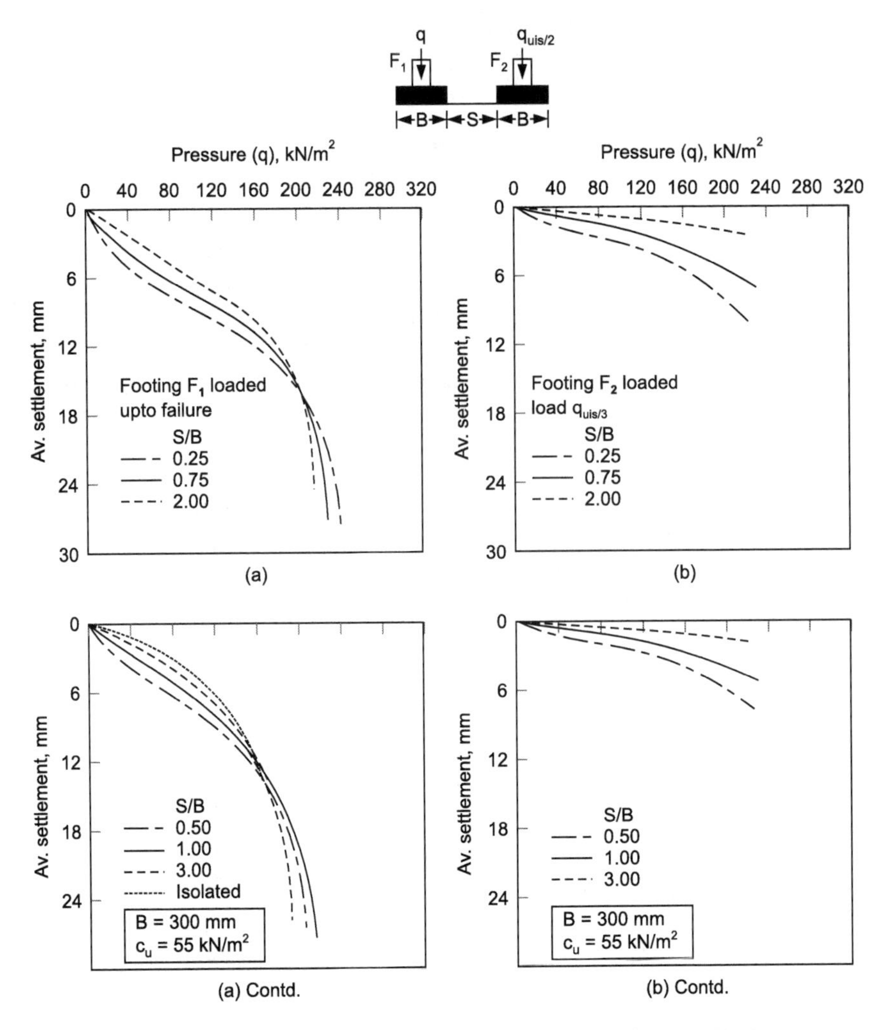

Fig. 8.10. Pressure versus settlement curves (Case III – Clay) (a) Footing F_1; (b) Footing F_2.

footings settle less. Tilts shown in Fig. 8.15 follow the same trends as given in Fig. 8.11. In Figs 8.16 and 8.17, pressure-settlement and pressure-tilt curves are shown for footings resting on sand. These curves also follow the same trend as in the case of clay (i.e. in Figs 8.14 and 8.15).

Case VII

Three strip footings of same widths, with the central footing (F_c) represent the existing foundation loaded half of the failure load of isolated footing and outer footings (F_0) loaded upto failure. In this case, due to symmetry,

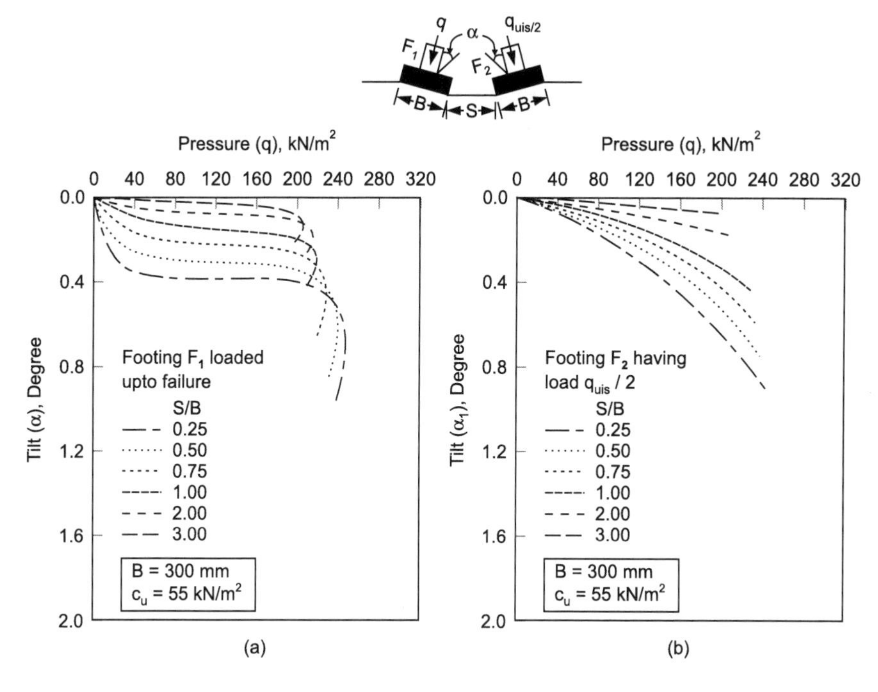

Fig. 8.11. Pressure versus tilt curves (Case III – Clay)
(a) Footing F_1; (b) Footing F_2.

central footing settles uniformly with no tilt. Pressure versus settlement and pressure versus tilt curves for footing resting on clay are shown in Figs 8.18 and 8.19 respectively. Similar curves for footing resting on sand are given in Figs 8.20 and 8.21. Settlement of all the three footings increases with decrease in spacing. For the same pressure intensity, side footings settle more in comparison to central footing.

Illustrative Example

Example 8.1

Two footings each of width equal to 1.0 m are spaced having centre to centre spacing as 2.0 m on the surface of soil. The new footing is subjected to a pressure intensity of 75 kN/m². The existing footing is having a pressure intensity equal to half of its ultimate bearing capacity. The properties of soil are: $c = 30$ kN/m²; $\phi = 30°$, $\gamma = 17$ kN/m³.

Kondner's hyperbola constants are: $\frac{1}{a} = 3.0 \times 10^3 (\sigma_3)^{0.4}$, $\frac{1}{b} = 100 + 2.0(\sigma_3)$, σ_3 being in kN/m².

Determine the vertical strain at a depth of 1.25 m along the vertical section passing through the right hand edge of the new footing.

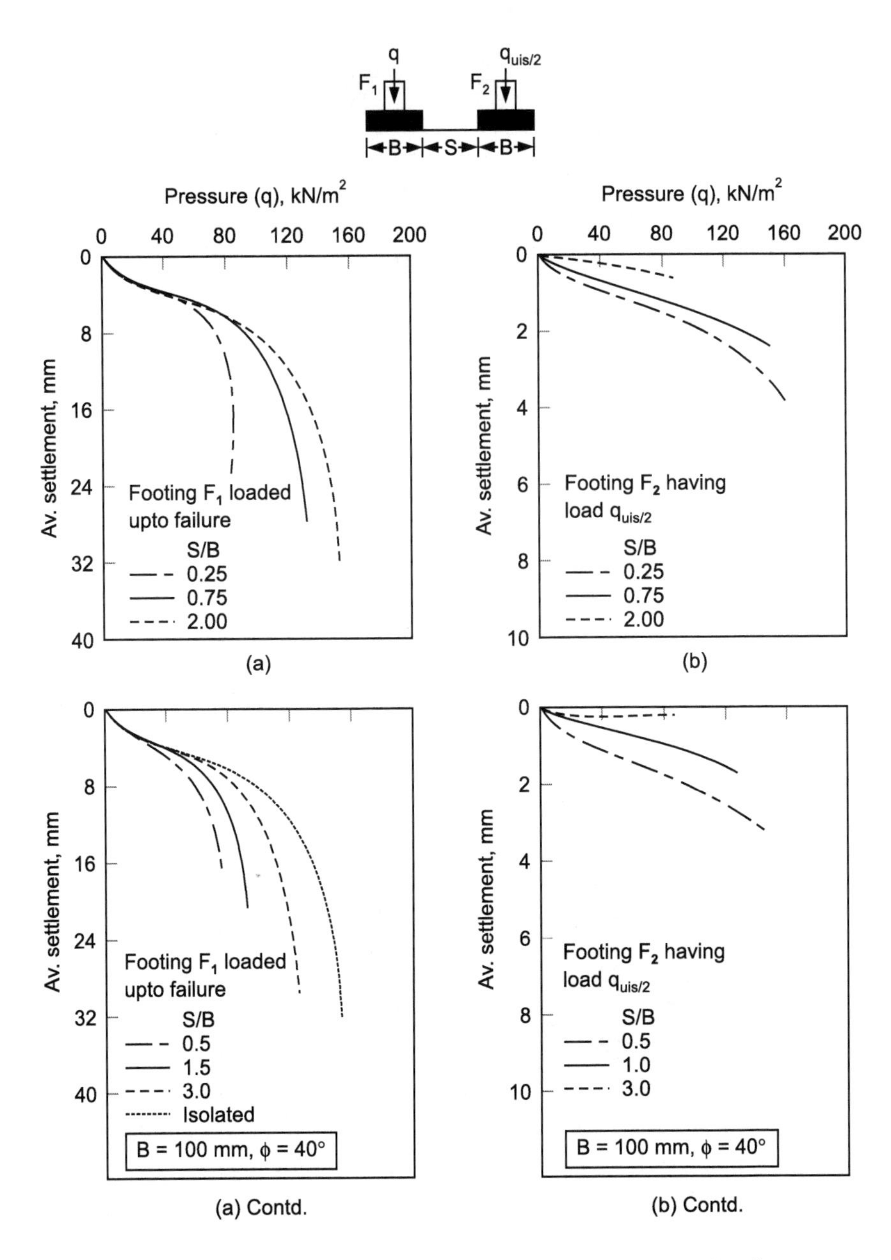

Fig. 8.12. Pressure versus settlement curves (Case III – Sand) (a) Footing F_1; (b) Footing F_2.

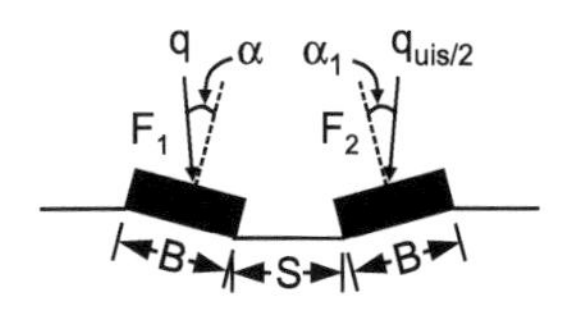

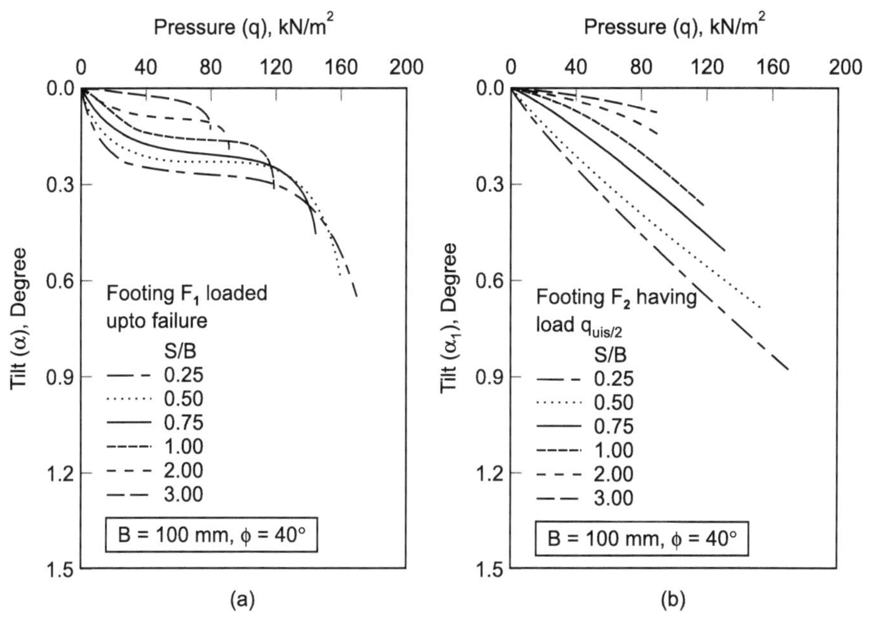

Fig. 8.13. Pressure versus tilt curves (Case III – Sand)
(a) Footing F_1; (b) Footing F_2.

Solution

(i) Refer to Fig. 8.22

$$B = 1.0 \text{ m}, b = \frac{1}{2} = 0.5 \text{ m}; a = \frac{1}{8} = 0.125 \text{ m}; \text{(Fig. 8.3d)}$$

$$q_{uis} = \left(cN_c + \frac{1}{2}\gamma BN_\gamma\right)$$

$$= 30 \times 30 + \frac{1}{2} \times 1.7 \times 1.0 \times 22.4$$

[See Appendix Table A4.1; for $\phi = 30°$, $N_c = 30$ and $N_\gamma = 22.4$]

$$= 90 + 190.4 = 280.4 \text{ kN/m}^2$$

Let the footing F_2 is subjected to a pressure intensity equal to

$$\frac{q_{uis}}{2} = \frac{280.4}{2} = 140.2 \text{ kN/m}^2 \text{ say } 140 \text{ kN/m}^2$$

Footing F_1 is subjected to a pressure intensity equal to 75 kN/m^2.

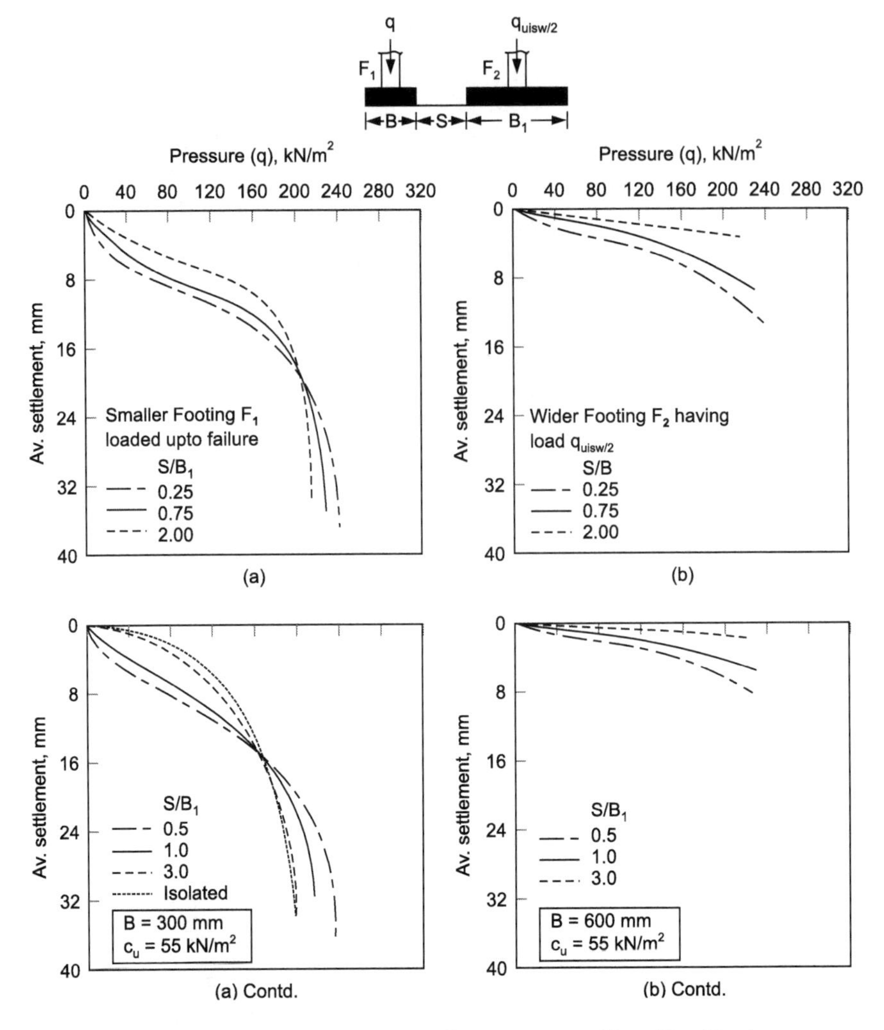

Fig. 8.14. Pressure versus settlement curves (Case V – Clay) (a) Smaller footing F_1; (b) Wider footing F_2.

(ii) For footing F_1

$$t_{a1} = \frac{q}{q_{uis}} c_a + q \tan \delta_m$$

$$c_a = \frac{2}{3} c = \frac{2}{3} \times 30 = 20 \text{ kN/m}^2$$

$$\tan \delta_m = \frac{q}{q_{uis}} \tan \delta = \frac{75}{280.4} \tan\left(\frac{2}{3} \times 30°\right) = 0.0973$$

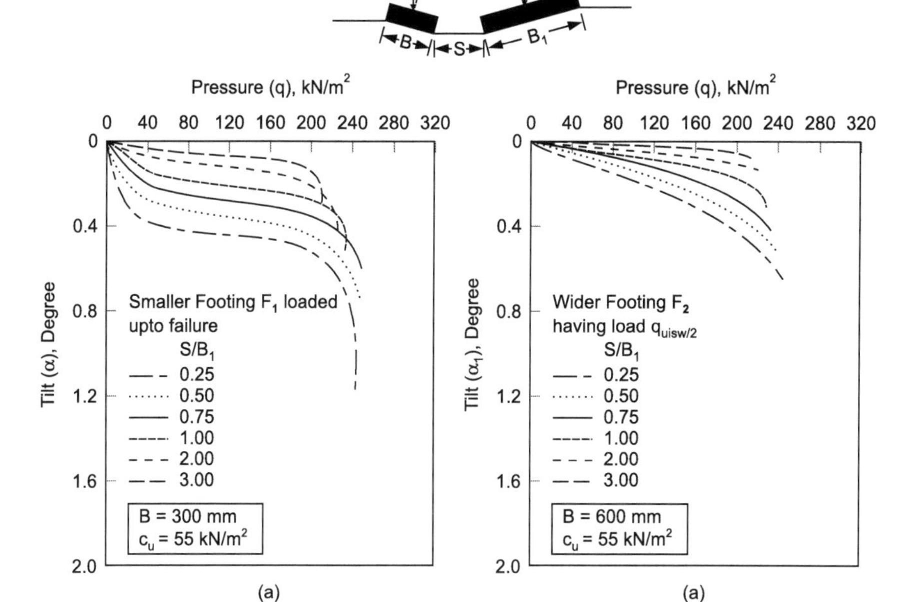

Fig. 8.15. Pressure versus tilt curves (Case V – Clay)
(a) Smaller footing F_1; (b) Wider footing F_2.

$$t_{a1} = \frac{75}{280.4} \times 20 + 75 \times 0.0973 = 5.35 + 7.3 = 12.65 \text{ kN/m}^2$$

For footing F_2

$$t_{a2} = \frac{c_a}{2} + \frac{q_{uis}}{2} \cdot \frac{\tan\delta}{2} = \frac{20}{2} + \frac{280.4}{2}\,\frac{\tan 20}{2} = 25.5 \text{ kN/m}^2$$

(iii) Stress at point 'B' (Fig. 8.22)

(a) Due to vertical pressure (Footing F_1)

$$b = 0.5 \text{ m};\ x = 0.5 \text{ m};\ z = 1.25 \text{ m and } q_1 = 75 \text{ kN/m}^2$$

$$\frac{\sigma_z}{q_1} = 0.370;\ \frac{\sigma_x}{q_1} = 0.059;\ \frac{\tau_{xz}}{q_1} = 0.124 \qquad \text{[Refer Table 3.1]}$$

$$\sigma_z = 27.75 \text{ kN/m}^2;\ \sigma_x = 4.425 \text{ kN/m}^2;\ \tau_{xz} = 9.3 \text{ kN/m}^2;$$

(b) Due to tangential (Footing F_1)

$$2b = 1.0 - 0.125 \text{ m};\ b = \frac{0.875}{2} = 0.4375 \text{ m};$$

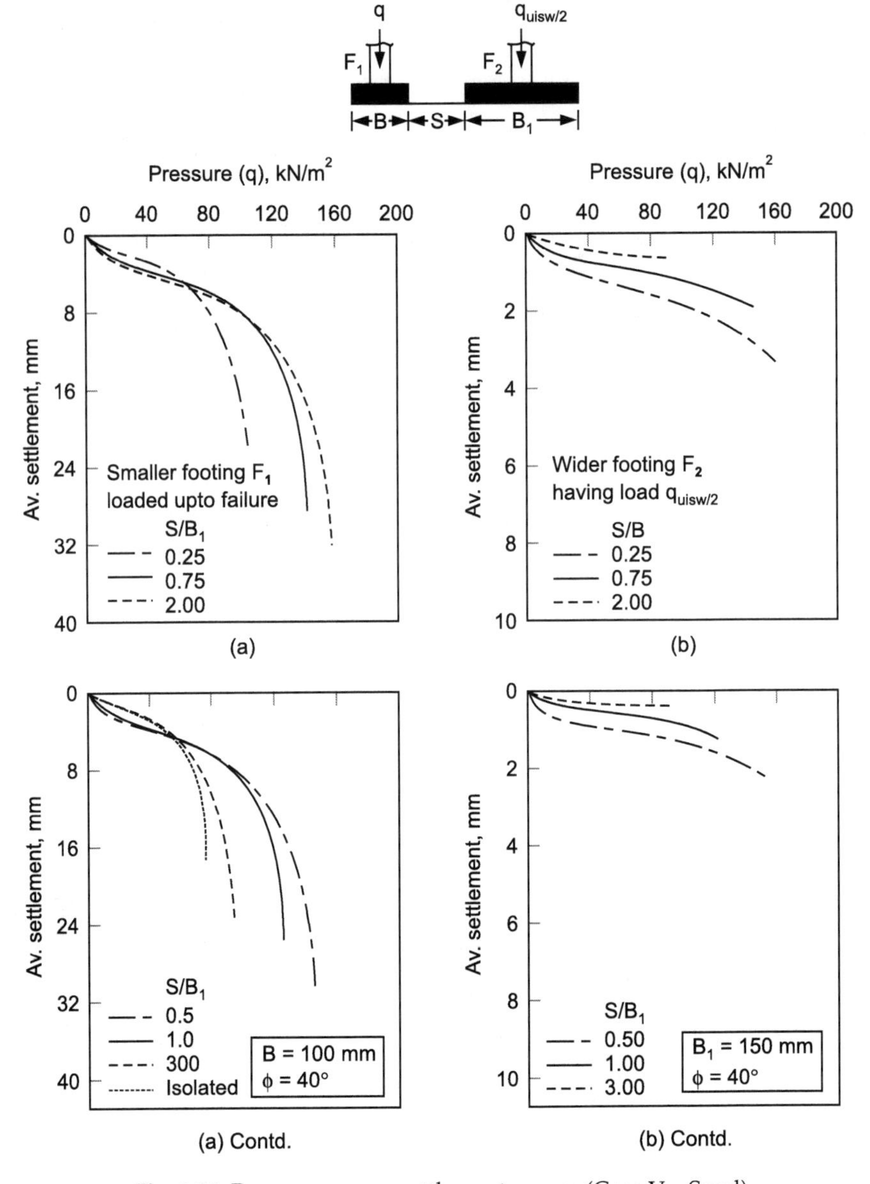

Fig. 8.16. Pressure versus settlement curves (Case V – Sand)
(a) Smaller footing F_1; (b) Wider footing F_2.

$x = 0.4375 + 0.125 = 0.5625$ m; $z = 1.25$ m and $q_{t1} = 12.65$ kN/m²;

$$\frac{\sigma_z}{q_{t1}} = 0.121; \quad \frac{\sigma_x}{q_{t1}} = 0.033; \quad \frac{\tau_{xz}}{q_{t1}} = 0.059 \quad \text{[Refer Table 3.2]}$$

σ_z = 1.53 kN/m^2; σ_x = 0.417 kN/m^2; τ_{xz} = 0.746 kN/m^2;

$$b = \frac{0.125}{2} = 0.0625 \text{ m}; x = -0.0625 \text{ m}; z = 1.25 \text{ m}$$

$$\frac{\sigma_z}{q_{t1}} = -0.0031; \frac{\sigma_x}{q_1} = -0.000016; \frac{\tau_{xz}}{q_1} = 0.00021 \text{ [Refer Table 3.2]}$$

σ_z = –0.039 kN/m^2; $\sigma_x \approx 0.0$; $\tau_{xz} \approx 0.0$

(c) Due to normal stress (Footing F_2)

b = 0.5 m; x = –(0.5 + 1)m = –1.5 m; z = 1.25 m and q_2 = 140.2 kN/m^2

$$\frac{\sigma_z}{q_1} = 0.095; \frac{\sigma_x}{q_1} = 0.119; \frac{\tau_{xz}}{q_1} = -0.1046 \quad \text{[Refer Table 3.1]}$$

σ_z = 13.3 kN/m^2; σ_x = 16.7 kN/m^2; τ_{xz} = –14.7 kN/m^2;

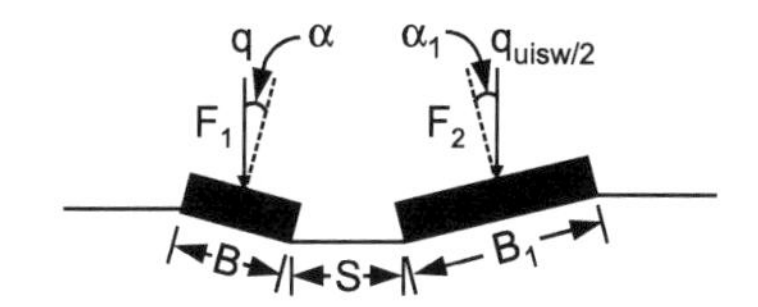

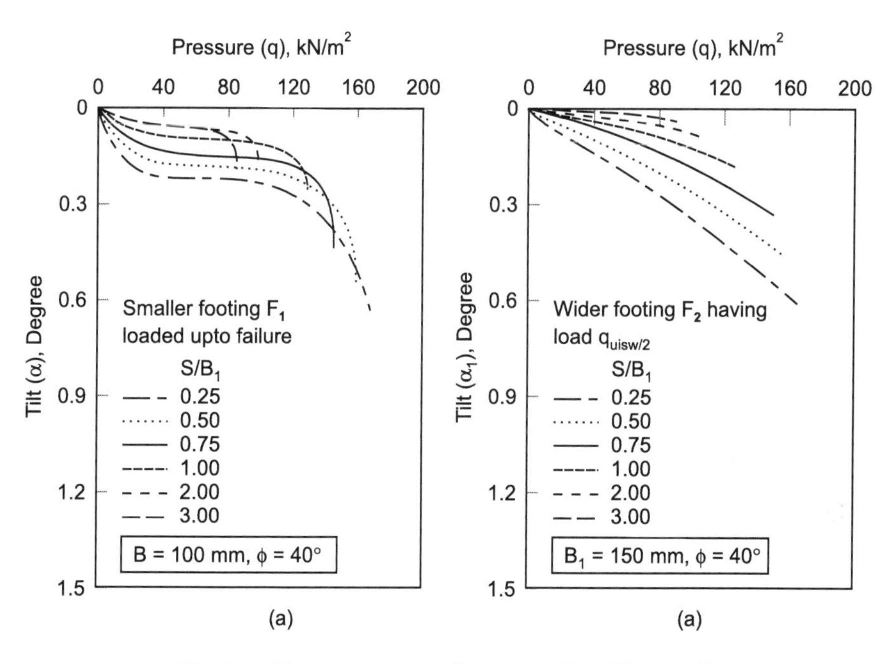

Fig. 8.17. Pressure versus tilt curves (Case V – Sand)
(a) Smaller footing F_1; (b) Wider footing F_2.

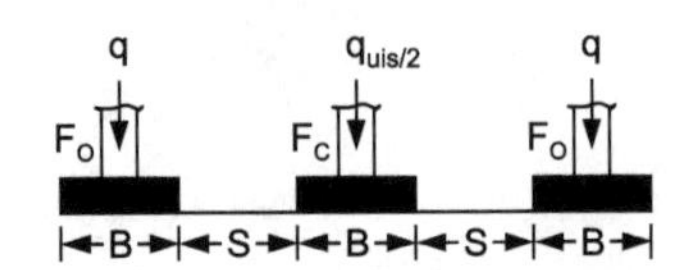

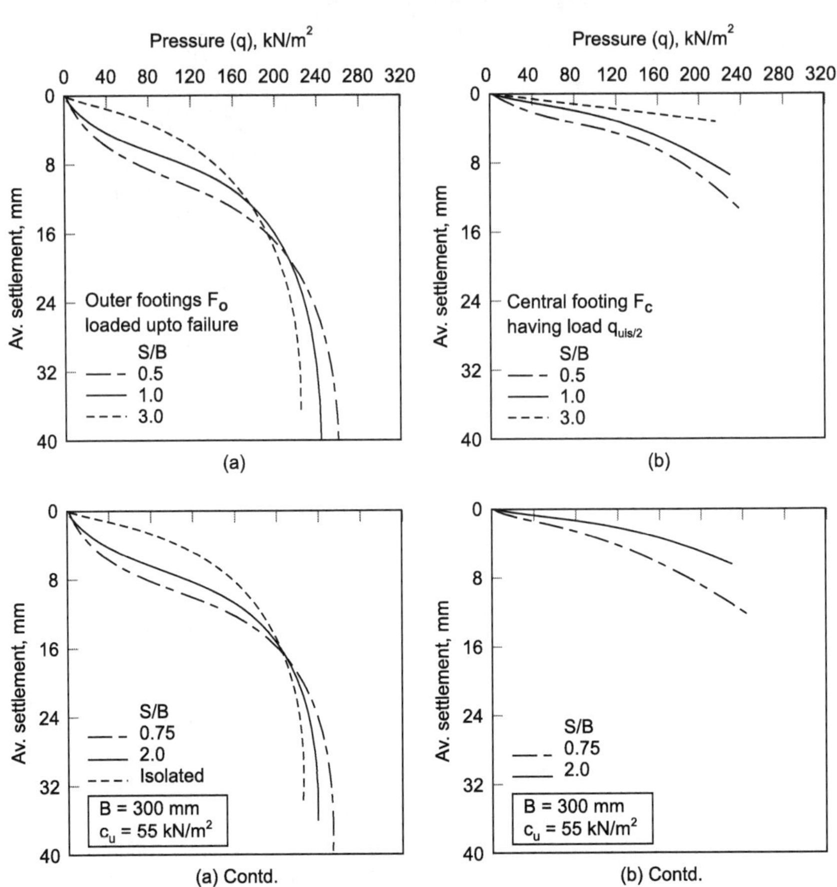

Fig. 8.18. Pressure versus settlement curves (Case VII – Clay)
(a) Outer footings F_O; (b) Central footing F_C.

(d) Due to tangential stress (Footing F_2)

$b = 0.4375$; $x = -(0.4375 + 1 + 0.125)$ m $= -1.5625$ m; $z = 1.25$ m and $q_{t2} = 25.5$ kN/m^2

$$\frac{\sigma_z}{q_{t2}} = 0.08645; \quad \frac{\sigma_x}{q_{t2}} = -0.1288; \quad \frac{\tau_{xz}}{q_{t2}} = 0.1041 \quad \text{[Refer Table 3.2]}$$

$\sigma_z = 2.2$ kN/m^2; $\sigma_x = 3.3$ kN/m^2; $\tau_{xz} = 2.65$ kN/m^2;

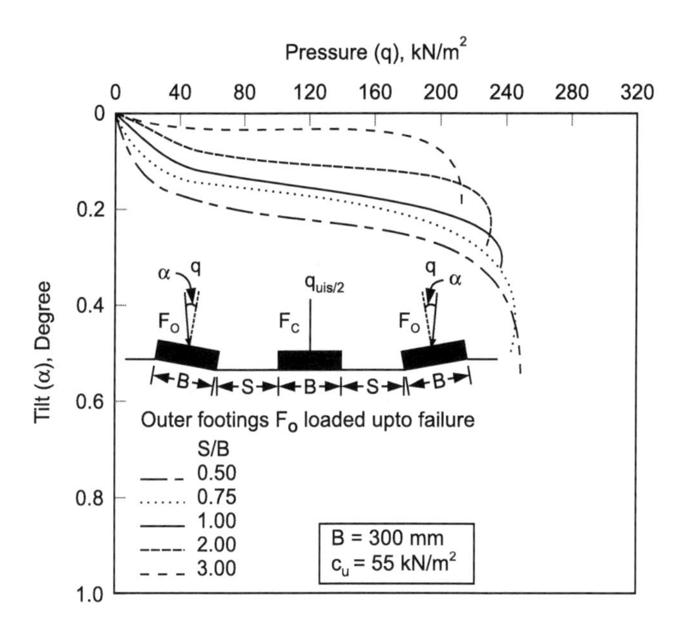

Fig. 8.19. Pressure versus tilt curves for outer footings F_O (Case VII – Clay).

$$b = 0.0625\text{m};\ x = -1.0625\text{m};\ z = 1.25\text{m and } q_{t2} = 25.5\ \text{kN/m}^2$$

$$\frac{\sigma_z}{q_{t2}} = -0.018;\ \frac{\sigma_x}{q_{t2}} = 0.0131;\ \frac{\tau_{xz}}{q_{t2}} = 0.0154 \quad \text{[Refer Table 3.2]}$$

$$\sigma_z = -0.46\ \text{kN/m}^2;\ \sigma_x = -0.33\ \text{kN/m}^2;\ \tau_{xz} = 0.39\ \text{kN/m}^2$$

Total stresses:

$$\sigma_z = 27.75 + 1.53 - 0.039 + 13.3 + 2.2 - 0.46 = 44.3\ \text{kN/m}^2$$

$$\sigma_x = 4.42 + 0.42 - 0.0 + 16.7 + 3.3 - 0.33 = 24.5\ \text{kN/m}^2$$

$$\tau_{xz} = 9.3 + 0.746 + 0.0 - 14.7 + 2.65 + 0.39 = -1.6\ \text{kN/m}^2$$

$$\sigma_1 = 44.43\ \text{kN/m}^2;\ \sigma_3 = 24.37\ \text{kN/m}^2;\ \theta_1 = -4.59;\ \theta_3 = 85.40;$$

$$\frac{1}{a} = 3.0 \times 10^3 (\sigma_3)^{0.4} = 3.0 \times 10^3 (24.37)^{0.4} = 9.59 \times 10^3\ \text{kN/m}^2$$

$$\frac{1}{b} = 100 + 2.0\,(\sigma_3) = 100 + 2.0 \times 24.37 = 148.74\ \text{kN/m}^2$$

$$a = 0.104 \times 10^{-3}\ \text{m}^2/\text{kN}$$

$$b = 0.0067\ \text{m}^2/\text{kN}$$

$$F = \frac{q_u}{q} = \frac{280.4}{75} = 3.74$$

$$E_s = \frac{(1-1/F)}{a} = \frac{(1-1/3.74)}{0.104\times10^{-3}} = 7.04\times10^{3}\ \text{kN/m}^2$$

$$\varepsilon_1 = \frac{(\sigma_1-\sigma_3)}{E_s} = \frac{44.43-24.37}{7.04\times10^{-3}} = 2.84\times10^{-3}$$

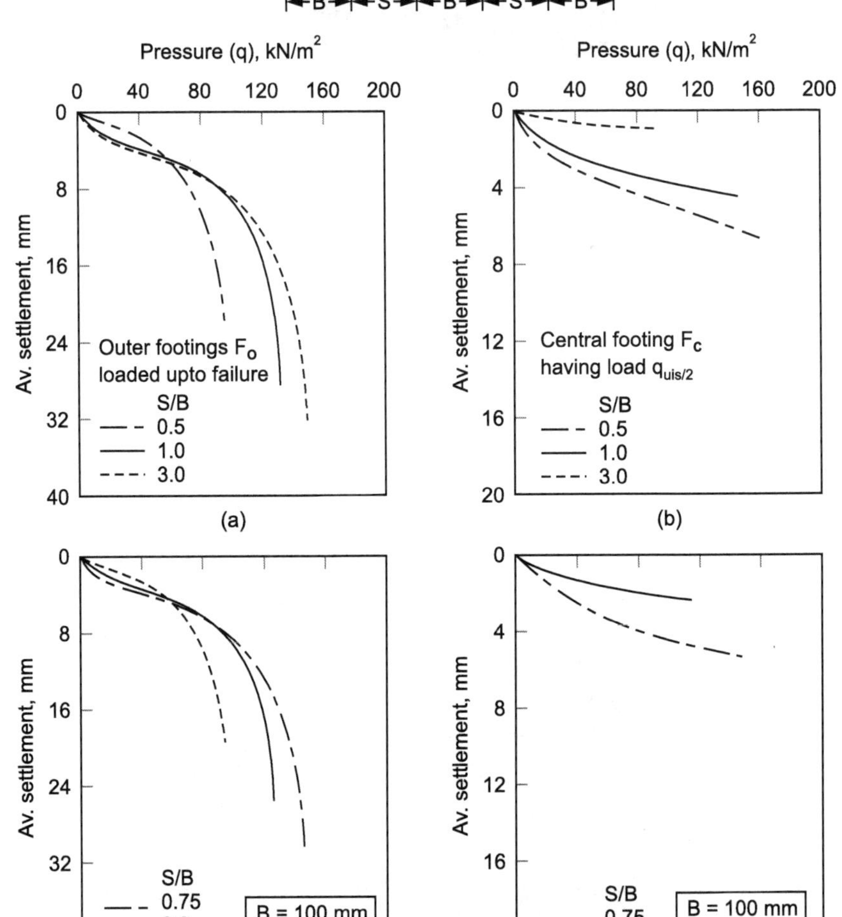

Fig. 8.20. Pressure versus settlement curves (Case VII – Sand) (a) Outer footings F_O; (b) Central footing F_C.

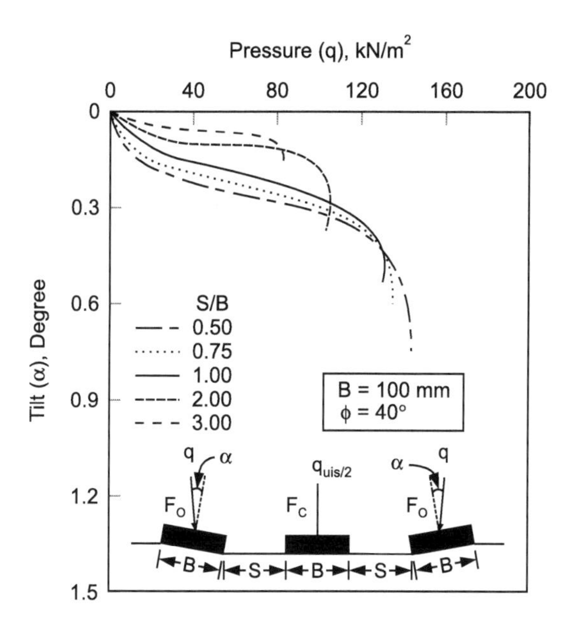

Fig. 8.21. Pressure versus tilt curves for outer footings (F_O) (Case VII – Sand).

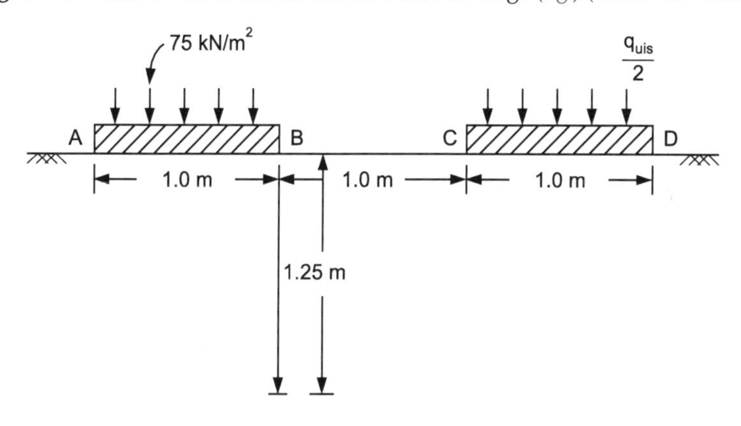

Fig. 8.22. Example 8.1.

$$\mu = 0.35; \mu_1 = \frac{\mu}{1-\mu} = \frac{0.35}{1-0.35} = 0.538$$

$$\mu_2 = \frac{\sigma_3 - \mu_1\sigma_1}{\sigma_1 - \mu_1\sigma_3}$$

$$= \frac{24.37 - 0.538 \times 44.43}{44.43 - 0.538 \times 24.37} = \frac{0.467}{31.32} = 0.015$$

$$\varepsilon_3 = -\mu_2.\varepsilon_1 = -0.015 \times 2.84 \times 10^{-3} = -0.0423 \times 10^{-3}$$

$$\varepsilon_z = \varepsilon_1\cos^2\theta_1 + \varepsilon_3\cos^2\theta_3$$

$$= 2.84 \times 10^{-3}\cos^2(-4.59) + (-0.0423 \times 10^{-3})\cos^2(85.4)$$

$$= 2.82 \times 10^{-3} - 2.72 \times 10^{-7} \approx 2.8 \times 10^{-3}$$

Practice Problems

1. How can the interference between two footings affect their behavior? Explain. Give some examples of structures in which interference between footing is likely to occur.
2. Explain stepwise the procedure of obtaining pressure-settlement of two interfering footings resting on clay while mentioning the loads on the footings.
3. Explain stepwise the procedure of obtaining pressure-settlement of two interfering footings resting on sand and mention the loads on the footings.
4. Two footings each of width equal to 1.0 m are spaced having centre to centre spacing as 2.0 m on the surface of soil. The properties of soil are: $c = 35$ kN/m^2; $\phi = 32°$, $\gamma = 20$ kN/m^3. Kondner's hyperbola constants are: $\frac{1}{a} = 4.0 \times 10^3(\sigma_3)$ $\frac{1}{b} = 100+3.0(\sigma_3)$, σ_3 being in kN/m^2.

REFERENCES

Amir, A.A.A. (1992). Interference Effect on the Behavior of Footing. Ph.D. Thesis, IIT, Roorkee.

Amir, J.M. (1967). Interaction of adjacent footings. *Proc. 3rd Asian Regional Conf. on Soil Mechanics and Foundation Engineering*. Haifa. **1(5):** 189-192.

Carothers, S.D. (1920). Direct Determination of Stresses. *Proc. of the Royal Society of London*, Series A, **97(682):** 110-123.

Das, B.M. and Labri-Cherif, S. (1983). Bearing capacity of two closely-spaced shallow foundations on sand. *Soils and Foundation*, Japanese Society of Soil Mechanics and Foundation Engrg., **23(1):** 1-7.

Dash, P.K. (1981). Interference between surface footings on purely cohesive soil. *Indian Geotech. J.*, **11(4):** 397-402.

Dash, P.K. (1990). Load carrying capacity of surface footings subjected to the effect of interference. *Indian Geotech. Conf.*, Visakhapatnam, **6(1):** 359-361.

Dembicki, E., Odrobinski, W. and Morzek, W. (1971). Bearing capacity of subsoil under strip foundations. *Proc. 10th Int. Conf. on Soil Mechanics and Foundation Engrg.*, Stockholm, **2:** 91-94.

Deshmukh, A.M. (1979). Interference of different types of footings on sand. *Indian Geotech. J.*, **8(4):** 193-204.

Graham, J., Raymond, G.P. and Supplah, A. (1984). Bearing capacity of three closely-spaced footings on sand. *Geotechnique*, **34(2):** 173-182.

Kumar, J. and Bhattacharya, P. (2013). Bearing capacity of two interfering strip footings from lower bound finite elements limit analysis, **37(5):** 441-452.

Khadilkar, B.S. and Varma (1977). Analysis of interference of strip footings by FEM. *Proc. 9th Int. Conf. on Soil Mechanics and Foundation Engrg.*, Tokyo, Japan, **1:** 597-600.

Kolosov, G.B. (1935). Application of complex diagrams and the theory of functions of complex variables to the theory of elasticity. ONTI.

Kouzer, K.M. and Kumar, J. (2008). Ultimate bearing capacity of equally spaced multiple strip footings on cohesion-less soils without surcharge. *International Journal for Numerical and Analytical Methods in Geomechanics*, **32(11):** 1417-1426.

Kouzer, K.M. and Kumar, J. (2010). Ultimate bearing capacity of a footing considering the interference of an existing footing on sand. *Geotech and Geol Eng.*, **28(4):** 457-470.

Kumar, J. and Kouzer, K.M. (2008). Bearing capacity of two interfering footings. *International Journal for Numerical and Analytical Methods in Geomechanics*, **32(3):** 251-264.

Kumar, J. and Ghosh, P. (2007). Ultimate Bearing Capacity of Two Interfering Rough Strip Footings. *Int. J. Geomech.*, **7(1):** 53-62.

Mandel, J. (1963). Interaction plastique de foundations super-ficielles. *Proc. Int. Conf. on Soil Mechanics and Foundation Engrg.*, Budapest, 267-270.

Murthy, S.S.N. (1970). Interference in surface footings on clear sands. *Proc. Symposium on Shallow Foundations*, Bombay, **1:** 109-115.

Myslivec, A. and Kysela. Z. (1973). Interaction of neighbouring foundations. *Proc. 8th Int. Conf. on Soil Mechanics and Foundation Engrg.*, Moscow, **1(3):** 181-184.

Patankar, M.V. and Khadllkar, B.S. (1981). Nonlinear analysis of interference of three surface strip footings by FEM. *Indian Geotech. J.*, **11(4):** 327-344.

Saran, S. and Agarwal, V.C. (1974). Interference of surface footings in sand. *Indian Geotech. J.*, **4(2):** 129-139.

Selvadurai, A.P.S. and Rabbaa, S.A.A. (1983). Some experimental studies concerning the contact stresses beneath interfering rigid strip foundations resting on a granular stratum. *Can. Geotech. J.*, **20:** 406-415.

Singh, A., Punmia, B.C. and Ohri, M.C. (1973). Interference between adjacent square footings on cohesion-less soil. *Indian Geotech. J.*, **3(4):** 275-284.

Siva Reddy, A. and Mogallah, G. (1976). Inteference between surface strip foundations on soil exhibiting anisotropy and non-homogeneity in cohesion. *J. Inst. of Engrs.*, **57:** 7-13.

Stuart, J.G.(1962). Interference between foundations with special reference to surface footings in sand. *Geotechnique*, **12(1):** 15-23.

CHAPTER 9

Ring Footings

9.1 General

Shallow foundations are a common type of foundation usually provided for various important structures resting on good soil. Annular foundations, however, are generally used for self-supporting vertical cantilevers (such as chimneys, silos, T.V. towers, etc.) and for tower-like structures (including elevated tanks, bins, vessels, etc.). Such foundations are usually subjected to eccentric-inclined loads due to moments and horizontal thrusts in conjunction with the vertical loads. Ring foundations, besides being economical, are the only solution for the aforementioned structures, when the dual conditions of full utilization of soil capacity and no tension under the foundation are to be satisfied.

Figure 9.1 illustrates some engineering application areas for the use of ring footings.

9.2 Brief Review

9.2.1 Ring Footing Subjected to Central Vertical Load

Various analytical solutions, based on the theory of elasticity, have been proposed by Egorov, 1965; Kovalev et al., 1966; Egorov and Shilova, 1968; Borodacheva, 1972; Popova, 1972; Snarskii, 1974; Zinov'ev, 1979 and Madhav, 1980 for computing settlement, contact pressure or stresses for ring or circular footings. Non-linear F.E. anlayses for predicting the bearing capacity and pressure-settlement relations have been developed by Desai et al. (1970); Duncan et al. (1970); Carrier et al. (1973); Hooper (1983); Murthy et al. (1995) for circular footings and by Patankar (1982) for ring footings. While, linear F.E. analyses to predict the same have been developed by Milovic (1973); Varadarajan et al. (1984); Kumar (1986);

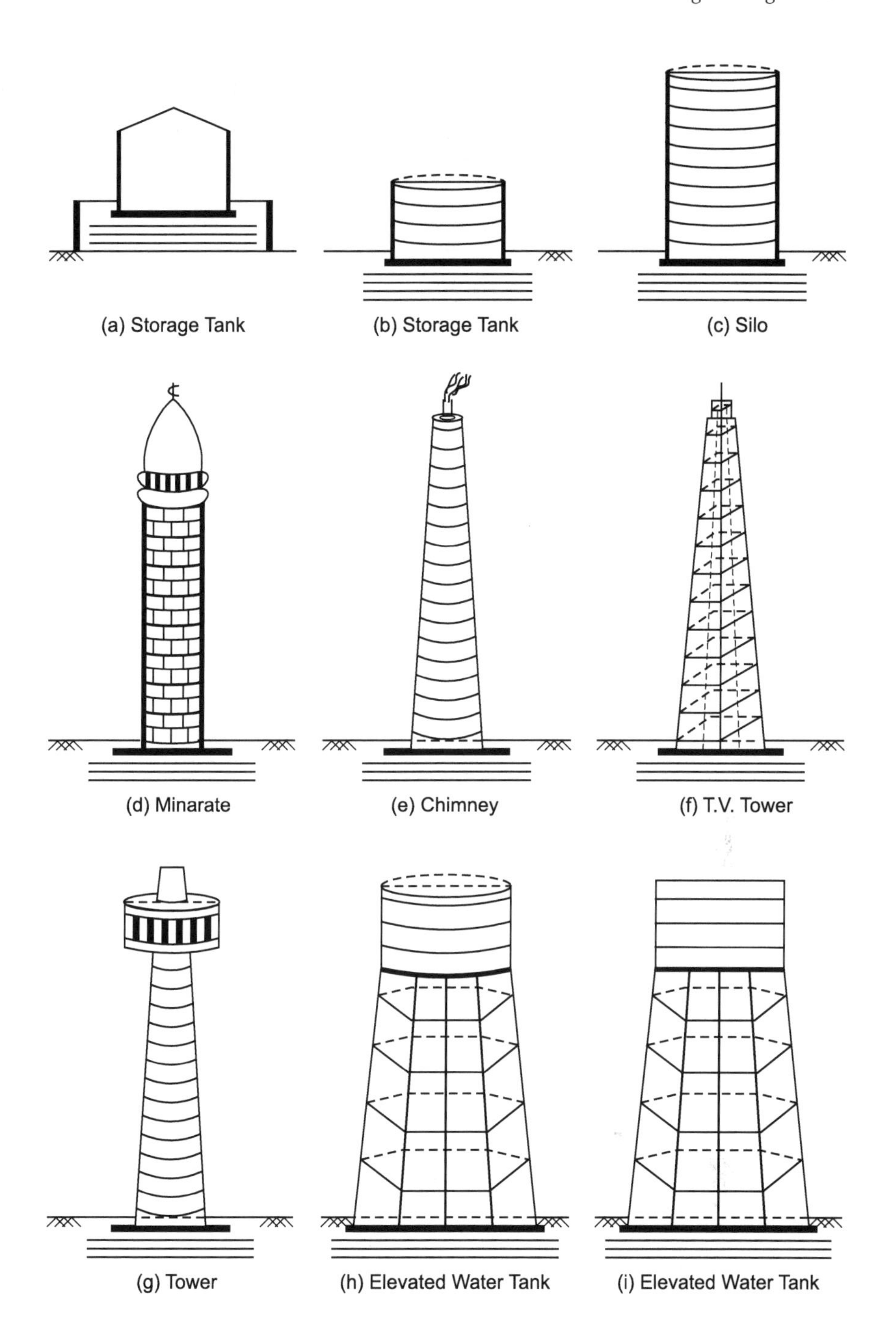

Fig. 9.1. Application areas for the use of ring footings and earth reinforcement.

Hataf and Razavi (2003); Snodi (2010) and Hosseininia (2016) for ring footings and by Dewaikar et al. (1992) for circular footings.

Choobbasti et al. (2010) carried out a numerical analysis using Plaxis software for calculating bearing capacity and settlement of a ring footing.

Kumar and Ghosh (2005) obtained $N\gamma$ factor for ring footings using the method of characteristics. Kumar and Chakraborty (2015) computed the bearing capacity factor for smooth and rough ring footings, using lower and upper bound theorems of limit analysis in conjunction with finite elements and linear optimization.

Experimental studies were conducted by Desai (1970), Saha (1978), Lo and Becker (1979), Haroon et al. (1980), Kakroo, (1985), Punmia (1985), Kumar (1986), Ranjan et al. (1987) and Al-Sanad et al. (1993) to investigate the behaviour of ring footings on sand or clay.

9.2.2 Ring Footing Subjected to Eccentric-vertical Load

Several investigators have developed equations based on the theory of elasticity to compute the settlement, tilt or contact pressure (Frohlich, 1953; Tettinek, 1953; Yegorov et al., 1961; Chandra et al., 1965; Arya, 1966; Zweig, 1966; Laletin, 1969; Glazer et al., 1975; Gusev, 1975; Dave, 1977 and Egorov et al., 1977) for circular or ring footings, while Meyerhof (1953); Karsenskii (1965); Gusev (1969) based their analyses on the plastic theory, complex resistance formula and Winkler's theory, respectively.

A few research workers have experimentally studied the behaviour of ring or circular footings (Meyerhof, 1953; Gusev, 1967; Tetior, 1971; Chaturvedi, 1982; Singh, 1982; Dhatrak and Gawande, 2016) on sand or clay.

9.2.3 Ring Footing Subjected to Eccentric-inclined Load

Al-Smadi (1998) developed an analytical analysis for getting pressure-settlement and pressure-tilt characteristic of ring footings subjected to eccentric-inclined loads resting on either clay or sand. He also performed model tests on solid circular footing and ring footings subjected toeccentric-inclined load and resting on sand. This work was summarized subsequently.

9.3 Footing on Clay

9.3.1 Assumptions

Assumptions made in the analysis are as given below:

- The soil mass is an isotropic and semi-infinite medium
- The footing base is smooth and resting on the surface of the ground

- The footing is initially considered to be flexible. The contact pressure distribution is assumed to be linear as shown in Figs 9.2 and 9.3
- The footing has been divided into n-small part (Fig. 9.5), so that each part is assumed to be acted upon by a concentrated vertical load and a concentrated horizontal load
- The soil mass supporting the footings is divided into a large number of thin horizontal layers, to a depth beyond which the stresses become very small, as shown in Fig. 9.4
- Stresses due to the vertical component of the applied load at the centre of each layer are computed using Bousinesq's theory (1885), since all stress equations are available
- Stresses due to the horizontal component of the applied load at the centres of each layer are computed using Cerruti's theory (1888), since all stress equations are available
- Strains have been computed from the known stress conditions using the constitutive law of soil
- There is no slippage at the interface of layers of the soil mass

9.3.2 Pressure-settlement and Pressure-tilt Characteristics

The proposed analytical procedure for predicting the pressure-settlement and pressure-tilt characteristics of ring footings, subjected to eccentric-inclined load, is described in the following steps:

Step 1

The whole area of the footing is divided into n-equal rings and N-equal sectors, as shown in Fig. 9.5. The resulting elementary areas are small; hence, the total load in each area has been considered as a point load, acting at the centre of each area.

Step 2

The contact pressure distribution diagrams, at the interface of the footing base and the soil mass, due to the eccentric-inclined load, are assumed to be linear, as shown in Figs 9.2 and 9.3, depending on the amount of eccentricity. The resultant of the vertical and horizontal contact pressure diagrams is equal to the applied eccentric-inclined load and its line of action coincides with the line of action of the applied eccentric-inclined load. Two cases have been considered, depending on the extent of eccentricity.

Case 1: $e/B \leq \dfrac{(1+n^2)}{8}$ (Appendix 9.1)

As seen in Fig. 9.2b, the maximum and minimum contact pressures due to the vertical component of the eccentric-inclined load (Roark, 1954; Teng, 1977) are given by:

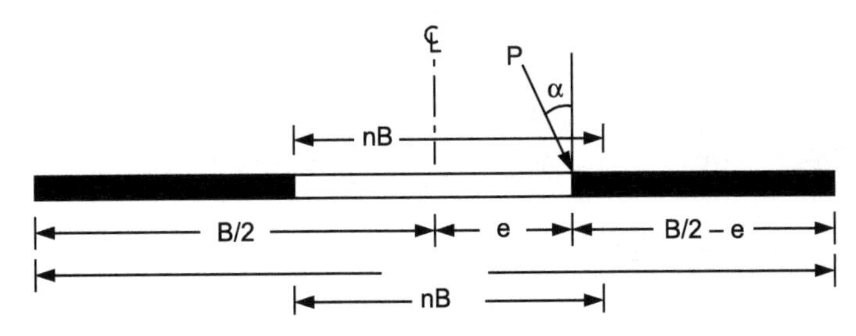

(a) Eccentric-inclined load on footing

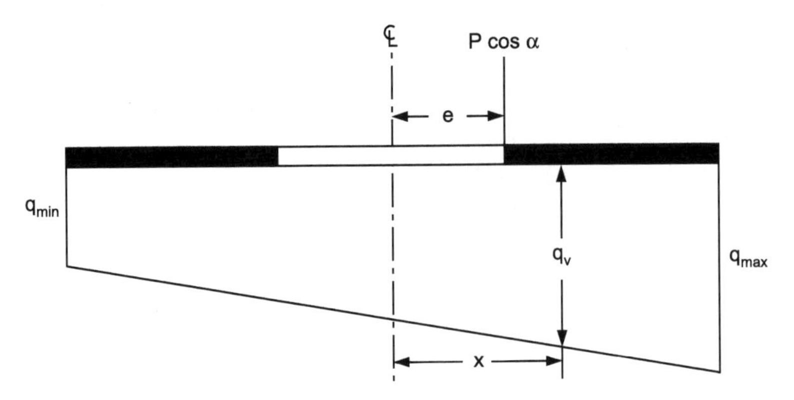

(b) Contact pressure diagram due to vertical component of eccentric-inclined load

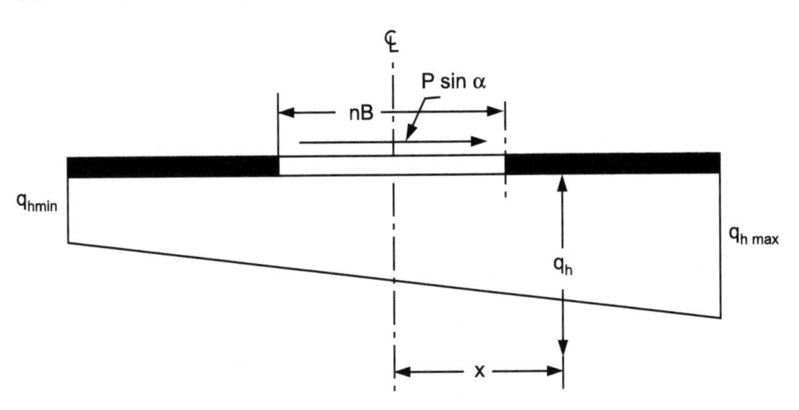

(c) Contact pressure diagram due to horizontal component of eccentric-inclined load

Fig. 9.2. Contact pressure distribution due to eccentric-inclined load, for $e/B \le (1 + n^2)/8$.

$$q = \frac{P}{A}\cos\alpha\left[1 \pm \left(\frac{8}{1+n^2}\right)\left(\frac{e}{B}\right)\right] \tag{9.1}$$

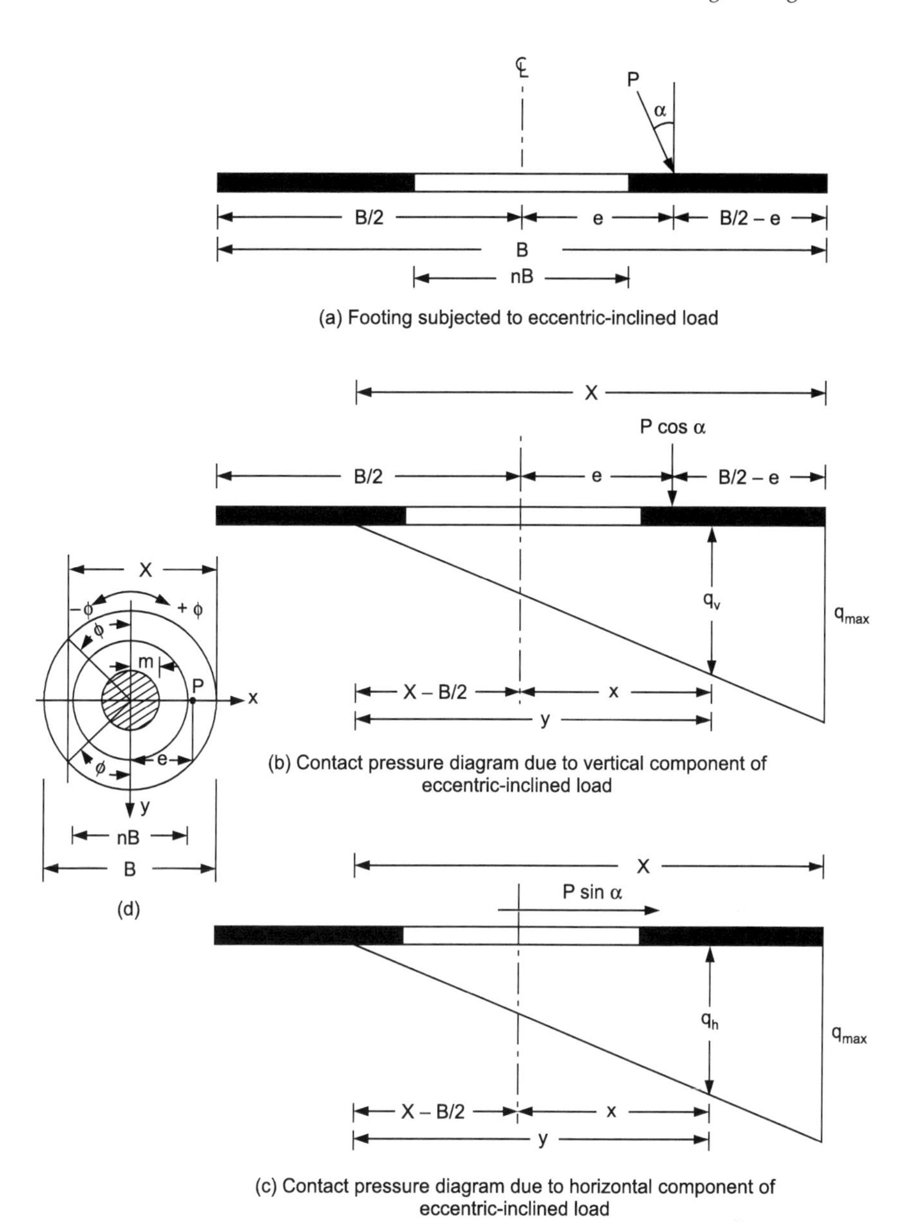

Fig. 9.3. Contact pressure distribution due to eccentric-inclined load, for $e/B > (1 + n^2)/8$.

The contact pressure at distance x from the centre is (Apendix 9.1)

$$q_v = \frac{P}{A}\cos\alpha\left[1+\left(\frac{8}{1+n^2}\right)\left(\frac{e}{B}\right)\left(\frac{x}{R}\right)\right] \tag{9.2}$$

$$\therefore\ \Delta P_v = q_v.\Delta A \tag{9.3}$$

where P = Applied eccentric-inclined load
e = Eccentricity of the load
B = Diameter of the footing (=2R)
q = Maximum or minimum contact pressure
α = Inclination of the applied load
A = Net area of the footing
n = Annular ratio
q_v = Contact pressure at point x
x = Coordinate of the point at which the contact pressure is desired
ΔA = Area of elementary strip
ΔP_V = Vertical point load acting in ΔA

However, the maximum and minimum contact pressures due to the horizontal component of the eccentric-inclined load 'P', as shown in Fig. 9.2c are assumed as follows:

$$q_{h\ \max,\min} = \frac{P}{A}\sin\alpha\left[1 \pm \left(\frac{8}{1+n^2}\right)\left(\frac{e}{B}\right)\right] \tag{9.4}$$

The contact pressure at distance x from the centre is given by

$$q_h = \frac{P}{A}\sin\alpha\left[1 + \left(\frac{8}{1+n^2}\right)\left(\frac{e}{B}\right)\left(\frac{x}{R}\right)\right] \tag{9.5}$$

$$\Delta P_h = q_h.\Delta A \tag{9.6}$$

where ΔP_h = Horizontal point load acting in small area ΔA.

Case 2: $e/B > \dfrac{(1+n^2)}{8}$

As seen in Fig. 9.3b, the maximum pressure due the vertical component of the eccentric-inclined load (Roark, 1954; Teng, 1977; Young, 1989) is given by

$$q_{max} = k\frac{P}{A}\cos\alpha \tag{9.7}$$

As shown in Fig. 9.3b, the contact pressure, due to the vertical component of the eccentric-inclined load, at a point x from the centre of the footing is given by

$$q_v = \frac{y}{X}q_{max},\ \text{where } y = x + \left(X - \frac{B}{2}\right) \tag{9.8}$$

$$\therefore \quad \Delta P_v = q_v . \Delta A \tag{9.9}$$

$$X = R(1 + \sin\phi) \tag{9.10}$$

where X and ϕ are as shown in Figs 9.3b and 9.3c, and ϕ satisfies Eq. (9.11).

$$e/R = \frac{(\pi/8) - 2.5\phi - (5/12)\sin\phi\cos\phi + (1/6\sin^3\phi\cos\phi)}{\cos\phi - (1/3)\cos^3\phi - (\pi/2)\sin\phi + \phi\sin\phi} \tag{9.11}$$

$$y = x + \left(X - \frac{B}{2}\right) \tag{9.12}$$

Values of k and X/R were calculated by Young (1989) for different n and e/R and presented in Table 9.1.

Step 3

The soil mass supporting the footing is divided into a large number of thin horizontal layers (i.e. 1-layer) upto a depth, beyond which, the stresses become negligible, as shown in Fig. 9.4.

Step 4

Stress components at the center of each layer along a vertical section, due to each vertical point load, were computed by using Boussinesq's equations (Eqs. 7.1 to 7.6) given in chapter 7, while the stress components at the centre of each layer along a vertical section, due to each horizontal point load (Fig. 9.6), are computed by using Cerruti's equations (Eqs. 9.13 to 9.18 as given below):

$$\sigma_z = \frac{3Hxz^2}{2\pi R^5} \tag{9.13}$$

$$\sigma_x = \frac{3H}{2\pi}\left\{\frac{x^3}{R^5} - \left(\frac{m-2}{3m}\right)\frac{x}{R^3}\left[1 - \frac{3R^2}{(R+z)^2} + \frac{x^2(3R+z)}{(R+z)^3}\right]\right\} \tag{9.14}$$

$$\sigma_y = \frac{3H}{2\pi}\left\{\frac{xy}{R^5} - \left(\frac{m-2}{3m}\right)\frac{x}{R^3}\left[1 - \frac{R^2}{(R+z)^2} + \frac{y^2(3R+z)}{(R+z)^3}\right]\right\} \tag{9.15}$$

$$\sigma_{xz} = \frac{3Hx^2z}{2\pi R^5} \tag{9.16}$$

$$\sigma_{yz} = \frac{3Hxyz}{2\pi R^5} \tag{9.17}$$

$$\tau_{xy} = \frac{3H}{2\pi}\left\{\frac{yx^2}{R^5} + \left(\frac{(m-2)}{3m}\right)\left(\frac{y}{R(R+z)^2}\right)\left[1 - \left(\frac{x^2}{R^2}\right) + \frac{(3R+z)}{(R+z)}\right]\right\} \tag{9.18}$$

Table 9.1. Values of k and X/R

n		e/R															
		0.25	*0.29*	*0.3*	*0.34*	*0.35*	*0.4*	*0.41*	*0.45*	*0.5*	*0.55*	*0.6*	*0.65*	*0.7*	*0.75*	*0.8*	*0.85*
0.0	$\frac{X}{R}$	2.00		1.82		1.66	1.51		1.37	1.23	1.10	0.97	0.84	0.72	0.60	0.47	0.35
	k	2.00		2.21		2.46	2.75		3.11	3.56	4.14	4.9	5.94	7.43	9.69	13.4	20.5
0.4	$\frac{X}{R}$		2.00	1.97		1.81	1.67		1.53	1.38	1.22	1.05	0.88	0.73	0.60	0.48	0.35
	k		2.00	2.03		2.22	2.43		2.68	2.99	3.42	4.03	4.9	6.19	8.14	11.3	17.3
0.6	$\frac{X}{R}$				2.00	1.97	1.84		1.71	1.56	1.39	1.21	1.02	0.82	0.64	0.48	0.35
	k				2.00	2.03	2.18		2.36	2.58	2.86	3.24	3.79	4.64	6.04	8.54	13.2
0.8	$\frac{X}{R}$							2.00	1.91	1.78	1.62	1.45	1.26	1.05	0.84	0.63	0.41
	k							2.00	2.10	2.24	2.42	2.65	2.94	3.34	3.95	4.98	7.16

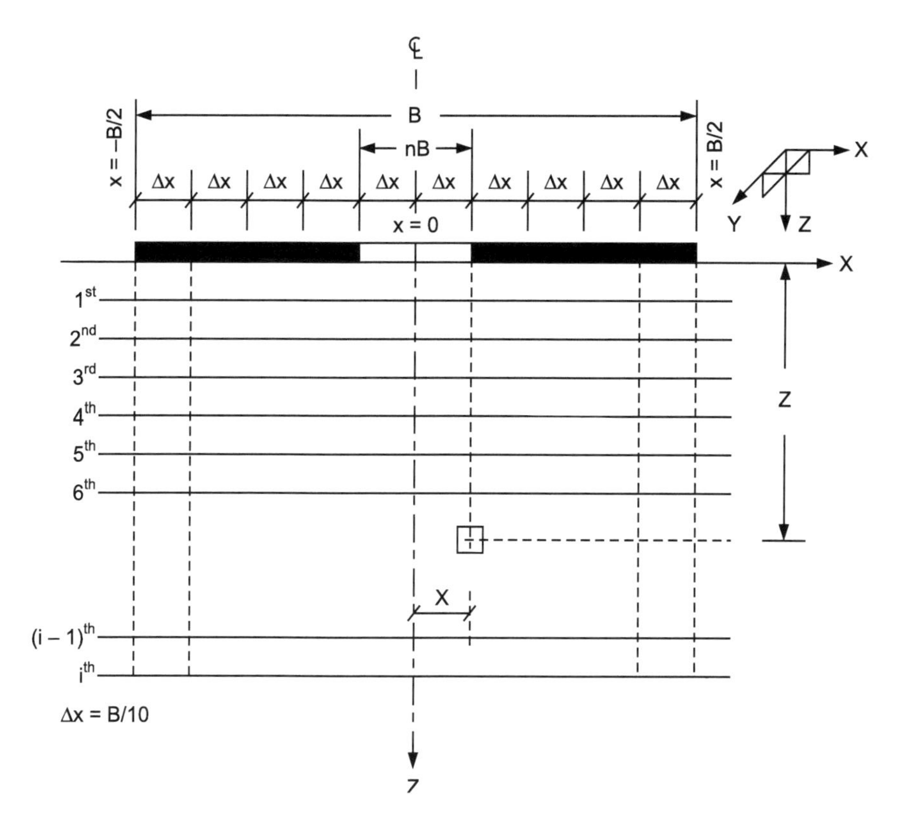

Fig. 9.4. Soil mass supporting a footing divided equally into large number of horizontal layers.

where H = Horizontal load; x, y, z, and R are as shown in Fig. 9.6; m is equal to $1/\mu$, μ being the Poisson's ratio.

Cerruti's theory has the same limitations as Boussinesq's theory.

Normal and shear stresses at the centre of each horizontal layer along a vertical section, due to the applied eccentric-inclined load P, are obtained by applying the method of super-position for all stresses resulting from vertical and horizontal load components in every small area.

Step 5

Principal stresses and their directions, with respect to the vertical z-axis as are computed, using equations of the theory of elasticity (Durielli et al., 1958; Harr, 1966; Poulos and Davis, 1973). The procedure is similar to that described in Chapter 7 for the rectangular footings. However, these steps have been repeated here for ready reference.

$$I_1 = \sigma_x + \sigma_y + \sigma_z \tag{9.19}$$

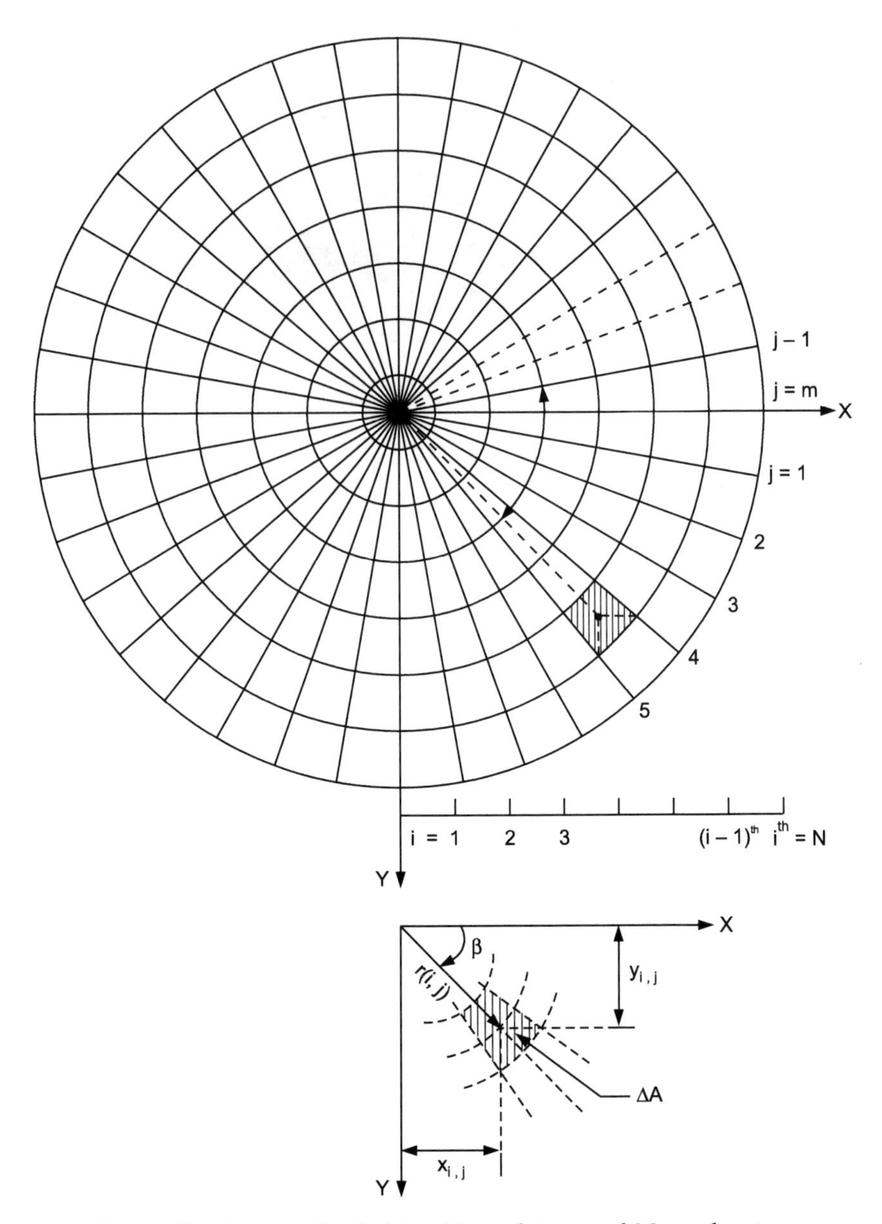

Fig. 9.5. Footing area divided into N equal rings and M equal sectors.

$$I_2 = \sigma_x.\sigma_y + \sigma_y.\sigma_z + \sigma_z.\sigma_x - \tau_{xy}^2 - \tau_{yz}^2 - \tau_{zx}^2 \tag{9.20}$$

$$I_3 = \sigma_x.\sigma_y.\sigma_z - \sigma_x.\tau_{yz}^2 - \sigma_y.\tau_{zx}^2 - \sigma_z.\tau_{xy}^2 + 2\,\tau_{xy}.\tau_{yz}.\tau_{zx} \tag{9.21}$$

Also

$$I_1 = \sigma_1 + \sigma_2 + \sigma_3 \tag{9.22}$$

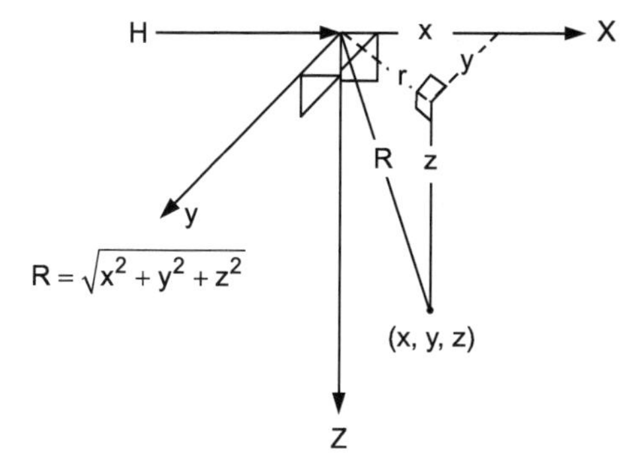

Fig. 9.6. Horizontal point load acting along the surface.

$$I_2 = \sigma_1.\sigma_2 + \sigma_2.\sigma_3 + \sigma_3.\sigma_1 \quad (9.23)$$

$$I_3 = \sigma_1.\sigma_2.\sigma_3 \quad (9.24)$$

Solving for σ_1, σ_2 and σ_3 we have

$$\sigma_1^3 - I_1\sigma_1^2 + I_2\sigma_1 - I_3 = 0 \quad (9.25)$$

$$\sigma_2^3 - I_1\sigma_2^2 + I_2\,\sigma_2 - I_3 = 0 \quad (9.26)$$

$$\sigma_3^3 - I_1\,\sigma_3^2 + I_2\,\sigma_3 - I_3 = 0 \quad (9.27)$$

Thus σ_1, σ_2 and σ_3 are the roots of equation

$$\sigma^3 + p\sigma^2 + q\sigma + r = 0 \quad (9.28)$$

where $p = -I_1$, $q = I_2$, and $r = -I_3$

Values of I_1, I_2 and I_3 are taken from Eqs. (9.19), (9.20) and (9.21) respectively. Solving Eq. (9.28), three values of σ i.e σ_1, σ_2 and σ_3 are obtained.

Solving for direction cosines with respect to z axis, we have (Sokolnikoff, 1956)

$$A_1 = (\sigma_y - \sigma_1)(\sigma_z - \sigma_1) - \tau_{yz}.\tau_{zy} \quad (9.29)$$

$$B_1 = -\tau_{xy}(\sigma_z - \sigma_1) + \tau_{yz}.\tau_{xz} \quad (9.30)$$

$$C_1 = \tau_{xy}.\tau_{yz} - (\sigma_y - \sigma_1) \quad (9.31)$$

$$A_2 = (\sigma_y - \sigma_2)(\sigma_z - \sigma_2) - \tau_{zy}.\tau_{xz} \quad (9.32)$$

$$B_2 = -\tau_{xy}(\sigma_z - \sigma_2) + \tau_{yz}.\tau_{xz} \quad (9.33)$$

$$C_2 = \tau_{xy}.\tau_{yz} - (\sigma_y - \sigma_2)\,.\,\tau_{xz} \quad (9.34)$$

$$A_3 = (\sigma_y - \sigma_3)(\sigma_z - \sigma_3) - \tau_{yz}.\tau_{zy} \quad (9.35)$$

$$B_3 = -\tau_{xy}(\sigma_z - \sigma_3) + \tau_{zy}.\tau_{xz} \tag{9.36}$$

$$C_3 = \tau_{xy}.\tau_{yz} - (\sigma_y - \sigma_3)\tau_{xz} \tag{9.37}$$

and

$$\cos\theta_1 = \frac{C_1}{\sqrt{A_1^2 + B_1^2 + C_1^2}} \tag{9.38}$$

$$\cos\theta_2 = \frac{C_2}{\sqrt{A_2^2 + B_2^2 + C_2^2}} \tag{9.39}$$

$$\cos\theta_2 = \frac{C_3}{\sqrt{A_3^2 + B_3^2 + C_3^2}} \tag{9.40}$$

Step 6

Evaluation of vertical strain

$$\varepsilon_z = \varepsilon_1\cos^2\theta_1 + \varepsilon_2\cos^2\theta_2 + \varepsilon_3\cos^2\theta_3 \tag{9.41}$$

The principal strains ε_1, ε_2 and ε_3 were evaluated, using the following equations:

$$\varepsilon_1 = \frac{a(\sigma_1 - \sigma_3)}{1 - b(\sigma_1 - \sigma_3)} \tag{9.42}$$

$$\varepsilon_2 = \frac{\sigma_2 - \mu(\sigma_3 - \sigma_1)}{\sigma_1 - \mu(\sigma_2 - \sigma_3)} \tag{9.43}$$

$$\varepsilon_3 = \frac{\sigma_3 - \mu(\sigma_1 - \sigma_2)}{\sigma_1 - \mu(\sigma_2 - \sigma_3)} \tag{9.44}$$

a and b are Kondner's hyperbola constants; μ is the Poisson's ratio.

Vertical strain ε_z at the point under consideration is then given by

$$\varepsilon_z = \varepsilon_1\cos^2\theta_1 + \varepsilon_2\cos^2\theta_2 + \varepsilon_3\cos^2\theta_3 \tag{9.45}$$

Step 7

The total settlement 'S' along any vertical section is computed by numerically integrating the expression

$$S = \int_0^L \varepsilon_z dz \tag{9.46}$$

The total settlements are computed along all vertical sections, namely, $x = -B/2, -4B/10, -3B/10, -2B/10, -B/10, 0, B/10, 2B/10, 3B/10, 4B/10, B/2$, for the given pressure intensity, eccentricity, angle of inclination and the annular ratio. Settlement diagrams as shown in Figs 9.7a and 9.8a were obtained.

Step 8

Evaluation of settlements and tilts of rigid ring footings are obtained for the following two cases using the following procedure:

Case 1: When $e/B \leq \dfrac{(1+n^2)}{8}$

As shown in Figs 9.7a and 9.7b, the values of S_{max} and S_{min} are obtained by equating:

- The area of settlement diagram of Fig. 9.7a to the area of settlement diagram of Fig. 9.7b.
- The distance of the centre of settlement diagram of Fig. 9.7a from the left edge of the footing, i.e. point A', to the distance of centre of settlement diagram of Fig. 9.7b from the left edge of the footing, point A'.

i.e.

$$A_s = \frac{y_2}{2}\,[(S_{\text{max}} + S_2) + (S_1 + S_{\text{min}})] \tag{9.47}$$

$$C_g = \frac{\left[A_3\left(\dfrac{(2S_1 + S_{\text{min}})}{(S_1 + S_{\text{min}})}\dfrac{y_2}{3}\right) + A_4\left(\dfrac{(2S_{\text{max}} + S_2)}{(S_{\text{max}} + S_2)}\dfrac{y_2}{3} + D\right)\right]}{(A_3 + A_4)} \tag{9.48}$$

Solving Eqs (9.47) and (9.48),

$$S_{\text{min}} = A_S \frac{\left[B(4 + n(1+n)) - 6C_g\right]}{\left[B^2(1-n)(1+n+n^2)\right]} \tag{9.49}$$

And

$$S_{\text{max}} = A_s \frac{\left[6C_g - B(1-n)(2+n)\right]}{B^2(1-n)(1+n+n^2)} \tag{9.50}$$

where A_s = area of the settlement diagram Fig. 9.6a

$y_2 = \dfrac{B}{2}(1-n)$

S_{min} = minimum settlement of footing

S_{max} = maximum settlement of footing

B = width (= external diamter) of footing

C_g = centre of gravity of the settlement diagram of Fig. 9.7a from point A'

n = annular ratio = internal diameter/external diameter

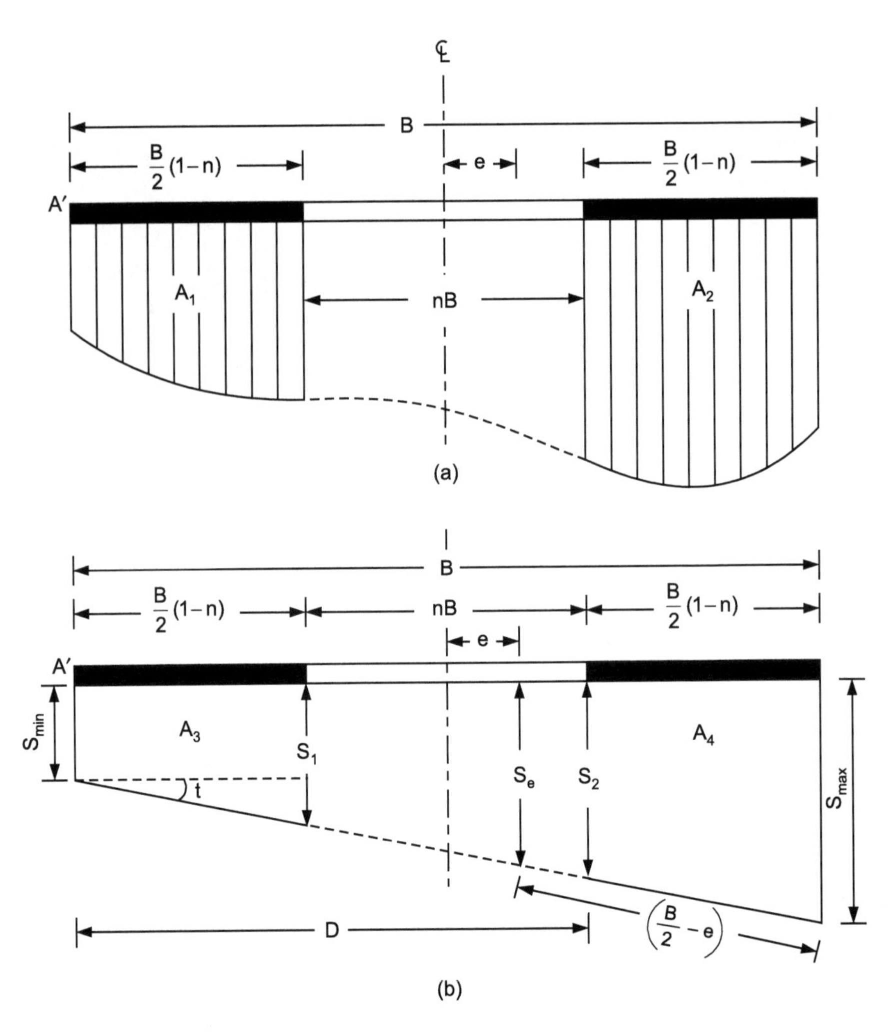

Fig. 9.7. Settlement diagrams under ring footing when $e/B \leq (1 + n^2)/8$.

$$D = \frac{B}{2}(1+n)$$

S_1 and S_2 are as defined in Fig. 9.7b.

Therefore, knowing the values of S_{max} and S_{min}, settlement of the point of load application (S_e) and tilt angle (t) of the rigid footing, as shown in Fig. 9.7b, are computed using the following expressions:

$$t = \frac{(S_{max} - S_{min})}{B} \tag{9.51}$$

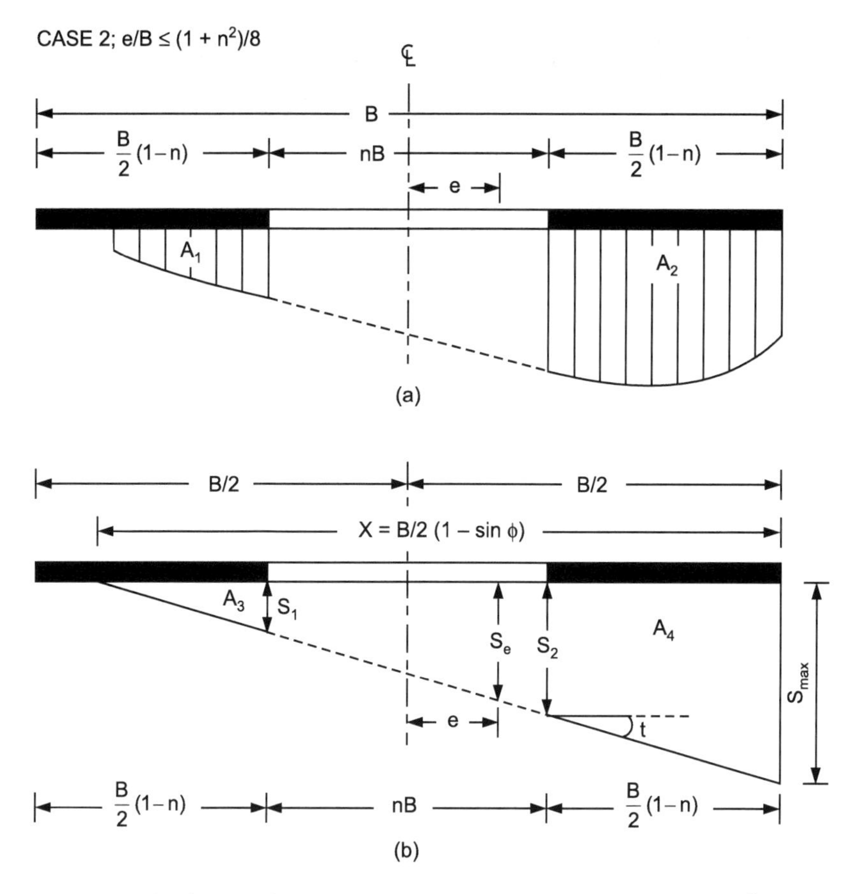

Fig. 9.8. Settlement diagrams under ring footing when $e/B > (1 + n^2)/8$.

and

$$S_e = S_{max} - (B/2 - e) \sin t \tag{9.52}$$

Case 2: When $e/B > \dfrac{(1+n^2)}{8}$

As shown in Figs 9.8a and 9.8b the value of S_{max} is obtained by equating

(i) The area of settlement diagram of Fig. 9.8a to the area of settlement diagram of Fig. 9.8b.

$$S_{max} = A_s\left[\frac{2X}{X(X-2Bn)+B^2 n}\right] \tag{9.53}$$

where X is defmed in Eq. (9.10). Values of X/R are given in Table 9.1 in terms of e/B and n.

Knowing the value of S_{max}, the settlement of the point of load application (S_e) and tilt (t) for the rigid ring footing subjected to eccentric-inclined load, as shown in Fig. 9.7b. are computed as follows:

$$t = \frac{S_{max}}{X} \tag{9.54}$$

and

$$S_e = S_{max} - (B/2 - e) \sin t \tag{9.55}$$

Step 9

Settlements (S_e), (S_{max}) and tilts (t) for a given set of e/B, a, n and B, are computed for different applied load intensities by repeating Step 1 through Step 8. Consequently, the pressure versus maximum settlement (S_{max}), pressure versus settlement (S_e) and pressure versus tilt (t) curves were obtained.

Step 10

Pressure-maximum settlement (S_{max}), pressure-settlement (S_e) and pressure-tilt (t) curves for other sets of e/B, a, n and B are obtained by repeating Step 1 through Step 9.

9.3.3 Results and Interpretation

Results have been obtained by the procedure described above for Buckshot clay using the constitutive laws. The properties and hyperbolic constants (a and b) for Buckshot clay are given in Section 2.4.1 of Chapter 2.

Pressure versus maximum settlement (S_{max}) and pressure versus tilt (t) curves for solid circular and ring footings (n = 0.0, 0.4) of size 1000 mm resting on Buckshot clay have been obtained for load inclination $\alpha = 0°, 10°$, $20°$ and eccentricity width ratio $e/B = 0.0, 0.1, 0.2$. Typical curves are shown in Figs 9.9 to 9.12. A close examination of these figures indicates that the proposed methodology predicts the pressure-maximum settlement and pressure-tilt curves as usually occurs in practice. If the bearing capacity is obtained by intersection tangent method, on the basis of studying the total curves (not shown), it was found that it decreases with increase in the e/B ratio and load inclination (α) for both solid circular footing ($n = 0$) and annular ring footing ($n = 0.4$). Further, for the same pressure intensity, settlement and tilt increase with increase in e/B and α for both $n = 0$ and 0.4 cases. However, for the same pressure intensity, both settlement and tilt are more in the case $n = 0$ in comparison to $n = 0.4$ case. It may be noted that when the pressure is same in solid ($n = 0$) and annular ($n = 0.4$) footing, the load acting on solid circular footing is more because the area of footing in $n = 0$ case is more than in $n = 0.4$ case.

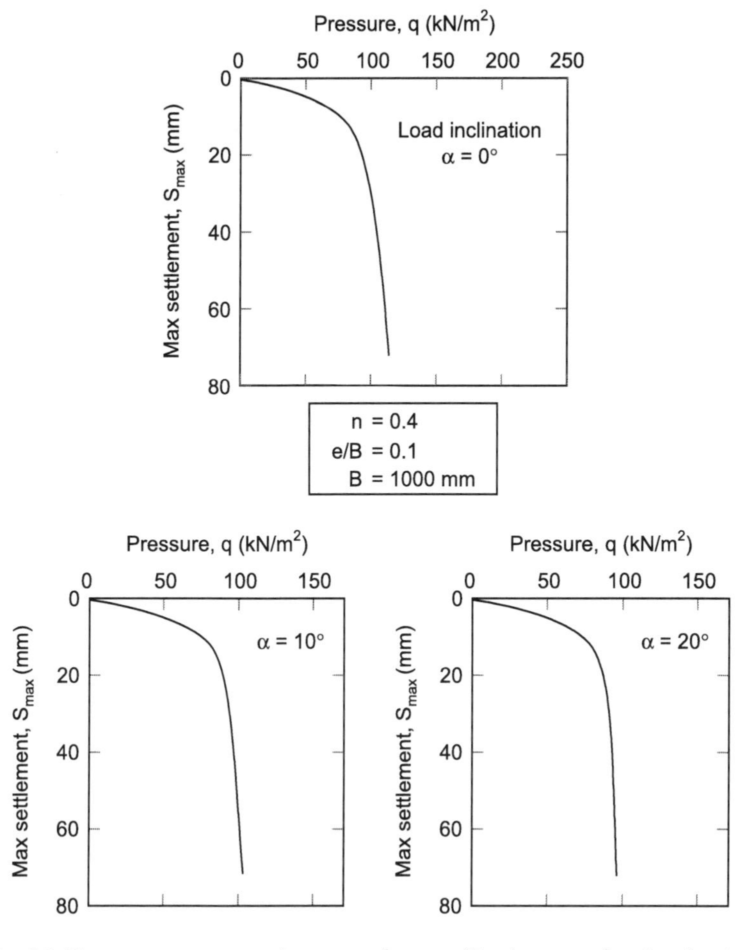

Fig. 9.9. Pressure versus maximum settlement (S_{max}) curves for ring footing (n = 0.4) on Buckshot clay, for e/B = 0.1.

9.4 Footing on Sand

9.4.1 Assumptions

- Contact pressure distribution beneath the footing is assumed as in the case of clay.
- The effect of the soil weight is taken into consideration for determination of stresses in the soil mass. The vertical stress component due the weight of soil is taken as γz, where γ is the unit weight of soil and z is the depth to the centre of soil layer. Whereas the horizontal stress components (in the x- and y-directions) due to the weight of the soil

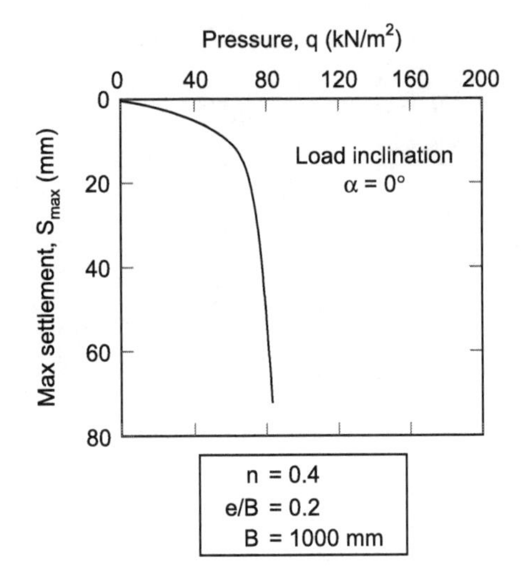

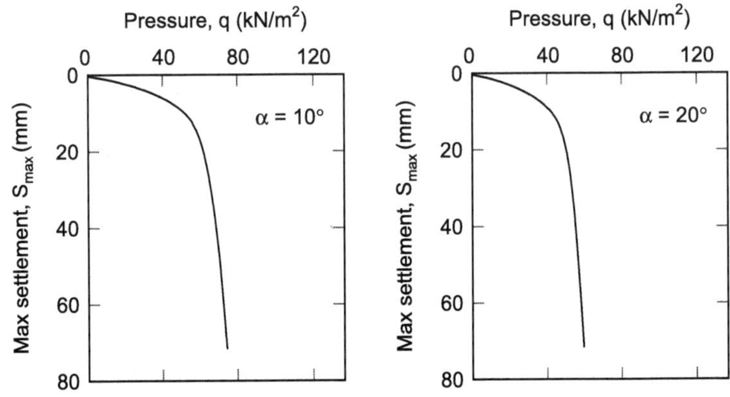

Fig. 9.10. Pressure versus maximum settlement (S_{max}) curves for ring footing (n = 0.4) on Buckshot clay, for e/B = 0.2.

are taken equal to $k_0\gamma z$, where k_0 is the coefficient of earth pressure at rest ($k_0 = 1 - \sin\phi$), ϕ being the angle of internal friction.

- The footing base is assumed to be smooth
- The ultimate bearing pressure (q_{urie}) for ring footing under eccentric-inclined load on sand, is computed as follows:

$$q_{urie} = R_{e\gamma} R_{i\gamma} q_{uro} \tag{9.56}$$

$R_{e\gamma}, R_{i\gamma}$ = Reduction factors due to eccentricity (e) and inclination (α) as proposed by Meyerhof (1956) are given below;

$$R_{e\gamma} = (1 - 2e/B)^2 \tag{9.57}$$

$$R_{i\gamma} = (1 - \alpha/\phi)^2 \tag{9.58}$$

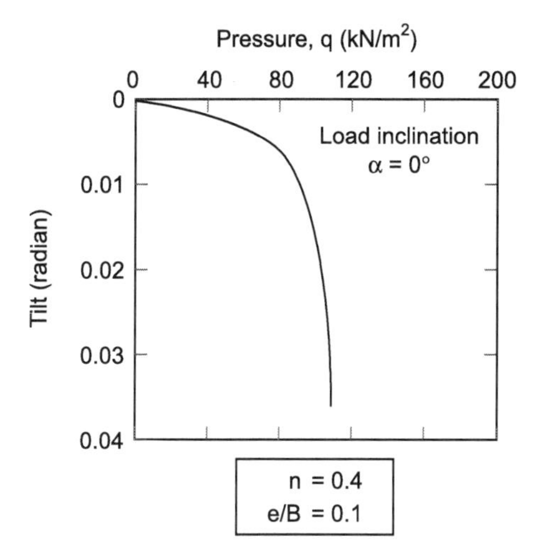

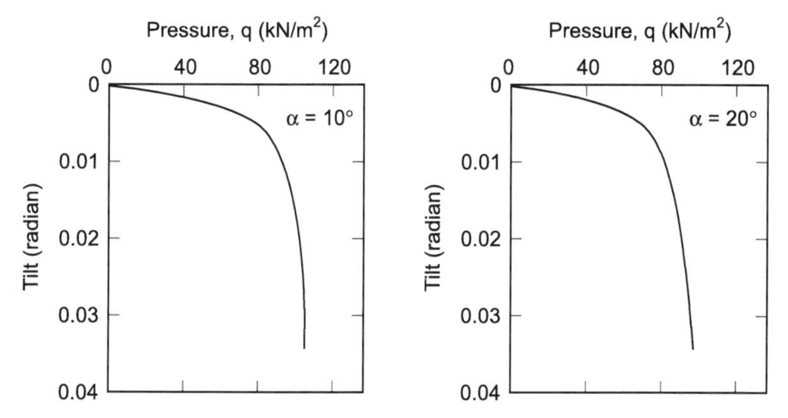

Fig. 9.11. Pressure versus tilt curves for ring footing (n = 0.4) on Buckshot clay, for e/B = 0.1.

q_{uro} = Bearing capacity of a centrally-loaded ring footing. For this case Kakroo (1985) gave the following equation for computing q_{uro}:

$$q_{uro} = \frac{\gamma B}{2} D_r \tan \phi (236 + 465n - 1420n^2 + 754n^3) \qquad (9.59)$$

where D_r is relative density of sand. The above equation is used for a surface ring footing.

- A factor of safety 'F' has been introduced, so that at all stress levels, the following relationship is satisfied:

$$\frac{q_{urie}}{q} = \frac{\sigma_u}{\sigma_1 - \sigma_{3a}} = F \qquad (9.60)$$

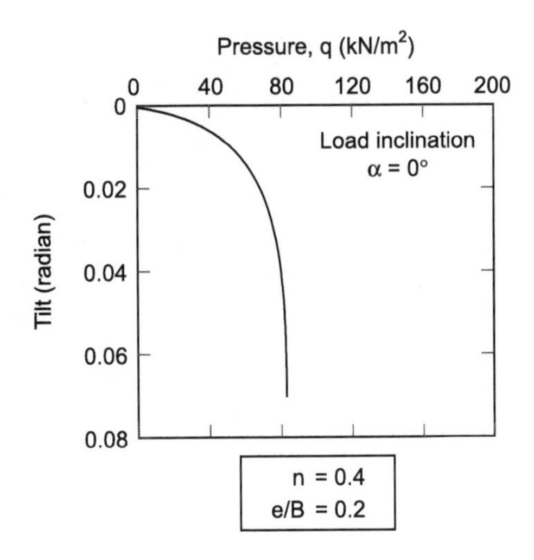

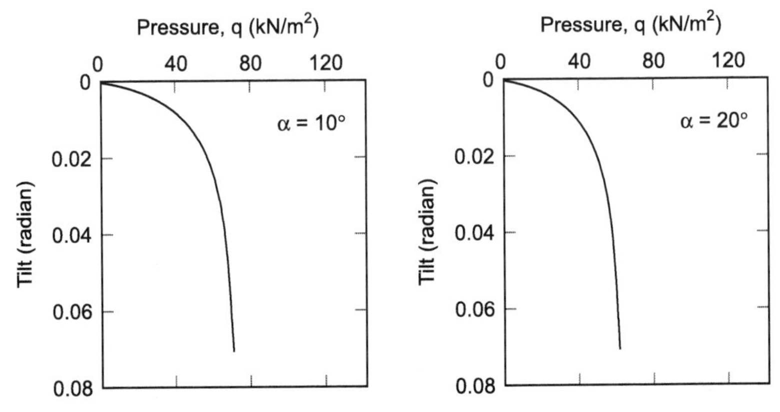

Fig. 9.12. Pressure versus tilt curves for ring footing ($n = 0.4$) on Buckshot clay, for $e/B = 0.2$.

where q = Intensity of surface eccentric-inclined load

σ_u = Ultimate stress from hyperbolic relation $(= 1/b)$

$\sigma_{3\underline{a}}$ = The average of σ_2 and σ_3

σ_1, σ_2 and σ_3 = The major, intermediate and minor principal stresses due to q and the weight of the soil.

- Modulus of elasticity, E_s has been obtained by the procedure given below. The strain at stress level of (σ_u/F) is given by:

$$\text{Strain} = \frac{a(\sigma_u / F)}{1 - b(\sigma_u / F)} \tag{9.61}$$

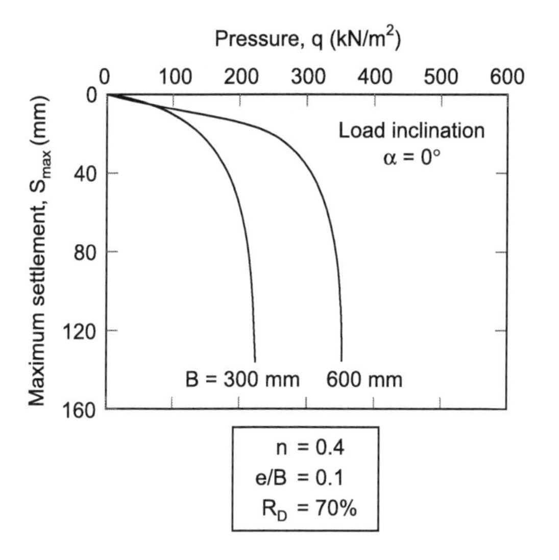

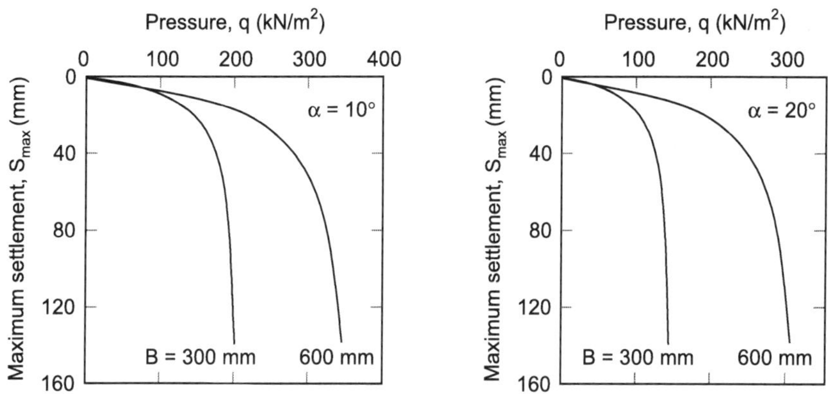

Fig. 9.13. Pressure versus maximum settlement (S_{max}) curves for ring footing (n = 0.4) on Amanatgarh sand (R_D = 70%), for e/B = 0.1.

$$E_s = \frac{\text{stress}}{\text{strain}} = \frac{1 - b(\sigma_u / F)}{a} \tag{9.62}$$

Parameters $1/a$ and $1/b$ are dependent on the confining pressure. Other assumptions made for footings in clay have also been assumed for the analysis of footings in sand.

9.4.2 Pressure-settlement and Pressure-tilt Characteristics

The proposed analytical procedure for predicting the pressure-settlement and pressure-tilt characteristics of ring footings subjected to eccentric-inclined loads resting on sand is given in the under-mentioned steps.

Step 1

Contact pressure distribution is adopted similar to that described in Step 2 of Sec. 9.3.2 for ring footings on clay.

Step 2

The value of Poisson's ratio 'μ' and the coefficient of earth pressure at rest 'k_0' are obtained from the following relationships:

$$k_0 = 1 - \sin \phi \tag{9.63}$$

$$\mu = k_0/(1 + k_0) \tag{9.64}$$

where ϕ = Angle of internal friction of sand

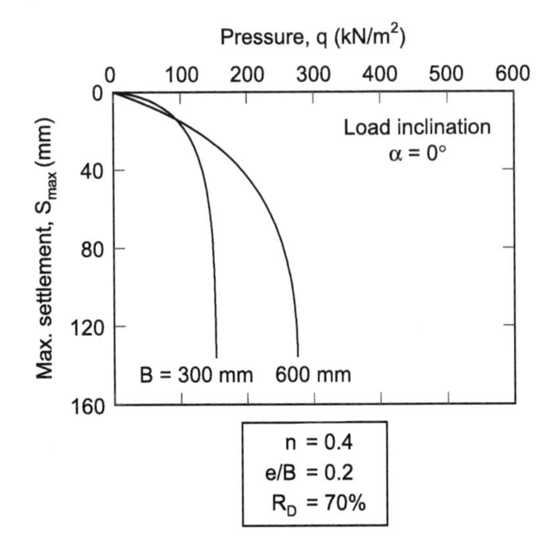

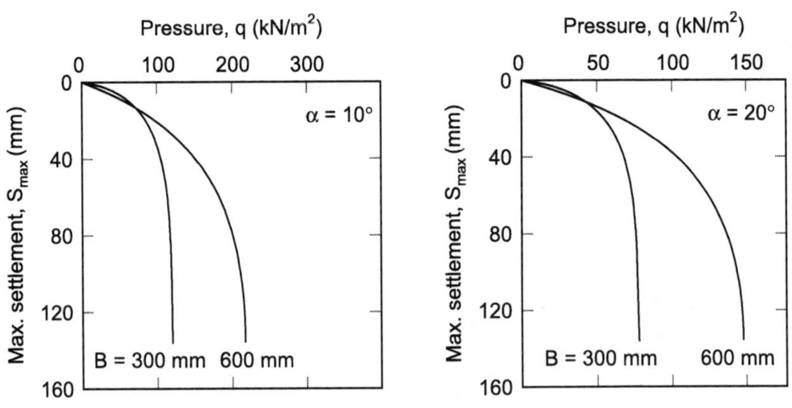

Fig. 9.14. Pressure versus maximum settlement (S_{max}) curves for ring footing (n = 0.4) on Amanatgarh sand (R_D = 70%), for e/B = 0.2.

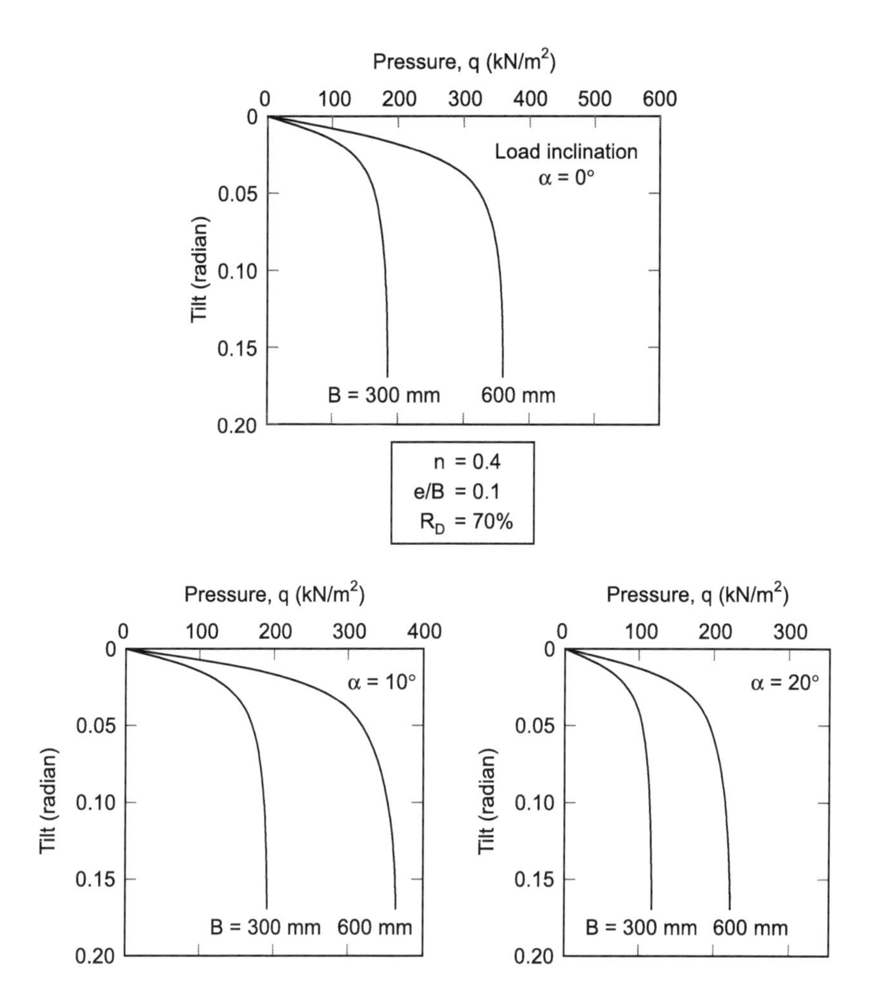

Fig. 9.15. Pressure versus tilt curves for ring footing ($n = 0.4$) on Amanatgarh sand ($R_D = 70\%$), for $e/B = 0.1$.

Step 3

Stresses, principal stresses and their directions in each layer have been determined according to Step 4 and Step 5 given for ring footings on clay.

Step 4

The ultimate bearing pressure 'q_{urie}' is computed from Eq. (9.65). This equation has been developed by Kakroo (1985) on the basis of extensive model testing.

$$q_{urie} = R_{e\gamma} R_{i\gamma} q_{uro} \tag{9.65}$$

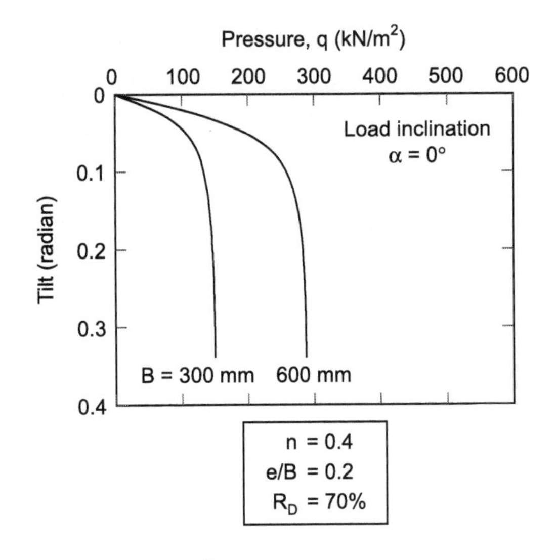

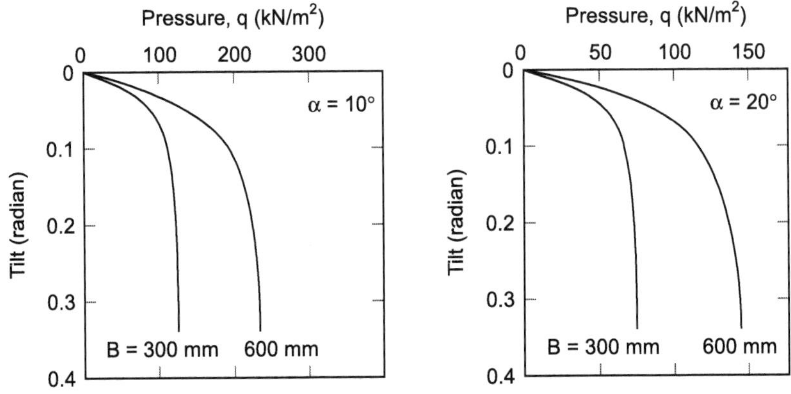

Fig. 9.16. Pressure versus tilt curves for ring footing ($n = 0.4$) on Amanatgarh sand ($R_D = 70\%$), for $e/B = 0.2$.

where
$$R_{e\gamma} = (1 - 2e/B)^2 \tag{9.66}$$
$$R_{i\gamma} = (1 - \alpha/\phi)^2 \tag{9.67}$$

and
$$q_{uro} = \frac{\gamma B}{2} D_r \tan \phi (236 + 465\,\mathrm{n} - 1420\,\mathrm{n}^2 + 754\,\mathrm{n}^3) \tag{9.68}$$

Step 5

The factor 'F' for the given surface load intensity 'q' is obtained from the relation

$$F = \frac{q_{urie}}{q} \tag{9.69}$$

q is the intensity of eccentric-inclined load (i.e. P/A).

Step 6

Modulus of elasticity 'E_s' is determined by using the following equation:

$$E_s = \frac{1 - b(\sigma_u / F)}{a} \tag{9.70}$$

where a and b are the constants of the hyperbola, the values of which are dependent on the confining pressure.

Step 7

Strains ε_1, ε_2 and ε_3 in each layer in the direction of the major, intermediate and minor principal stresses respectively, have been computed using the following formulae:

$$\varepsilon_1 = \frac{(\sigma_1 - \sigma_{3a})}{E_s} \tag{9.71}$$

$$\varepsilon_2 = \frac{\sigma_2 - \mu(\sigma_3 + \sigma_1)}{\sigma_1 - \mu(\sigma_2 + \sigma_3)}.\varepsilon_1 \tag{9.72}$$

$$\varepsilon_3 = \frac{\sigma_3 - \mu(\sigma_1 + \sigma_2)}{\sigma_1 - \mu(\sigma_2 + \sigma_3)}.\varepsilon_1 \tag{9.73}$$

where σ_1, σ_2 and σ_3 are the principal stresses due to load intensity q and weight of soil.

Step 8

Strain in the vertical direction (ε_z) of each layer, the vertical settlement (S) of each layer and the total settlement (S) along any vertical section are computed as for footings on clay.

Step 9

Maximum settlement (S_{max}), settlement of the point of load application (S_e) and tilt (t) are computed as for footings on clay.

Step 10

Values of (S_{max}), (S_e) and (t) are computed for various load intensities by repeating Step 1 to Step 9.

Step 11

Pressure-maximum settlement (S_{max}), pressure-settlement (S_e) and pressure tilt (t) curves are obtained for various sets of e/B, a, n and B by repeating Step 1 through Step 10.

9.5 Results and Interpretation

Results have been obtained using the aforementioned procedure for Amanatgarh sand by using its constitutive law. The properties and hyperbolic constants (a and b) for Amanatgarh sand are given in Chapter 2.

The pressure-maximum settlement (S_{max}) and pressure-tilt (t) curves for ring footings resting on Amanatgarh sand are obtained for B = 300 mm, 600 mm; n = 0.0, 0.4; α = 0°, 10°, 20° and e/B = 0.0, 0.1, 0.2 and are presented in Figs 9.13 to 9.16.

It is evident from these figures that for the same pressure intensity, the settlement of a ring footing increases with increase in width of footing (B), e/B ratio and load inclination α.

Illustrative Examples

Example 9.1

Determine the value of vertical strain at a point lying on the vertical section passing through the centre of a ring footing (external diameter = 6.0 m; annular ratio = 0.4) located at depth 2.0 m below the ground surface. The ring footing is subjected to:

Vertical load = 1400 kN

Horizontal load = 245 kN

Moment = 840 kN-m

Assume the value of Poisson's ratio, μ as 0.35. Adopt Kodner's hyperbola constants as: $a = 9.83 \times 10^{-5}$ m²/kN and b = 0.014 m²/kN.

Solution

1. The footing size, B (i.e. external diameter) is 6.0 m with annular ratio (n) equal to 0.4.

2. $A = \frac{\pi}{4}(1 - n^2)B^2 = \frac{\pi}{4} \times (1 - 0.4^2) \times 6^2 = 22.68 \text{ m}^2$

$$e = \frac{M}{P_v} = \frac{840}{1400} = 0.6 \text{ m}; \quad \frac{e}{B} = \frac{0.6}{6.0} = 0.1$$

$$\frac{(1+n^2)}{8} = \frac{(1+0.4^2)}{8} = \frac{1.16}{8} = 0.145$$

$$\therefore \quad e/B < \frac{(1+n^2)}{8}$$

Contact pressure at a distance x from the centre.

$$q_{vx} = \frac{V}{A}\left[1 + \left(\frac{8}{1+n^2}\right)\left(\frac{e}{B}\right)\left(\frac{x}{R}\right)\right]$$

$$= \frac{1400}{22.68}\left[1+\left(\frac{8}{0.145}\right)(0.1)\left(\frac{x}{3.0}\right)\right]; R = B/2 = \frac{6.0}{2} = 3.0 \text{ m}$$

$$= 61.73[1 + 0.23x]$$

3. Divide the ring into 8 equal parts (Fig. 9.17)

   Area of each ring $\frac{22.68}{8} = 2.835\ m^2$

   Distance of *c.g* of sector *oab* from point *o*

$$= \frac{2R\sin\alpha}{3\alpha}; \alpha = \frac{2\pi}{8\times 2} = 0.3925; \sin\alpha = 0.383$$

$$= \frac{2\times 3\times 0.383}{3\times 0.3925} = 1.95 \text{ m}$$

   Distance of *c.g* of sector *ocd* from point *o*

$$= \frac{1.2}{3.0} \times 1.95 = 0.78 \text{ m}$$

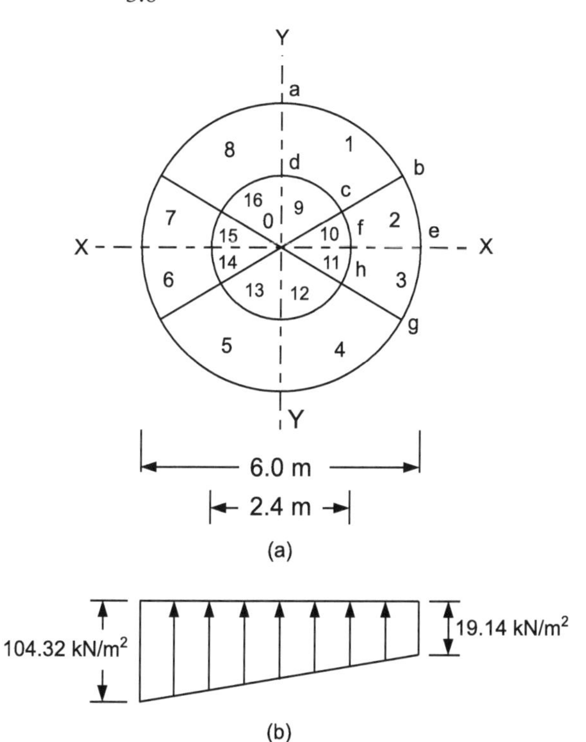

Fig. 9.17. (a) Ring footing divided in eight equal parts (b) Base pressure diagram (Example 9.1).

Therefore distance of portion *abcd* (i.e. Part 1) from point *o*

$$C_{gl} = \left[\frac{\pi \times 3^2 \times 1.95 - \pi \times 1.2^2 \times 0.78}{\pi \times 3^2 - \pi \times 1.2^2}\right]$$

$$\left[\frac{55.11 - 3.53}{28.26 - 4.52}\right] = \left[\frac{51.58}{23.74}\right] = 2.17 \text{ m}$$

Due to symmetry distance of *c.g.* of parts 2, 3,…, 8 will also be same, i.e. 2.17 m

For parts 1, 4, 5 and 8:

$$x = 2.17 \cos\left(\frac{180}{8} + \frac{180}{16}\right) = 1.804 \text{ m}$$

$$y = 2.17 \sin\left(\frac{180}{8} + \frac{180}{16}\right) = 1.205$$

For parts 2, 3, 6 and 7:

$$x = 2.17 \cos\left(\frac{180}{16}\right) = 2.128 \text{ m}$$

$$y = 2.17 \sin\left(\frac{180}{16}\right) = 0.423 \text{ m}$$

Pressure intensities acting on different parts:

$$q_{V1,4} = 61.73(1 + 0.23 \times 1.804) = 87.34 \text{ kN/m}^2$$

$$q_{V2,3} = 61.73(1 + 0.23 \times 2.128) = 91.94 \text{ kN/m}^2$$

$$q_{V5,8} = 61.73(1 - 0.23 \times 1.804) = 36.12 \text{ kN/m}^2$$

$$q_{V6,7} = 61.73(1 - 0.23 \times 2.128) = 31.52 \text{ kN/m}^2$$

Therefore

$$P_1 = P_2 = 87.34 \times 2.835 = 247.61 \text{ kN}$$

$$P_2 = P_3 = 91.94 \times 2.835 = 260.65 \text{ kN}$$

$$P_1 = P_2 = 36.12 \times 2.835 = 102.40 \text{ kN}$$

$$P_1 = P_2 = 31.52 \times 2.835 = 89.36 \text{ kN}$$

$$700.02 \text{ kN}$$

$\therefore$ Total load = 700.02 × 2 = 1400 kN (matches well with given data)

Proceeding exactly in the same way as illustrated above.

$$H_1 = H_4 = \frac{245}{1400} \times 247.61 = 43.33 \text{ kN}$$

$$H_2 = H_3 = \frac{245}{1400} \times 260.65 = 45.61 \text{ kN}$$

$$H_5 = H_8 = \frac{245}{1400} \times 102.40 = 17.92 \text{ kN}$$

$$H_6 = H_7 = \frac{245}{1400} \times 89.36 = 17.92 \text{ kN}$$

$$\overline{\qquad 122.498 \text{ kN} \qquad}$$

$\therefore$ Total load = 122.498 × 2 = 245 kN (matches well with given data)

Stresses caused by loads ($P_1, P_2, \ldots, P_8$) are given in Table 9.2. Similarly stresses due to ($H_1, H_2, \ldots, H_8$) are given in Table 9.3.

Therefore, total stress due to vertical and horizontal loads will be:

$$\sigma_z = 23.91 + 1.88 = 25.79 \text{ kN/m}^2$$

$$\sigma_x = 19.36 + 1.75 = 21.11 \text{ kN/m}^2$$

$$\sigma_y = 3.56 + 0.044 = 3.60 \text{ kN/m}^2$$

$$\tau_{zx} = 10.70 + 4.06 = 14.76 \text{ kN/m}^2$$

$$\tau_{yz} = 0$$

$$\tau_{xy} = 0$$

4. $I_1 = \sigma_x + \sigma_y + \sigma_z = 25.79 + 21.11 + 3.60 = 50.50 \text{ kN/m}^2$

$$I_2 = \sigma_x \sigma_y + \sigma_y \sigma_z + \sigma_x \sigma_z - \tau_{xy}^2 - \tau_{yz}^2 - \tau_{zx}^2$$

$$= 21.11 \times 3.6 + 3.6 \times 25.79 + 21.11 \times 25.79 - 0 - 0 - 14.76^2$$

$$= 76.00 + 92.84 + 544.43 - 217.86 = 495.41 \text{ (kN/m}^2\text{)}$$

$$I_3 = \sigma_x \sigma_y \sigma_z - \sigma_x \sigma_{yz}^2 - \sigma_y \tau_{zx}^2 - \tau_z \tau_{xy}^2 + 2\tau_{xy} . \tau_{yz} . \tau_{zx}$$

$$= 21.11 \times 3.60 \times 25.74 - 0 - 3.6 \times 14.76^2 - 0 + 0$$

$$= 1956.13 - 784.29 = 1171.84 \text{ (kN/m}^2\text{)}^3$$

Now the equation:

$$\sigma^3 - I_1 \sigma^2 + I_2 \sigma - I_3 = 0$$

$$\sigma^3 - 50.5\, \sigma^2 - 495.41\, \sigma - 1171.84 = 0$$

$$\sigma_1 = 38.39 \text{ kN/m}^2$$

Table 9.2. Computation of stresses due to vertical loads P

$P(kN)$	$z(m)$	x	y	r	R	μ	m	σ_z (kN/m^2)	σ_x (kN/m^2)	σ_y (kN/m^2)	τ_{xz} (kN/m^2)	τ_{yz} (kN/m^2)	t_{xy} (kN/m^2)
247.61	2.00	1.80	1.21	2.17	2.95	0.35	2.86	4.23	2.85	1.21	3.81	2.55	1.98
260.65	2.00	2.13	0.42	2.17	2.95	0.35	2.86	4.45	4.21	0.05	4.74	0.94	0.86
260.65	2.00	2.13	-0.42	2.17	2.95	0.35	2.86	4.45	4.21	0.05	4.74	–0.94	–0.86
247.61	2.00	1.80	–1.21	2.17	2.95	0.35	2.86	4.23	2.85	1.21	3.81	–2.55	–1.98
102.4	2.00	–1.80	–1.21	2.17	2.95	0.35	2.86	1.75	1.18	0.50	–1.58	–1.05	0.82
89.36	2.00	–2.13	–0.42	2.17	2.95	0.35	2.86	1.53	1.44	0.02	–1.62	–0.32	0.30
89.36	2.00	–2.13	0.42	2.17	2.95	0.35	2.86	1.53	1.44	0.02	–1.62	0.32	–0.30
102.4	2.00	–1.80	1.21	2.17	2.95	0.35	2.86	1.75	1.18	0.50	–1.58	1.05	–0.82
				Total				**23.91**	**19.37**	**3.56**	**10.70**	**0.00**	**0.00**

Table 9.3. Computation of stresses due to horizontal loads H

$H(kN)$	$z(m)$	x	y	r	R	μ	m	$\sigma_z(kN/m^2)$	$\sigma_x(kN/m^2)$	$\sigma_y(kN/m^2)$	$\tau_{xz}(kN/m^2)$	$\tau_{Yz}(kN/m^2)$	$\tau_{XY}(kN/m^2)$
43.33	2.00	1.80	1.21	2.17	2.95	0.35	2.86	0.67	0.51	0.13	0.60	0.40	0.37
45.61	2.00	2.13	0.42	2.17	2.95	0.35	2.86	0.83	0.88	–0.08	0.88	0.18	0.18
45.61	2.00	2.13	–0.42	2.17	2.95	0.35	2.86	0.83	0.88	–0.08	0.88	–0.18	–0.18
43.33	2.00	1.80	–1.21	2.17	2.95	0.35	2.86	0.67	0.51	0.13	0.60	–0.40	–0.37
17.92	2.00	–1.80	–1.21	2.17	2.95	0.35	2.86	–0.28	–0.21	–0.05	0.25	0.17	–0.15
15.64	2.00	–2.13	–0.42	2.17	2.95	0.35	2.86	–0.28	–0.30	0.03	0.30	0.06	–0.06
15.64	2.00	–2.13	0.42	2.17	2.95	0.35	2.86	–0.28	–0.30	0.03	0.30	–0.06	0.06
17.92	2.00	–1.80	1.21	2.17	2.95	0.35	2.86	–0.28	–0.21	–0.05	0.25	–0.17	0.15
				Total				**1.87**	**1.75**	**0.044**	**4.07**	**0.00**	**0.00**

$$\sigma_2 = 8.53 \text{ kN/m}^2$$

$$\sigma_3 = 3.58 \text{ kN/m}^2$$

5. $$\varepsilon_1 = \frac{a(\sigma_1 - \sigma_3)}{1 - b(\sigma_1 - \sigma_3)}$$

$$= \frac{9.86 \times 10^{-5}(38.39 - 3.58)}{1 - 0.014(38.39 - 3.58)} = \frac{9.86 \times 10^{-5} \times 34.81}{0.5127}$$

$$= 669.44 \times 10^{-5}$$

$$\varepsilon_2 = \frac{\sigma_2 - \mu(\sigma_3 + \sigma_1)}{\sigma_1 - \mu(\sigma_2 + \sigma_3)}.\varepsilon_1$$

$$= \frac{8.53 - 0.35(38.39 + 3.58)}{38.39 - 0.35(8.53 + 3.58)} \times 669.44 \times 10^{-5}$$

$$= \frac{8.53 - 14.69}{38.39 - 4.24} \times 669.44 \times 10^{-5} = 120.75 \times 10^{-5}$$

$$\varepsilon_3 = \frac{\sigma_3 - \mu(\sigma_1 + \sigma_2)}{\sigma_1 - \mu(\sigma_2 + \sigma_3)}.\varepsilon_1$$

$$= \frac{3.58 - 0.35(38.39 + 8.53)}{38.39 - 0.35(8.53 + 3.58)} \times 669.44 \times 10^{-5} = 251.74 \times 10^{-5}$$

6. $$A_1 = (\sigma_y - \sigma_1)(\sigma_z - \sigma_1) - \tau_{yz}.\tau_{zy}$$

$$= (3.60 - 38.39)(25.79 - 38.39) - 0 = 438.35$$

$$B_1 = -\tau_{xy} \times (\sigma_z - \sigma_1) + \tau_{xz}.\tau_{zy}$$

$$= 0 + 0 = 0$$

$$C_1 = \tau_{yz}.\tau_{xy} - (\sigma_y - \sigma_1)\,\tau_{xz}$$

$$= 0 - (3.6 - 38.39) \times 14.76 = 513.5$$

$$\cos\theta_1 = \frac{C_1}{\sqrt{A_1^2 + B_1^2 + C_1^2}} = \frac{513.5}{\sqrt{438.35^2 + 0^2 + 513.5^2}}$$

Or, $\cos\theta_1 = 0.761$

Similarly

$$A_2 = -85.09$$

$$B_2 = 0$$

And $C_2 = 72.77$

$$\cos\theta_2 = 0.65$$

$$A_3 = 0.444$$

$$B_3 = 0$$

$$\text{and } C_3 = -0.295$$

$$\cos\theta_3 = -0.553$$

7. $$\varepsilon_z = \varepsilon_1\cos^2\theta_1 + \varepsilon_2\cos^2\theta_2 + \varepsilon_3\cos^2\theta_3$$

$$= 669.44\times10^{-5}\times(0.761)^2 + 120.75\times10^{-5}\times(0.65)^2 - 251.74\times10^{-5}\times(-0.553)^2$$

Or, $$\varepsilon_z = 259\times10^{-5}$$

Example 9.2

Solve Example 9.1, assuming that the ring footing rests on sand having the following properties to obtain the pressure-settlement characteristics of a ring footing ($B = 7.0$ m, $n = 0.45$) resting on sand. Kondner's hyperbola constants are:

$$\gamma = 16.5 \text{ kN/m}^2$$

$$\varphi = 37°$$

$$\frac{1}{a} = K_1(\sigma_3)^n \text{ and } \frac{1}{b} = K_2 + K_3\sigma_3$$

$$K_1 = 4.0\times10^3$$

$$K_2 = 100$$

$$K_3 = 3.6 \text{ and } n = 0.6$$

σ_3 is in kN/m²

Solution

1. Computations of stresses (σ_z, σ_x, σ_y, σ_{xz}, σ_{xy}, σ_{yz}) will be same as described in steps 1, 2 and 3 of Example 9.1.
2. $K_0 = 1 - \sin\phi = 1 - \sin 37 = 1 - 0.602 = 0.398$

$$\gamma z = 16.5\times2.0 = 33 \text{ kN/m}^2$$

$$K_0\gamma z = 0.398\times33 = 13.13 \text{ kN/m}^2$$

$$\mu = \frac{K_0}{1+K_0} = \frac{0.398}{1+0.398} = 0.285$$

Therefore

$$\sigma_z = 25.79 + 33 = 58.79 \text{ kN/m}^2$$

$$\sigma_x = 21.11 + 13.13 = 34.24 \text{ kN/m}^2$$

$$\sigma_y = 3.60 + 13.13 = 16.73 \text{ kN/m}^2$$

$$\tau_{zx} = 14.76 \text{ kN/m}^2$$

$$\tau_{yz} = 0$$

$$\tau_{xy} = 0$$

Using the above normal and shear stress as principal stresses, their directions are obtained. These are given below:

$$I_1 = \sigma_x + \sigma_y + \sigma_z = 34.24 + 16.73 + 58.79 = 109.76 \text{ kN/m}^2$$

$$I_2 = \sigma_x \sigma_y + \sigma_y \sigma_z + \sigma_x\sigma_z - \tau_{xy}^2 - \tau_{yz}^2 - \tau_{zx}^2$$

$$= 21.11 \times 3.6 + 3.6 \times 25.79 + 21.11 \times 25.79 - 0 - 0 - 14.76^2$$

$$= 3351.504 \text{ (kN/m}^2)$$

$$I_3 = \sigma_x \sigma_y \sigma_z - \sigma_x \tau_{yz}^2 - \sigma_y \tau_{zx}^2 - \sigma_z \tau_{xy}^2 + 2\tau_{xy} . \tau_{yz} . \tau_{zx}$$

$$= 34.24 \times 16.73 \times 58.79 - 0 - 16.73 \times 14.76^2 - 0 + 0$$

$$= 30032.22 \text{ (kN/m}^2)$$

Now the equation:

$$\sigma^3 - I_1 \sigma^2 + I_2 \sigma - I_3 = 0$$

$$\sigma^3 - 109.76 \sigma^2 + 3351 \sigma - 30032.22 = 0$$

$$\sigma_1 = 65.71 \text{ kN/m}^2$$

$$\sigma_2 = 27.31 \text{ kN/m}^2$$

$$\sigma_3 = 16.73 \text{ kN/m}^2$$

$$\sigma_{3a} = \frac{27.31 + 16.73}{2} = 22.02 \text{ kN/m}^2$$

$$\frac{1}{a} = 4.0 \times 10^3 \times (22.02)^{0.6} = 25.57 \times 10^3 \text{ kN/m}^2$$

$$a = 0.0391 \times 10^{-3} \text{ m}^2\text{/kN}$$

$$\frac{1}{b} = 100 + 3.6 \times 22.02 = 179.27 \text{ kN/m}^2$$

$$b = 5.58 \times 10^{-3} \text{ m}^2\text{/kN}$$

3. $$q_{urie} = R_{e\gamma} R_{i\gamma} q_{uro}$$

$$R_{e\gamma} = (1 - 2e/B)^2 = (1 - 2 \times 0.1)^2 = 0.64$$

$$\alpha = \tan^{-1}\left(\frac{245}{1400}\right) = 9.92^0$$

$$R_{i\gamma} = (1 - \alpha/\phi)^2 = \left(1 - \frac{9.92}{37}\right) = 0.535$$

$$q_{uro} = \frac{\gamma B}{2} D_r \tan\phi(236 + 465\,n - 1420\,n^2 + 754\,n^3)$$

$$= \frac{16.5 \times 6.0}{2} \times 0.70 \tan 37^0 (236 + 465 \times 0.4 - 1420$$

$$\times 0.4^2 + 754 \times 0.4^3)$$

$$= 26.11\,(236 + 186 - 227.2 + 48.26) = 6346 \text{ kN/m}^2$$

$$q_{urie} = 6346 \times 0.535 \times 0.64 = 2172 \text{ kN/m}^2$$

Eccentric-inclined load

$$= \sqrt{1400^2 + 245^2} = 142 \text{ kN}$$

$$q = \frac{1421}{22.68} = 62.65 \text{ kN/m}^2$$

$$F = \frac{2172}{62.65} = 34.7$$

A higher value factor of safety means that the capacity of the assumed ring footing is very large. It can take an eccentric-inclined load of magnitude 2172 × 22.68 = 49261 kN instead of 1421 kN. This example is prepared for illustration. Readers are advised to solve this problem assuming less value of external diameter or higher value of external loads.

$$E_s = \frac{1 - b(\sigma_u / F)}{a}$$

$$= \frac{1 - (1/F)}{a}\left[\text{as } b \approx \frac{1}{\sigma_u}\right]$$

$$= \frac{1 - (1/34.7)}{0.0391 \times 10^{-3}}$$

$$= 24.84 \times 10^3 \text{ kN/m}^2$$

$$\varepsilon_1 = \frac{(\sigma_1 - \sigma_{3a})}{E_s} = \frac{65.71 - 22.02}{24.84 \times 10^3} = 1.76 \times 10^{-3}$$

$$\varepsilon_2 = \frac{\sigma_2 - \mu(\sigma_3 + \sigma_1)}{\sigma_1 - \mu(\sigma_2 + \sigma_3)}.\varepsilon_1$$

$$= \frac{1.76 \times 10^{-3}[27.31 - 0.285\,(65.71 + 16.73)]}{65.71 - 0.285\,(6.73 + 27.31)}$$

$$= \frac{1.76\times10^{-3}[3.815]}{53.16} = 0.126 \times 10^{-3}$$

$$\varepsilon_3 = \frac{\sigma_3 - \mu(\sigma_1 + \sigma_2)}{\sigma_1 - \mu(\sigma_2 + \sigma_3)}.\varepsilon_1$$

$$= \frac{1.76\times10^{-3}[16.73 - 0.285(65.71 + 27.31)]}{53.16} = -0.324 \times 10^{-3}$$

$$A_1 = (\sigma_y - \sigma_1)(\sigma_z - \sigma_1) - \tau_{yz}\cdot\tau_{zy}$$

$$= (16.76 - 65.71)(58.79 - 65.71) - 0 = 338.94$$

$$B_1 = -\tau_{xy} \times (\sigma_z - \sigma_1) + \tau_{xz}.\tau_{zy}$$

$$= 0 + 0 = 0$$

$$C_1 = \tau_{yz}.\tau_{xy} - (\sigma_y - \sigma_1)\,\tau_{xz}$$

$$= 0 - (16.73 - 65.71) \times 14.76 = 722.94$$

$$\cos\theta_1 = \frac{C1}{\sqrt{A_1^2 + B_1^2 + C_1^2}} = \frac{722.94}{\sqrt{338.94^2 + 0^2 + 722.94^2}}$$

Or, $\cos\theta_1 = 0.905$

Similarly

$$A_2 = -333.06$$

$$B_2 = 0$$

And $C_2 = 156.16$

$$\cos\theta_2 = 0.424$$

$$A_3 = 0$$

$$B_3 = 0$$

and $C_3 = 0$

$$\cos\theta_3 = 0$$

$$\varepsilon_z = \varepsilon_1\cos^2\theta_1 + \varepsilon_2\cos^2\theta_2 + \varepsilon_3\cos^2\theta_3$$

$$= 1.76 \times 10^{-3} \times (0.905)^2 + 0.126 \times 10^{-3} \times (0.424)^2 + 0$$

Or, $\varepsilon_z = 146 \times 10^{-5}$

Practice Problems

1. In what situations, ring footings are provided? Give their merits. Describe briefly the steps to get pressure-settlement footing of a ring footing resting on clay.
2. Obtain the pressure-settlement characteristics of a ring footing having external diameter equal to 5.0 m and annular ratio 0.3 m. It is resting on clay which has the following properties: $a = 8 \times 10^{-5}$ m^2/kN; $b = 0.013$ m^2/kN.
3. Obtain the pressure-settlement characteristics of a ring footing ($B = 7.0$ m, $n = 0.45$) resting on sand. Kondner's hyperbola constants are:

$$\frac{1}{a} = K_1 (\sigma_3)^n, \frac{1}{b} = K_2 + K_3\sigma_3$$

$$K_1 = 3.8 \times 10^3$$

$$K_2 = 90$$

$$K_3 = 3.35$$

σ_3 is in kN/m^2

REFERENCES

Al-Sanad, H.A., Ismael, N.F. and Brenner, R.P. (1993). Settlement of Circular and Ring Plates in Very Dense Calcareous Sands. *Journal of the Geotechnical Engineering Division, ASCE,* **119(4):** 622-637.

Al-Smadi, M.M. (1998). Behavior of Ring Foundation on Reinforced Soil, Ph.D Thesis, University of Roorkee, Roorkee, India.

Arya, A.S. (1966). Foundations of Tall Circular Structures. *The Indian Concrete Journal,* **40(4):** 142-147.

Borodacheva, F.N. (1972). Displacements and Stresses in the Base of a Rigid, Symmetrically Loaded Ring Foundation. *Soil Mechanics and Foundation Engineering,* **9(4):** 217-221.

Carrier, W.D. and Christian, J.T. (1973). Rigid Circular Plate Resting on a Non-Homogeneous Elastic Half-Space. *Geotechnique,* **23(1):** 67-84.

Chandra, B. and Chandrasekaran, A.R. (1965). The Design of Circular Footings for Water Towers, Chimneys and Similar Structures. *Indian Concrete Journal,* **39:** 304-310.

Chaturvedi, A. (1982). Behavior of Eccentrically-loaded Ring Footings on Sand. M.E. Thesis, University of Roorkee, Roorkee, India.

Choobbasti, A.J. Hesami, S., Najafi, A., Pirzadeh, S., Farrokhzad, F. and Zahmatkesh, A. (2010). Numerical Evaluation of Bearing Capacity and Settlement of Ring Footing: Case Study of Kazeroon Cooling Towers. *Int. J. Res. Rev. Appl. Sci.,* **4(2):** 263-271.

Dave, P.C. (1977). Soil Pressure Under Annular Foundation. *Journal of Institution of Engineers (India)*, Civil Engineering Div. **(58):** 56-61.

Desai, C.S. and Reese, L.C. (1970). Ultimate Capacity of Circular Footings on Layered Soils. *Journal of the Indian National Society of SMFE*, **9(1):** 41-50.

Dewaikar, D.M. and Prajapati, A.H. (1992). Finite Element Analysis of Embedded and Surface Circular Footings in Clay. *IGC-1992*, **1:** 132-136.

Dhatrak, A.I. and Gawande, Poonam (2016). Behavior of Eccentrically-loaded Ring Footing on Sand. *IJIRS*, **5(4)**.

Duncan, J.M. and Chang, C.Y. (1970). Nonlinear Analysis of Stress and Strain in Soils. *Journal of the Soil Mechanics and Foundations Division*, ASCE, **96(5):** 1629-1653.

Durielli, A.J., Phillips, E.A. and Tsao, C.H. (1958). Introduction to the Theoretical and Experimental Analysis of Stress and Strain. McGraw-Hill Book Company, New York.

Egorov, K.E. and Shilova, O.D. (1968). Determination of the Standard Pressure on the Supporting Soil under a Circular Foundation. *Soil Mechanics and Foundation Engineering*, **2:** 84-86.

Egorov, K.E. (1965). Calculation of Bed for Foundation with Ring Footing. *Proc. of the 6th International Conference on SMFE*, Montreal, **2:** 41-45.

Egorov, K.E., Konovalov, P.A., Kitaykina, O.V., Salnikov, L.F. and Zinovyev, A.V. (1977). Soil Deformations under Circular Footing. *Proc. of the 9th International Conference on SMFE*, Tokyo, **1:** 489-492.

Frohlich, O.K. (1953). On the Settling of Buildings Combined with Deviation from their Originally Vertical Position. *Proc. of the 3rd International Conference on SMFE*, Switzerland, **1:** 362-365.

Glazer, S.I. and Shkolnik, S. Sh. (1975). Analysis of Annular Foundations Partially Separated from the Soil. *Soil Mechanics and Foundation Engineering*, **12(2):** 108-110.

Gusev, Yu. M. (1967). Experimental Study of the Effect of Ring Foundation Width on Settling and Inclination Movements. *Soil Mechanics and Foundation Engineering*, **3:** 178-180.

Gusev, Yu. M. (1969). Tilting of Annular Footings Considering Specific Soil Deformations under Compression and Decompression. *Soil Mechanics and Foundation Engineering*, **2:** 126-130.

Gusev, Yu. M. (1975). Analysis of Tilt of Annular Foundation with Plastic Soil Deformations. *Soil Mechanics and Foundation Engineering*, **12(1):** 19-22.

Haroon, M. and Misra, S.K. (1980). A Study on the Behaviour of Annular Footings on Sand. *IGC-1980*, Bombay, **1:** 87-91.

Harr, M.E. (1966). Foundations of Theoretical Soil Mechanics. McGraw-Hill Book Company, New York.

Hataf, N. and Razavi, M.R. (2003). Behavior of ring footing on sand. *Iranian Journal of Science and Technology*, Transaction B, **27:** 47-56.

Hooper, J.A. (1983). Non-linear Analysis of a Circular Raft on Clay. *Geotechnique*, **33(1):** 1-20.

Hosseininia Ehsan Seyedi (2016). Bearing Capacity Factors of Ring Footings. *Iranian Journal of Science and Technology*, Transactions of Civil Engineering, **40(2):** 121-132.

Kakroo, A.K. (1985).Bearing Capacity of Rigid Annular Foundations under Vertical Loads. Ph.D. Thesis, University of Roorkee, Roorkee, India.

Kovalev, K.V. and Le, C.C. (1966). Calculation of a Ring on a Compressible Layer. *Soil Mechanics and Foundation Engineering,* **6:** 371-373.

Kumar, J. and Ghosh, P. (2005). Bearing Capacity Factor Nγ for Ring Footings Using the Method of Characteristics. *Can Geotech J.,* **40(3):** 1474-1484.

Kumar, J. and Chakraborty, M. (2015). Bearing Capacity Factors for Ring Foundations. *J. Geotech. Geoenviron. Eng.,* 10.1061/(ASCE)GT.1943-5606.0001345, 06015007.

Kumar, P. (1986). Ultimate Bearing Capacity of Rigid Ring Footings on Clay. M.E. Thesis, University of Roorkee, Roorkee, India.

Laletin, N.V. (1969). Calculation of the Tilt of a Circular Rigid Foundation. *Soil Mechanics and Foundation Engineering,* **6(4):** 289.

Lo, K.Y. and Becker, D.E. (1979). Pore-Pressure Response Beneath a Ring Foundation on Clay. *Canadian Geotechnical Journal,* **16(3):** 551-566.

Madhav, M.R. (1980). Settlement and Allowable Pressure for Ring or Annular Footings. *Indian Geotechnical Journal,* **10(3):** 267-271.

Meyerhof, G.G. (1953). The Bearing Capacity of Foundations Under Eccentric and Inclined Loads. *Proc. of 3rd International Conference on SMFE,* Zurich, **1:** 440-445.

Milovic, D.M. (1973). Stresses and Displacements Produced by a Ring Foundation. *Proc. of 8th International Conference on Soil Mechanics and Foundation Engineering,* **3:** 167-171.

Murthy. B.R.S., Sridharan, A. and Vinod, P. (1995). Settlement of Circular Footings on Layered Soils. *Indian Geotechnical Journal,* **25(2):** 249-266.

Patankar, M.V. (1982) Non-Linear Analysis of Annular and Circular Foundations by Finite Element Method. *Journal of Institution of Engineering, Civil Engineering Div.,* **62(CI6):** 347-352.

Popova, O.V. (1972). Stress and Displacement Distributions in a Homogeneous Half-Space Below a Circular Foundation. *Soil Mechanics and Foundation Engineering,* **9(2):** 86-89.

Poulos, H.G. and Davis, E.H. (1973). Elastic Solutions for Soil and Rock Mechanics. John Wiley and Sons, Inc., New York.

Punmia, B.C., Ohri, M.L. and Chowdhary, G.R. (1985). Behavior of Annular Footings on Dune Sand. *Indian Geotechnical Conference, IGC-85,* Roorkee, **1:** 89-93.

Ranjan, G., Viladkar, M.N. and Kumar, P. (1987). Behavior of Rigid Ring Footings on Layered Cohesive Deposit. *Indian Geotechnical Conference, IGC-87,* Bangalore, **1:** 131-135.

Roark, R.J. (1954). Formulas for Stress and Strain, McGraw-Hill Book Co., Inc., New York.

Saha, M.C. (1978).Ultimate Bearing Capacity of Ring Footings on Sand, M.E. Thesis, University of Roorkee, Roorkee, India.

Singh, C. (1982). Behaviour of Circular Footings with Eccentric Vertical Loads. M.E. Thesis, University of Roorkee, Roorkee, India.

Snarskii, A.S. (1974). Stresses and Displacements in the Soil under a Circular Foundation. *Soil Mechanics and Foundation Engineering,* **11(5):** 339-342.

Snodi Lamyaa Najah (2010). Ultimate Bearing Capacity of Ring Footings on Sand. *Wasit Journal for Science & Medicine,* **3(1):** 71-80.

Tetior, A.N. and Litvinenko, A.G. (1971). Foundations for Tower-like Structures. *Proc. of the 4th European Conference on Soil Mechanics,* Budapest, 12-15 October, 819-827.

Tettinek, W. (1953). A Contribution to Calculating the Inclination of Eccentrically Loaded Foundations. *Proc. of the 3rd International Conference on SMFE,* **1(4):** 461-465.

Varadarajan, A. and Aliuja, R. (1984). Analysis of Ring Foundation by FEM. *1GC-1984,* Calcutta, **1:** 27-30.

Yegorov, K.E. and Nichiporovich, A.A. (1961). Research on the Deflexion of Foundations. *Proc. of the 5th International Conference on SMFE,* **1:** 861-866.

Zinov'cv, A.V. (1979). Determinations of the Deformation of a Base of Finite Thickness under an Annular Foundation. *Soil Mechanics and Foundation Engineering,* **16(3):** 152-154.

Zweig, A. (1966). Eccentrically-loaded Trapezoidal or Round Footings. *Journal of the Structural Division,* ASCE, **92(1):** 161-168.

Appendix 9.1

In a ring footing shown in Fig. A9.1.

$$A = \text{Net area of ring footing} = \frac{\pi}{4}B^2 - \frac{\pi n^2 B^2}{4} = \frac{\pi}{4} = (1 - n^2)B^2$$

I_{xx} = Moment of inertia of ring footing about X-X axis

I_{yy} = Moment of inertia of ring footing about Y-Y axis

Due to symmetry

$$I_{xx} = I_{yy} = \frac{\pi B^4}{64} - \frac{\pi n^4 B^4}{64} = \frac{\pi B^4}{64}(1 - n^2)$$

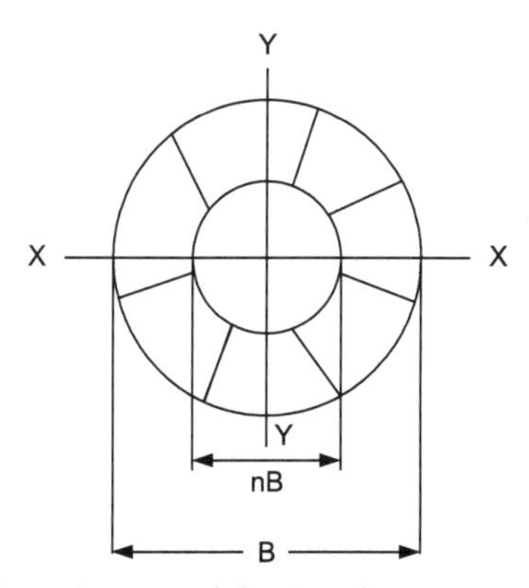

Fig. A9.1. A view of the ring footing in a plan.

If the ring footing is subjected to moment M_{yy} about Y-Y axis, then base pressure will be:

$$p_{1,2} = \frac{Q}{A} \pm \frac{M_{yy}.x}{I_{xx}} \;;$$

(Q is the central vertical load acting on the footing)

$$p_{\max} = \frac{Q}{A} \pm \frac{Q.e_x.B/2}{\frac{\pi B^4}{64}(1-n^2)} ; e_x = \frac{M_{yy}}{Q}$$

$$= \frac{Q}{\frac{\pi}{4}B^2(1-n^2)} \pm \frac{Q.e_x.B/2}{\frac{\pi B^3}{32}(1-n^4)}$$

$$= \frac{Q}{\frac{\pi}{4}B^2(1-n^2)}\left[1+\frac{8Q.e_x}{(1-n^2)B}\right]$$

or, $$p_{\max} = \frac{Q}{A}\left[1+\frac{8e_x}{(1+n^2)B}\right]$$

And $$p_{\min} = \frac{Q}{A}\left[1-\frac{8e_x}{(1+n^2)B}\right]$$

The above expressions hold good when $e_x \leq \frac{(1+n^2)}{8}$

Pressure at a distance x from centre of footing is:

$$p_x = \frac{Q}{A} \pm \frac{M_{yy}.x}{I_{xx}}$$

$$= \frac{Q}{A} \pm \frac{Q.e_x.x}{\frac{\pi B^4}{64}(1-n^4)}$$

$$= \frac{Q}{A}\left[1 \pm \frac{16e_x.x}{(1+n^2)B^2}\right]$$

$$= \frac{Q}{A}\left[1 \pm \frac{16e_x.x}{(1+n^2)B^2}\frac{1}{(1+n^2)}\right]$$

$$= \frac{Q}{A}\left[1 \pm \frac{Be_x.x}{B}\cdot\frac{x}{R}\cdot\frac{1}{(1+n^2)}\right]$$

where $R = B/2$.

CHAPTER 10

Strip Footing Located in Seismic Region

10.1 General

A foundation engineer frequently comes across the problem of designing footings for structures located in the seismic region. The seismic excitation induces inertial stresses within the soil mass due to the vibration induced in the soil mass which is participating in the event. These induced inertial stresses lead to a reduction in resistance of soil beneath the footing. Due to this, the bearing capacity of the footing gets reduced, while its settlement increases. These reduced bearing capacity and increased settlement are termed as seismic-bearing capacity and seismic settlement respectively. These two are needed for proportioning of the footing.

As in the static case (Chapter 4), so far seismic-bearing capacity and seismic settlement are obtained in two independent steps. In Table 10.1, a summary of the works done on seismic-bearing capacity evaluation has been presented.

Budhu, M and Al-Karni (1993) suggested a procedure for obtaining the seismic settlement. Using the method of characteristic (Kumar and Rao, 2002), Majidi and Mirghasemi (2008) by adopting discrete element method and Shafiee and Jahanandish (2010) by finite element method gave the solution for obtaining seismic settlement.

In this chapter, a simple pseudo-static analysis is presented to obtain the pressure-settlement and pressure-tilt characteristics of a footing subjected to seismic loads, using non-linear constitutive law of the soil. In pseudo-static analysis, these additional forces are taken in terms of horizontal and vertical coefficients (i.e. A_h and A_v).

The problem analyzed in this chapter has been pictorially represented in Fig. 10.1. Fig. 10.1a shows the forces acting on the

Table 10.1: Summary of literature review of seismic bearing capacity of footings resting on flat ground

Sr. No.	*Investigator (Year)*	*Method of analysis*	*Approach*	*Failure Pattern Modified from*	*Vertical Acceleration*	*Effect of Eccentricity*
1	Chen (1975)	Limit Analysis	Pseudo-Static	Prandtl (1921)	✓	×
2	Sarma and Lossifelis (1990)	Limit Equilibrium	Pseudo-Static	Meyerhof (1951)	×	×
3	Pecker and Salencon (1991)	Limit Analysis	Pseudo-Static	Prandtl (1921)	✓	✓
4	Richards et.al. (1993)	Limit Equilibrium	Pseudo-Static	Coulomb (1776)	✓	×
5	Budhu and Al-Karni (1993)	Limit Equilibrium	Pseudo-Static	Vesic (1975)	✓	×
6	Dormieux and Pecker (1995)	Limit Analysis	Pseudo-Static	Prandtl (1921)	✓	×
7	Paolucci and Pecker (1997)	Yield Design Theory	Kinematic	Pradtl (1921)	✓	✓
8	Soubra (1999)	Limit Analysis	Pseudo-Static	Prandtl (1921)	×	×
9	Kumar and Mohan Rao (2002)	Method of Characteristics	Pseudo-Static	Sokolovski (1960)	×	×
10	Choudhury and Rao (2005)	Limit Equilibrium	Pseudo-Static	Vesic (1975)	✓	×
11	Merlos and Romo (2006)	Limit Equilibrium	Pseudo-Static	Circular Surface	×	×
12	Chatzigogos et.al. (2007)	Yield Design Theory	Kinematic	Vesic (1975)	✓	×
13	Majidi and Mirghasemi (2008)	Discrete Element Method	Pseudo-Static	-	×	×
14	Ghosh (2008)	Limit Analysis	Pseudo-Dynamic	Coulomb (1776)	✓	×
15	Shafiee and Jahanandish (2010)	Finite Element Method	Pseudo-Static	-	×	×
16	Ghosh and Choudhury (2011)	Limit Equilibrium	Pseudo-Dynamic	Coulomb (1776)	✓	×
17	Saran and Rangwala (2011)	Limit Equilibrium	Pseudo-Static	Saran (1970)	✓	×
18	Rangwala, H.M. (2014)	Limit Equilibrium	Pseudo-Static	-	✓	✓

foundation system in terms of seismic coefficients (A_h and A_v) and total weight of the superstructure and footing. Due to these, the footing is subjected to a vertical load V', a horizontal load (H) and a moment (M), (Fig. 10.1b) (Biswas, 2017; Biswas et al. 2016)

where $V' = V(1 \pm A_v)$ = Total vertical load including due to seismicity

$H = V.A_h$ = Horizontal seismic force

$M = V.A_h$ = Moment at the base of the footing due to the horizontal seismic force

V = Vertical load due to superstructure and the footing. It passes through the centre of the footing

h = Height of point of application of horizontal seismic force above the base of the footing. Point C represents the combined superstructure and footing.

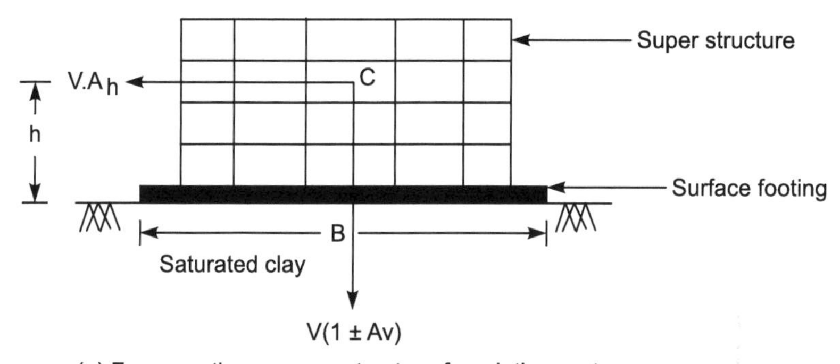

(a) Forces acting on superstructure-foundation system

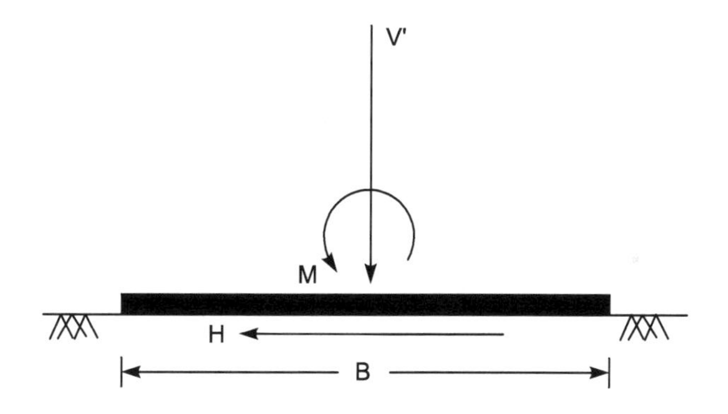

(b) Forces and moment acting on the footing

Fig. 10.1. The problem analyzed in chapter.

10.2 Analysis

The analysis has been carried out for saturated clay. Hyperbola parameter 'a' and 'b' are considered independent to confining pressure.

$$e = \frac{M}{V} = \frac{V.A_h.h}{V(1 \pm A_v)} = \frac{A_h.h}{(1 \pm A_v)} \tag{10.1}$$

At the first instance the footing is assumed flexible, due to which the base contact pressure will be linear as shown in Fig. 10.2. There will be two cases as given below:

(i) $e \leq \frac{B}{6}$ (Fig. 10.2a)

$$q_1 = \frac{V'}{B}\left(1 + \frac{6e}{B}\right) \tag{10.2a}$$

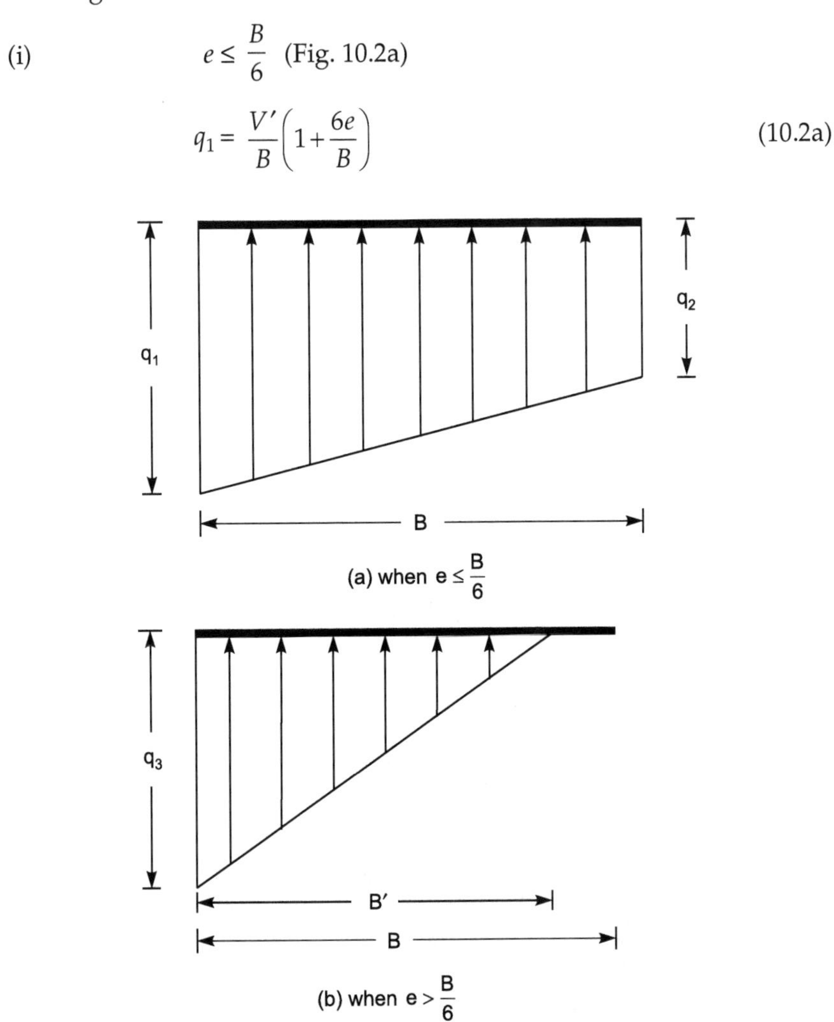

Fig. 10.2. Contact stresses at the base of the footings.

$$q_2 = \frac{V'}{B}\left(1 - \frac{6e}{B}\right) \tag{10.2b}$$

(ii) $e > \frac{B}{6}$ (Fig.10.2b)

$$q_3 = \frac{V'}{B}\left(\frac{4B}{3B - 6e}\right) \tag{10.3a}$$

$$B' = 3\left(\frac{B}{2} - e\right) \tag{10.3b}$$

Assumptions

The analysis is based on the following basic assumptions:

- The soil mass has been assumed as a semi-infinite and isotropic medium
- The whole soil mass supporting a footing is divided into a large number of thin horizontal strips (Fig. 10.3) in which stresses and strains have been obtained along any vertical section
- The stresses in each layer have been computed by using theory of elasticity as the stress equations for various types of loads are available
- The strains have been computed from the known stress condition using constitutive law of soil

Procedure

Procedure of the analysis is described in the following steps:

(i) Divide the soil strata in thin layers upto significant depth ($\approx$ 5.0 B). Thickness of each layer may be taken equal to $B/8$, B being the width of footing (Fig. 10.4).

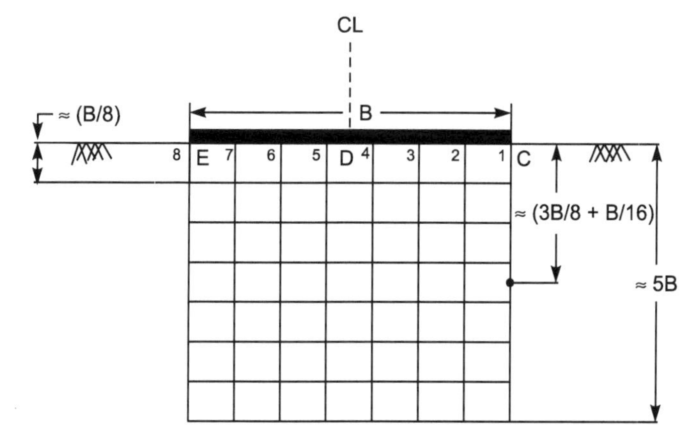

Fig. 10.3. Soil strata divided into n horizontal strip.

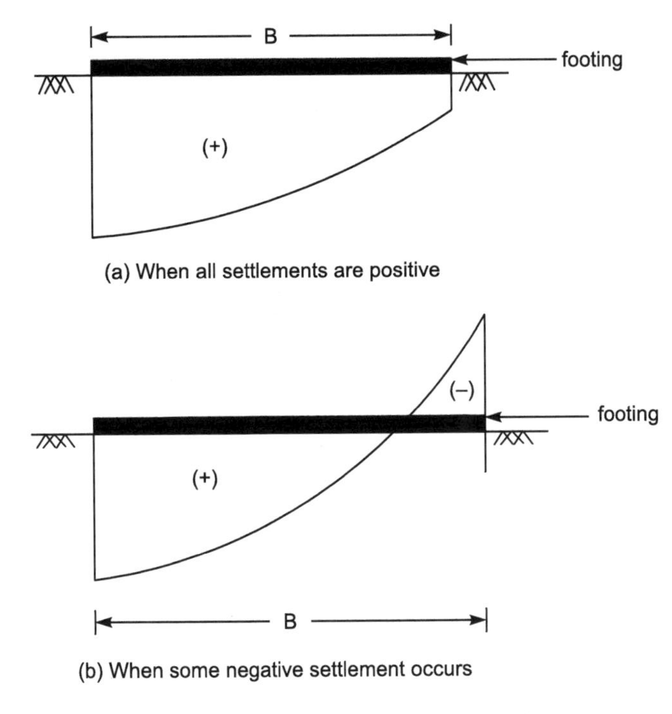

Fig. 10.4. Settlement diagrams in flexible footing.

(ii) Select about eight points on the base of footing including points C, D, and E. Further procedure is to determine the values of settlements of these points. For this, consider a vertical section passing through any of the selected points.

(iii) For determining the settlement of the selected point on the base of footing, determine the settlement of each layer along the section passing through at that point. For example, consider fourth layer and vertical section passing through point C. Depth of the centre of this layer below base of footing will be $\left(\frac{3B}{8}+\frac{B}{16}\right)$.

(iv) Determine the value of e/B ratio (Eq. 10.1). Depending on its value, determine the stresses (σ_z, σ_x, and τ_{xz}) at this point due to applied stresses as shown in Fig. 10.2a or Fig. 10.2b, using the theory of elasticity. Also determine the stresses (σ_z, σ_x, and τ_{xz}) due to horizontal pressure intensity (H/B) and add to the earlier obtained values. The stress equations are already given in Chapter 3. Using these values of stresses, obtain the values of principal stresses (σ_1 and σ_3), and their directions (θ_1 and θ_3) with vertical in the usual way. Steps (v) to (vii) are same as already described in Chapter 4 and reproduced here for ready reference.

(v) In case of clays, the value of principal strain (ε_1) in the direction major principal stress may be obtained as per Eq. (10.4)

$$\varepsilon_1 = \frac{a(\sigma_1 - \sigma_3)}{1 - b(\sigma_1 - \sigma_3)} \tag{10.4}$$

where a and b are Kondner's hyperbola constants, their values may be obtained by performing drained triaxial tests. With soil as saturated clay, these are independent to confining pressure σ_3

$$\varepsilon_3 = -\mu_2 . \varepsilon_1 \tag{10.5a}$$

$$\mu_2 = \frac{-\sigma_3 + \mu_1 \sigma}{\sigma_1 - \mu_1 \sigma_3} \tag{10.5b}$$

$$\mu_1 = \frac{\mu}{1 - \mu} \tag{10.5c}$$

μ is Poisson's ratio

(vi) Vertical strain at the selected point will then be:

$$\varepsilon_z = \varepsilon_1 \cos^2\theta_1 + \varepsilon_3 \cos^2\theta_3 \tag{10.6}$$

(vii) On multiplying ε_z with the thickness of the strip (i.e $\approx$ B/8), settlement of the strip at the point under consideration is obtained. The evaluation of the total settlement along any vertical section is done by numerically integrating the quantity, i.e.
where

$$S_t = \int_0^n \varepsilon_z dz \tag{10.7}$$

S_t = total settlement along i^{th} vertical section

ε_z = vertical strain at depth z below the base of footing

dz = thickness of strips at depth z

n = number of strips in which the soil strata upto significant depth is divided. In this way values of total settlements along other vertical sections passing through the base of footing are obtained.

Actually in flexible footing, settlement patterns will be non-linear and therefore, it will be as shown in Fig. 10.4a or Fig. 10.4b depending upon the value of e/B ratio. Values of e depends basically on A_h and h values, considering the base of the footing as rigid its settlement pattern will depend on the shapes of settlement diagrams of the flexible footing (i.e. Fig. 10.4a and Fig. 10.4b). If in flexible footing, settlement pattern is as shown in Fig. 10.4a, it is likely that in rigid footing will be as shown in Fig. 10.5a. Values of S_{max} and S_{min} may be obtained by solving the following equations:

$$\left(\frac{S_{max}+S_{min}}{2}\right)B = A \tag{10.8a}$$

$$\left(\frac{2S_{min}+S_{max}}{S_{max}+S_{min}}\right)\frac{B}{3} = C_g \tag{10.8b}$$

A = area of settlement diagram shown in Fig. 10.4a, and C_g = distance of C_g of settlement diagram shown in Fig. 10.4a from left edge of footing.

Therefore, knowing the values of S_{max} and S_{min} tilt angle (t) of the rigid footing is computed using the following expressions:

$$t = \tan^{-1}\left(\frac{S_{max}-S_{min}}{B}\right) \tag{10.9}$$

If settlement pattern of flexible footing is as shown in Fig. 10.4b, the settlement pattern of a rigid footing will be as shown in Fig. 10.5b. Then in this case, values of S_{max} are obtained by solving the following equation:

$$\frac{1}{2}S_{max}.B'' = A \tag{10.10a}$$

where A is area of positive portion of settlement diagram (Fig. 10.4b). Further in this case, $S_{min} = 0$.

B'' is the reduced width of footing. Also the centre of settlement diagram (Fig. 10.4b) should coincide the settlement (Fig. 10.5b), i.e. for the rigid footing. Therefore,

$$\frac{B''}{3} = C_g \text{ or } B'' = 3C_g \tag{10.10b}$$

Hence, in rigid footing S_{max} will be:

$$S_{max} = \frac{2A}{B''} = \frac{2A}{3C_g} \tag{10.10c}$$

Tilt will be given by the following expression:

$$t = \tan^{-1}\left(\frac{S_{max}}{B''}\right) \tag{10.11}$$

10.3 Parametric Study

Pressure versus maximum settlement, pressure versus minimum settlement and pressure versus tilt curves were obtained for the following parameters:

(i) Width of following, B: 0.5 m, 1.0 m, 2.0 m and 2.5 m
(ii) Horizontal seismic coefficient, A_h: 0, 0.5, 0.10 and 0.15

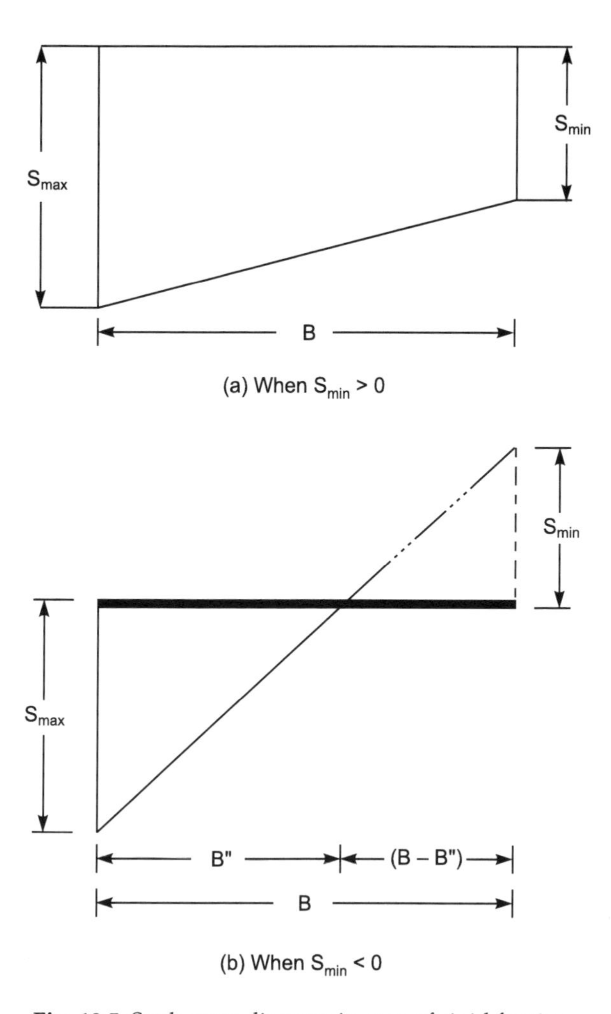

Fig. 10.5. Settlement diagram in case of rigid footing.

(iii) Vertical seismic coefficient, A_v: $A_h/2$

(iv) Hyperbola parameters $1/a$ and $1/b$ respectively: (5000 kN/m^2, 35 kN/m^2), (7000 kN/m^2, 50 kN/m^2), (9000 kN/m^2, 60kN/m^2), and (12000 kN/m^2, 80 kN/m^2)

(v) Height point of application of horizontal seismic force from base of footing, h:B, 1.5 B, 2.0 B, 2.5 B.

After studying the values of maximum settlement (S_{max}) and minimum settlement (S_{min}), it was found that at a particular pressure intensity (q), these increase in direct proportion to the width footing. This point has already been illustrated in Chapter 4 for uniform pressure intensity acting on the footing. Keeping these facts in view, plots have been prepared between q versus S_{max}/B, q versus S_{min}/B and q versus tilt for all the parameters listed above. It may be stressed again that these curves are independent to the width of footing, i.e. hold good for all the values of B ($\approx$ 0.5 m, 1.0 m, 2.0 m, 2.5 m).

For $1/a = 9000$ kN/m^2, $1/b = 60$ kN/m^2; some typical curves (q versus S_{max}/B, q versus S_{min}/B and q versus tilt) are shown in Figs 10.6, 10.7 and 10.8 respectively for $A_h = 0.05$, $A_h = 0.10$ and $A_h = 0.15$ respectively. In these variation with h/B ratio has also been shown. A study of these figures indicated that for a given pressure intensity:

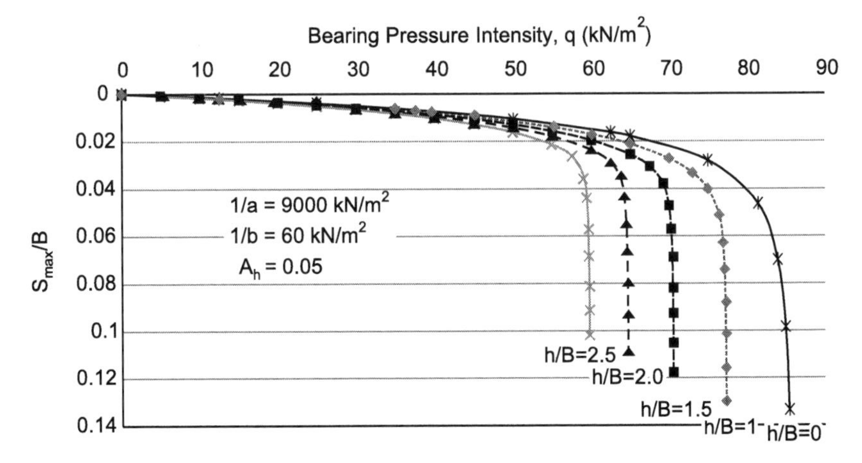

Fig. 10.6a. q versus (S_{max}/B) curves.

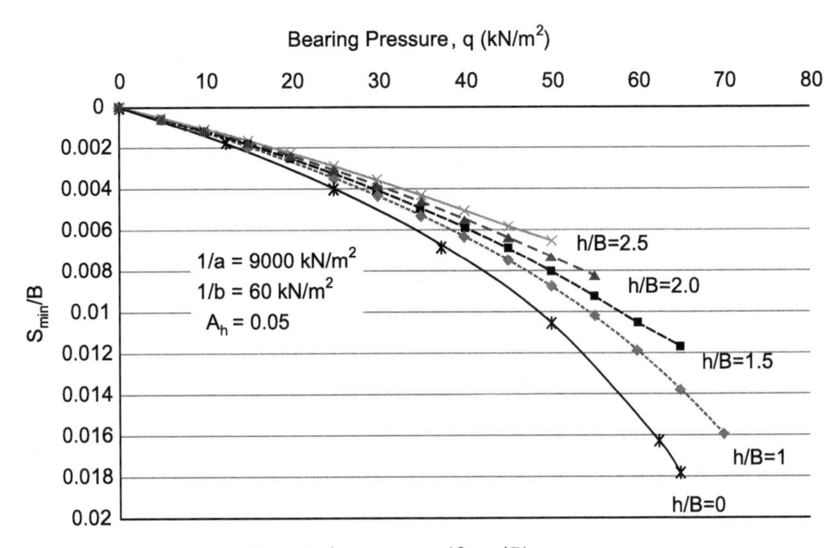

Fig. 10.6b. q versus (S_{min}/B) curves.

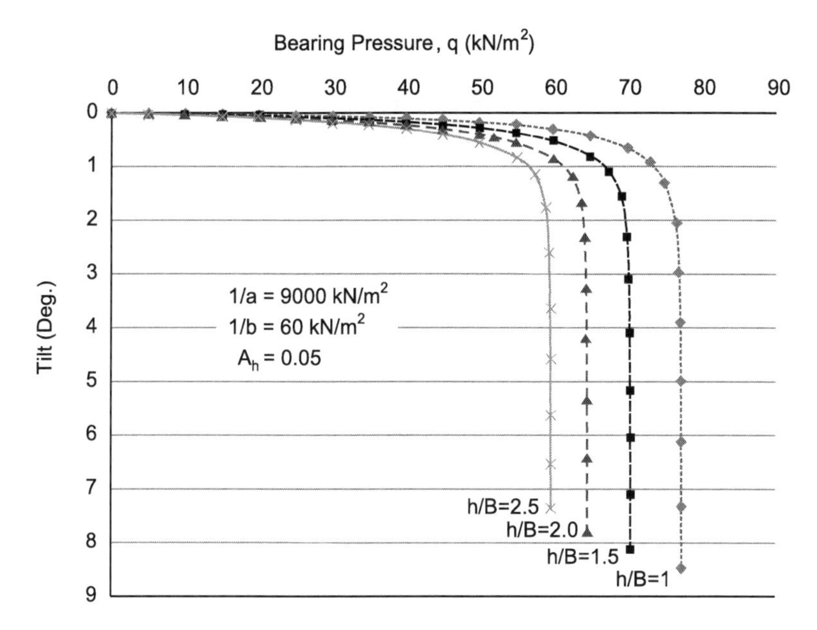

Fig. 10.6c. q versus tilt curves.

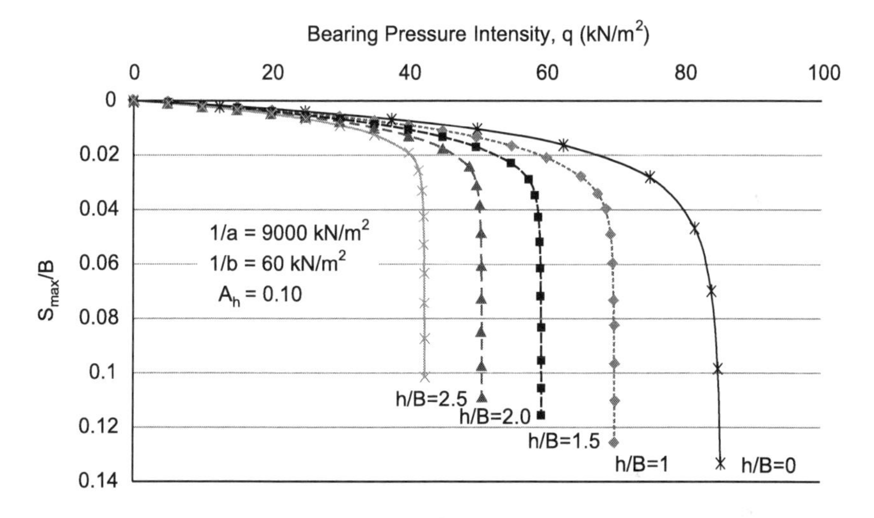

Fig. 10.7a. q versus (S_{max}/B) curves.

S_{max} increases with the increase in A_h and h/B ratio
S_{min} decreases with the increase in A_h and h/B ratio
Tilt increases with the increase in A_h and h/B ratio

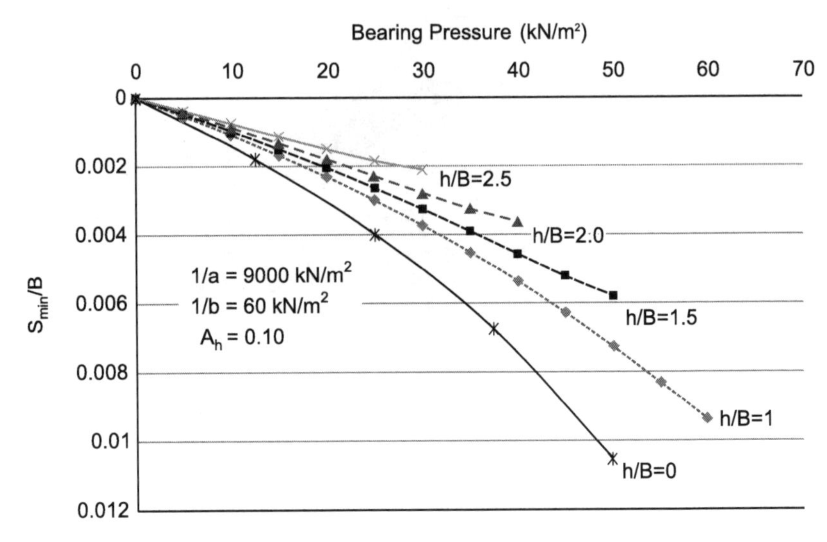

Fig. 10.7b. q versus (S_{min}/B) curve.

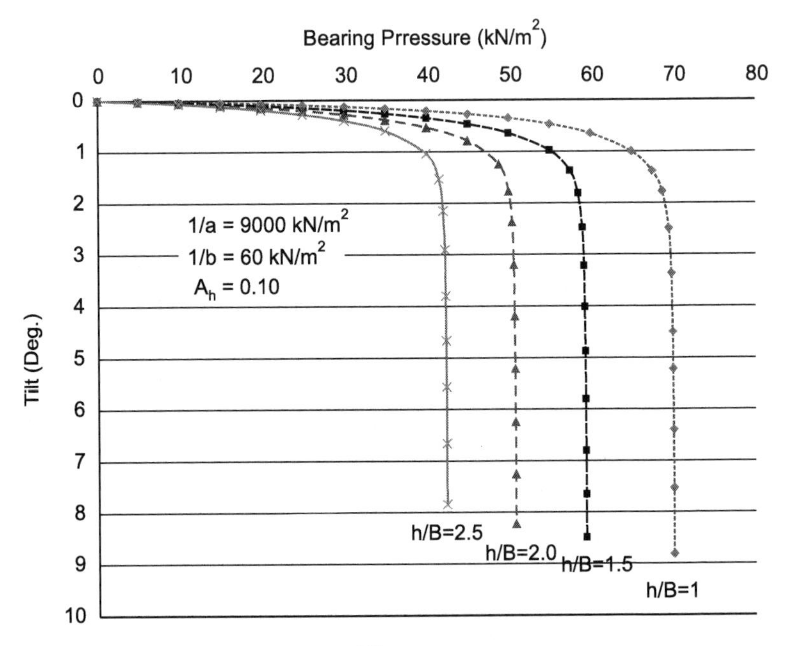

Fig. 10.7c. Tilt versus q curve.

This trend occurs with other values of $1/a$ and $1/b$ parameters also. In these figures, curve showing for $h/B = 0$, belongs to static case i.e. $A_h = 0$.

From the curves shown in Fig. 10.6, ultimate bearing capacities have been obtained by intersection tangent method. It was found that bearing capacity ratio (BCR) equal to $\frac{(q_u)_{A_h}}{(q_u)_{A_h=0}}$ is same for all the values $1/a$ and $1/b$ (Tables 10.2 to 10.5). $(q_u)_{A_h}$ and $(q_u)_{A_h=0}$ are respectively the bearing capacity values for a specified value of A_h and in static case, respectively. Therefore, B.C.R. is dependent only on h/B ratio and seismic coefficient A_h.

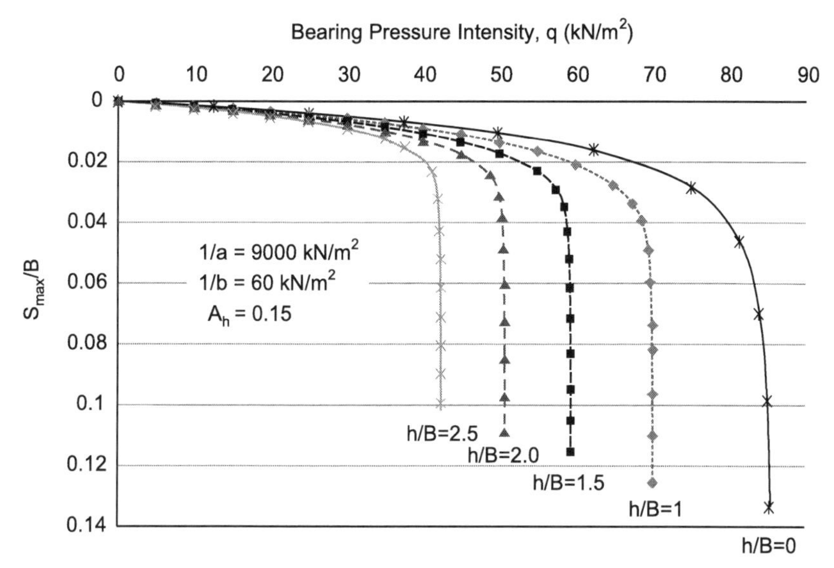

Fig. 10.8a. q versus (S_{max}/B) curves.

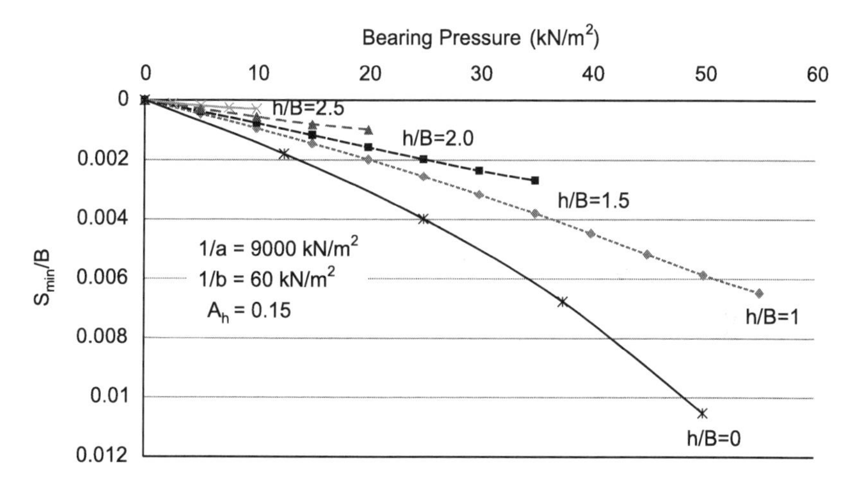

Fig. 10.8b. q versus (S_{min}/B) curves.

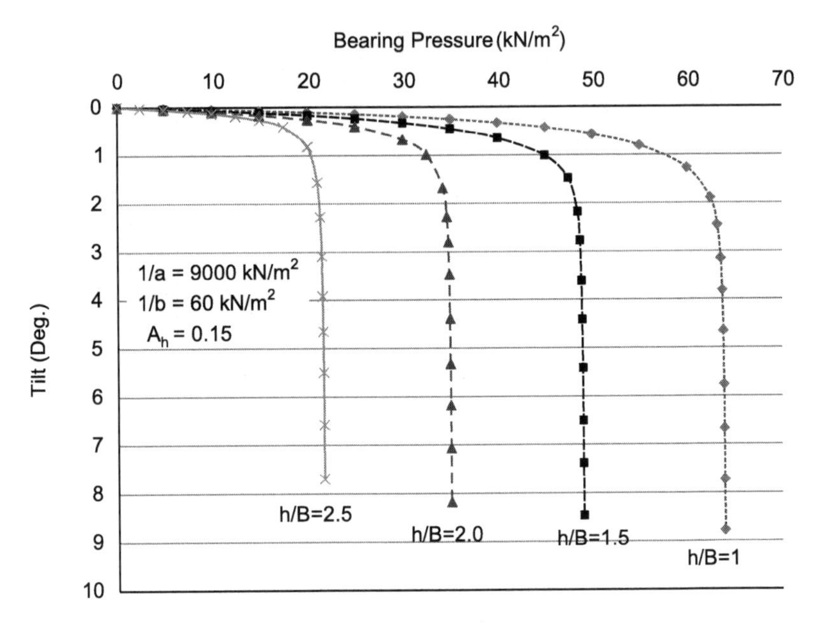

Fig. 10.8c. q versus tilt curve.

Table 10.2. Computation the ratio of ultimate bearing pressure for $A_h = 0.05$ ($1/a = 5000$ kN/m² and $1/b = 35$ kN/m²)

$B(m)$	h/B	A_h	$q_u(kN/m^2)$	q_u/q_{u0}
0.5	0.00	0.05	46.98	1.000
0.5	1.00	0.05	41.62	0.886
0.5	1.50	0.05	38.78	0.826
0.5	2.00	0.05	35.94	0.765
0.5	2.50	0.05	33.10	0.705

Table 10.3. Computation the ratio of ultimate bearing pressure for $A_h = 0.05$ ($1/a = 7000$ kN/m² and $1/b = 50$ kN/m²)

$B(m)$	h/B	A_h	$q_u(kN/m^2)$	q_u/q_{u0}
0.5	0.00	0.05	66.94	1.000
0.5	1.00	0.05	59.31	0.886
0.5	1.50	0.05	55.26	0.826
0.5	2.00	0.05	51.21	0.765
0.5	2.50	0.05	47.16	0.705

Table 10.4. Computation the ratio of ultimate bearing pressure for $A_h = 0.05$ ($1/a = 9000$ kN/m^2 and $1/b = 60$ kN/m^2)

$B(m)$	h/B	A_h	$q_u(kN/m^2)$	q_u/q_{u0}
0.5	0.00	0.05	81.01	1.000
0.5	1.00	0.05	71.78	0.886
0.5	1.50	0.05	66.88	0.826
0.5	2.00	0.05	61.98	0.765
0.5	2.50	0.05	57.07	0.705

Table 10.5. Computation the ratio of ultimate bearing pressure for $A_h = 0.05$ ($1/a = 12000$ kN/m^2 and $1/b = 80$ kN/m^2)

$B(m)$	h/B	A_h	$q_u(kN/m^2)$	q_u/q_{u0}
0.5	0.00	0.05	108.02	1.000
0.5	1.00	0.05	95.70	0.886
0.5	1.50	0.05	89.17	0.826
0.5	2.00	0.05	82.63	0.765
0.5	2.50	0.05	76.10	0.705

Figure 10.9 shows the trend of variation of BCR. It is evident from this figure that BCR decreases with increase in h/B ratio and A_h.

It is very common to obtain $(q_u)_{A_h=0}$ in a conventional way using Terzaghi's theory or by any other approach including field tests. Therefore, the seismic bearing capacity can be obtained using charts given in Fig. 10.9 for saturated clay. Doing regression analysis, Fig. 10.9 may be summed up by the following equation:

$$BCR = 1 - (2.0422 \times A_h + 0.006)h/B \tag{10.12}$$

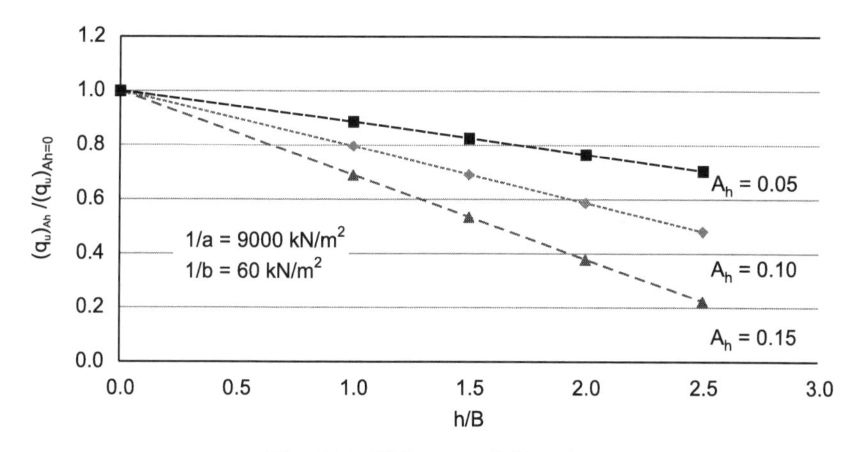

Fig. 10.9. BCR versus h/B ratio.

From the pressure versus maximum settlement curves Figs. 10.6(a), 10.7(a), 10.8(a), ratios of $\frac{(S_{\max})_{A_h}}{(S_{\text{avg}})A_{h=0}}$ were computed for factors of safety 2, 2.5 and 3.0 for all the values of A_h and h/B. $(S_{\max})A_h$ represents the values of maximum settlement for specified value of A_h and factor of safety; $(S_{\text{avg}})_{A_h=0}$ is the value of average settlement for static case i.e. A_h = 0 considering the same factor of safety. Factors of safety 2, 2.5 and 3 correspond to pressure intensities $q_u/2$, $q_u/2.5$ and $q_u/3$, where q_u is the ultimate bearing pressure which is obtained by intersection tangent method from a given pressure-settlement curve. It was found that the values of ratio $\frac{(S_{\max})A_h}{(S_{\text{avg}})A_{h=0}}$ are independent to the value of factor of safety. Typical plots are shown in Fig. 10.10. From this figure one can determine the value of $\frac{(S_{\max})A_h}{(S_{\text{avg}})A_{h=0}}$ for the given values of h/B and A_h. After doing regression analysis, these curves can be summed up by the following equation:

$$\frac{(S_{\max})_{A_h}}{(S_{\text{avg}})_{A_h=0}} = [-2.86 \times A_h^2 + 0.0039 \times A_h + 0.0039] \times (h/B)^2 + [-4.24 \times A_h^2 + 0.172 \times A_h - 0.0657] \times (h/B) + 1 \quad (10.13)$$

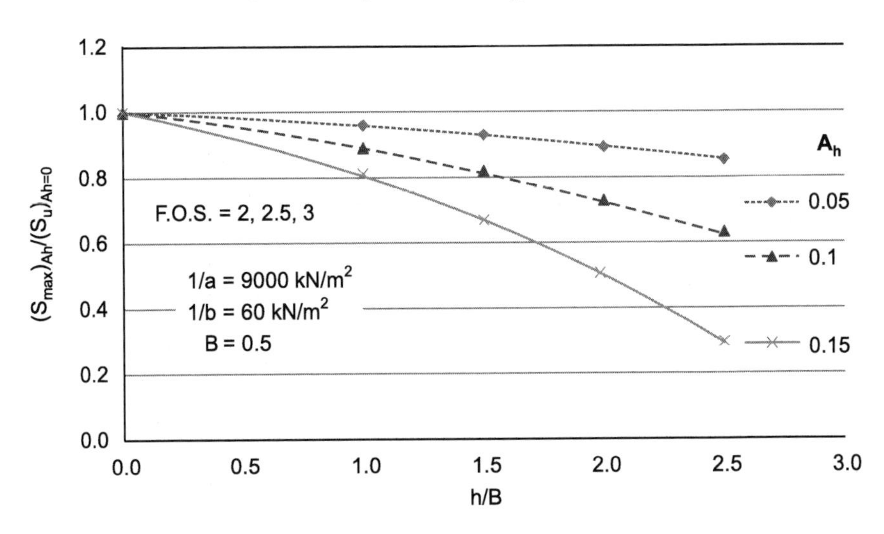

Fig. 10.10. $\frac{(S_{\max})}{(S_{\text{avg}})_{A_h=0}}$ versus $\frac{h}{B}$ for different values of A_h.

Similar exercise has been done for $\frac{(S_{\min})_{A_h}}{(S_{\text{avg}})_{A_h=0}}$ using the data shown in Figs 10.6b, 10.7b and 10.8b. The final curves are shown in Fig. 10.11 and the corresponding equation is as given below:

$$\frac{(S_{\min})_{A_h}}{(S_{\text{avg}})_{A_h=0}} = [2.36 \times A_h^2 + 0.5256 \times A_h - 0.0003] \times (h/B)^2$$

$$+ [8.66 \times A_h^2 - 5.9658 \times A_h - 0.003] \times (h/B) + 1 \qquad (10.14)$$

It may be noted that the above Eqs (10.13) and (10.14) are not dependent on the values of $1/a$ and $1/b$.

The procedure of obtaining the pressure-settlement of a rigid footing in static case has been explained in detail in Chapter 4. Once this is obtained one can get $(q_u)_{A_h=0}$ by intersection tangent method. Further the value of $(S_{\text{avg}})_{A_h=0}$ i.e. in static case may also be obtained for any pressure intensity. Usually settlements and tilts are obtained for pressure intensities lying between $\frac{(q_u)_{A_h=0}}{3}$ to $\frac{(q_u)_{A_h=0}}{2}$. Further the values of $(q_u)_{A_h=0}$ and $(S_{\text{avg}})_{A_h=0}$ may also be obtained, using classical approaches i.e Terzaghi's theory, one-dimensional consolidation approach or using plate load test data or penetration tests data. In this chapter, procedure has been given for obtaining the ultimate bearing capacity, settlement and tilt of a footing located in seismic region. It is further illustrated in Example 10.1.

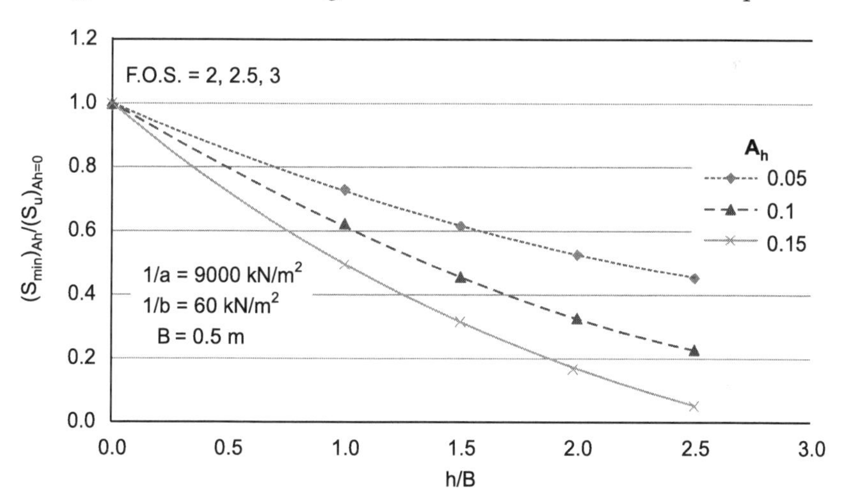

Fig. 10.11. $\frac{(S_{\min})_{A_h}}{(S_{\text{avg}})_{A_h=0}}$ versus $\frac{h}{B}$ for different values of A_h.

Illustrative Example

Example 10.1

Proportion a strip footing subjected to a vertical load of magnitude 100 kN/m and located in seismic region having horizontal seismic coefficient equal to 0.1. Values of unit cohesion and compression index are 40 kN/m^2 and 0.12 respectively. Height of point of application of the horizontal seismic force above the base of footing is 3.0 m. The unit weight of soil is 19 kN/m^3 and degree of saturation is 95 per cent. Permissible values of settlement and tilt are 25 mm and 0.5° respectively.

Solution

1. Assume the width of the footing as 1.5 m
2. Using Terzaghi's theory,

$$(q_u)_{A_h=0} = c.N_c$$

c is the unit cohesion and N_c is bearing capacity factor equal to 5.7.
Therefore $(q_u)_{A_h=0} = 40 \times 5.7 = 228$ kN/m^2
Load carrying capacity of the footing in static case

$$= \frac{228}{3} \times 1.5 = 114 \text{ kN/m}$$

3. From Eq. (10.12)

$$BCR = \frac{(q_u)_{A_h}}{(q_u)_{A_h=0}} = 1 - [2.0422A_h + 0.006](h/B)$$

$$= 1 - [2.0422 \times 0.1 + 0.006]\ (3.0/1.5)$$

$$= 1 - 0.42044 = 0.57956 \approx 0.58$$

$$\therefore \quad (q_u)_{A_h} = 228 \times 0.58 = 132.2 \text{ kN/m}^2$$

Load-carrying capacity of strip footing

$$= \frac{132.2}{3} \times 1.5 = 66.12 \text{ kN/m} < 100 \text{ kN/m (hence unsafe)}$$

4.
$$\left(\frac{G_s + Se}{1+e}\right)\gamma_w = 19 \text{ kN/m}^3$$

$$\left(\frac{2.65 + 0.95e}{1+e}\right)10.0 = 19$$

$$2.65 + 0.95\,e = 1.9 + 1.9\,e$$

or $$0.95\,e = 0.75$$

$\therefore$ $$e = 0.7895$$

Assuming that the soil stratum from the base of the footing to a depth equal to $2B$ is contributing to the settlement of the footing,

$$(S_{avg})_{A_h=0} = \frac{C_c H}{1+e_0}\log_{10}\frac{p_0+\Delta p}{p_0}$$

C_c = Compression index = 0.12

$H = 2B = 2 \times 1500 = 3000$ mm

e_0 = Initial void ratio = 0.7895

p_0 = Initial overburden pressure

Δp = Increase in pressure at the center of clay layer due to superimposed load

$p_0 = 19 \times 1.5 = 28.5$ kN/m^2

$$\Delta p = \frac{66}{(1.5+1.5)} = 22 \text{ kN/m}^2$$

$$(S_{avg})_{A_h=0} = \frac{0.12\times 3000}{(1+0.7895)}\log_{10}\frac{28.5+22}{28.5}$$

$$= \frac{360}{1.7895} \times 0.2484 = 50 \text{ mm}$$

From Eq. (10.13)

$$\frac{(S_{max})_{A_h}}{(S_{avg})_{A_h=0}} = [-2.86 \times A_h^2 + 0.0039 \times A_h + 0.0039] \times (h/B)^2$$

$$+ [4.24 \times A_h^2 + 0.172 \times A_h\, 0.0657] \times (h/B) + 1$$

$$= [-2.86 \times (0.1)^2 + 0.0039(0.1) + 0.0039](3/1.5)^2$$

$$= [4.25 \times (0.1)^2 + 0.172(0.1) - 0.00657](3/1.5)^2$$

$$= 0.735$$

Therefore,

$$(S_{max})_{A_h} = 50 \times 0.735 = 36.75 \text{ mm} > 25 \text{ mm (hence unsafe)}$$

From Eq. (10.14)

$$\frac{(S_{min})_{A_h}}{(S_{avg})_{A_h=0}} = [2.36 \times A_h^2 + 0.5256 \times A_h - 0.0003] \times (h/B)^2$$

$$+ [8.66 \times A_h^2 - 5.9658 \times A_h - 0.003] \times (h/B) + 1$$

$$= [2.36 \times (0.1)^2 + 0.5256\ (0.1) + 0.0003]\ (3/1.5)^2$$

$$+ [8.66 \times (0.1)^2 - 5.9658\ (0.1) - 0.003]\ (3/1.5) + 1 = 0.277$$

$$(S_{\min})_{A_h} = 50 \times 0.277 = 13.85 \text{ mm}$$

$$\text{Tilt} = \tan^{-1}\left[\frac{(S_{\max})_{A_h} - (S_{\min})_{A_h}}{B}\right]$$

$$= \tan^{-1}\left[\frac{(36.75 - 13.85)}{1500}\right] = 0.875° > 0.5° \text{ (hence unsafe)}$$

Less than 1° (hence safe)

Practice Problems

1. Considering pseudo-static approach, show the forces and moments acting on the base of footing. Draw the typical settlement patterns considering (i) $e/B \leq 1/6$ and (ii) $e/B > 1/6$, e being the eccentricity and B the width of footing.
2. Explain stepwise the procedure for obtaining maximum and minimum settlements of a smooth flexible footing resting on saturated clay and located in seismic region.
3. How is the settlements of a rigid footing obtained knowing the settlements of a flexible footing of same width.
4. Determine the amount of vertical load on a strip footing of width 1.5 m resting on saturated clay ($1/a = 9000$ kN/m^2, $1/b = 60$ kN/m^2) considering that located in seismic region having horizontal seismic coefficient equal to 0.15. Permissible value of settlement is 30 mm.

REFERENCES

Biswas, T., Saran, S. and Shanker, D. (2016). Behavior of a Strip Footing Resting on Clay Using Non-linear Costitutive Law. *Geosciences*, **6(3):** 75-77.

Biswas, T. (2017). Pseudo-static Analysis of a Strip Footing Using Constitutive Laws of Soils. Ph.D. Thesis, IIT Roorkee, Roorkee.

Budhu, M. and Al-Karni, A.A. (1993). Seismic Bearing Capacity of Soils. *Géotechnique*, **43:** 181-187.

Chatzigogos, C.T., Pecker, A. and Salencon, J. (2007). Seismic bearing capacity of a circular footing on a heterogeneous cohesive soil. *Soils and Foundations*, **47:** 786-797.

Chen, W.F.(1975). Limit Analysis and Soil Plasticity. Elsevier, New York.

Choudhury, D. and Subba Rao, K.S. (2005). Seismic bearing capacity of shallow strip footings. *Geotechnical and Geological Engineering*, **23:** 403-418.

Dormieux, L. and Pecker, A. (1995). Seismic bearing capacity of foundation on cohesionless soil. *Journal of Geotechnical Engineering*, **121:** 300-303.

Ghosh, P. (2008). Upper bound solutions of bearing capacity of strip footing by pseudo-dynamic approach. *Acta Geotechnica*, **3:** 115-123 .

Ghosh, P. and Choudhury, D. (2011). Seismic bearing capacity factors for shallow strip footings by pseudo-dynamic approach. *Disaster Advances*, **4(3):** 34-42.

Kumar, J. and Mohan Rao, V.B.K. (2002). Seismic bearing capacity factors for spread of foundations. *Géotechnique*, **52(2):** 79-88.

Majidi, A. and Mirghasemi, A. (2008). Seismic 3D-bearing capacity analysis of shallow foundations. *Iranian Journal of Science and Technology*, **32:** 107-124.

Merlos, J. and Romo, M. (2006). Fluctuant bearing capacity of shallow foundations during earthquakes. *Soil Dynamics and Earthquake Engg.*, **26:** 103-114.

Paolucci, R. and Pecker, A. (1997). Seismic bearing capacity of shallow strip foundations on dry soils. *Soils and Foundations*, **37:** 95-105.

Pecker, A. and Salencon, J. (1991). Seismic bearing capacity of shallow strip footing on clays. *Proceedings of International Workshop and Seismology and Earthquake Engineering*, Mexico, 287-304.

Rangwala, H. (2014). Seismic Bearing Capacity of a Strip Footing Adjacent to Stable Slopes, Ph.D. Thesis, IIT Roorkee, Roorkee.

Richards, R., Elms, D.G. and Budhu, M. (1993). Seismic bearing capacity and settlements of foundations. *Journal of Geotechnical Engineering*, **119:** 662-674.

Saran, S. and Rangwala, H. (2011). Seismic bearing capacity of footings. *International Journal of Geotechnical Engineering*, **5(4):** 447-455.

Sarma, S.K. and Lossifelis, I.S. (1990). Seismic bearing capacity factors of shallow strip footings. *Géotechnique*, **40(2):** 265-273.

Shafiee, A.H. and Jahanandish, M. (2010). Seismic bearing capacity factors for strip footings. *Proceedings of 5th National Congress on Civil Engineering*, Mashhad, May 4-6, 1-8.

Soubra, A.-H. (1999). Upper-bound solutions for bearing capacity of foundations. *Journal of Geotechnical and Geoenvironmental Engineering*, ASCE, **125(1):** 59-68.

CHAPTER 11

Strip Footing Located Below Ground Surface

11.1 General

The base of a footing is always located at a certain depth d below the ground surface. Usually this depth ranges between 0.25 B and 1.0 B, where B is the width of footing (Terzaghi, 1943; Terzaghi and Peck, 1967; Saran, 2006). In shallow foundations, $d/B > 0$, for the same pressure intensity, the values of settlement and tilt are lesser in comparison to the surface footing. This means that the pressure-settlement and pressure-tilt characteristics improve, i.e. shift upwards. Hence, neglecting the effect of depth of the footing will lead to a safe design, but it is uneconomical.

In this chapter, the methodology has been developed for analyzing the behavior of a strip footing having its base located below the ground surface, using non-linear constitutive law of the soil. Since stress equations are available only for vertical and horizontal point loads, solutions are based on the principle of numerical addition.

11.2 Stress Equations

The solution for determining the stresses σ_z, σ_x and τ_{xz} was given by Melan (1919, 1932) for both vertical and horizontal line loads acting at a depth d below the surface of semi-infinite mass. The equations of stresses in convenient form were reproduced by Harr (1966). By using superimposition, the stresses can be obtained for any orientation of the line load.

For a vertical line load (Fig. 11.1a), he found

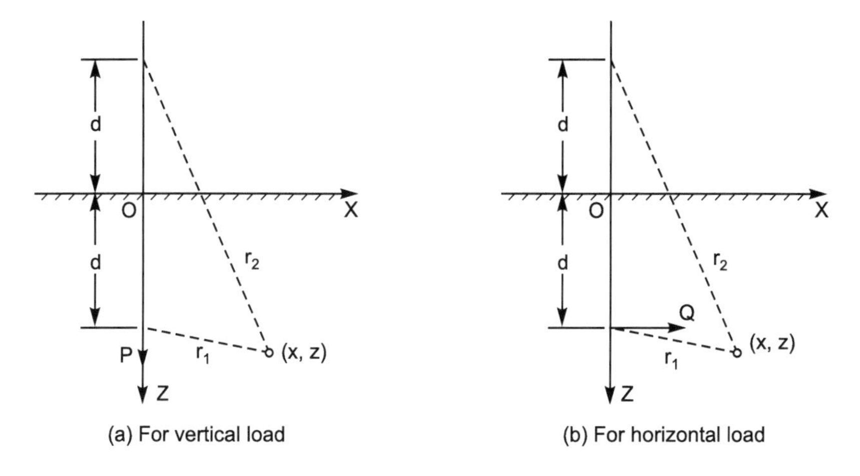

Fig. 11.1. Geometry with respect to vertical and horizontal loads is shown.

$$\sigma_z = \frac{P}{\pi}\left[\frac{m+1}{2m}\left\{\frac{(z-d)^3}{r_1^4}+\frac{(z+d)\left[(z+d)^2+2dz\right]}{r_2^4}-\frac{8dz(d+z)x^2}{r_2^6}\right\}\right.$$

$$\left.+\frac{m-1}{4m}\left\{\frac{(z-d)}{r_1^2}+\frac{(3z+d)}{r_2^2}-\frac{4zx^2}{r_2^4}\right\}\right] \tag{11.1}$$

$$\sigma_x = \frac{P}{\pi}\left[\frac{m+1}{2m}\left\{\frac{(z-d)x^2}{r_1^4}+\frac{(z+d)(x^2+2d^2)-2dx^2}{r_2^4}+\frac{8dz(d+z)x^2}{r_2^6}\right\}\right.$$

$$\left.+\frac{m-1}{4m}\left\{-\frac{(z-d)}{r_1^2}+\frac{(z+3d)}{r_2^2}+\frac{4zx^2}{r_2^4}\right\}\right] \tag{11.2}$$

$$\tau_{xz} = \frac{Px}{\pi}\left[\frac{m+1}{2m}\left\{\frac{(z-d)^2}{r_1^4}+\frac{z^2-2dz-d^2}{r_2^4}+\frac{8dz(d+z)^2}{r_2^6}\right\}\right.$$

$$\left.+\frac{m-1}{4m}\left\{\frac{1}{r_1^2}-\frac{1}{r_2^2}+\frac{4z(d+z)}{r_2^4}\right\}\right] \tag{11.3}$$

For a horizontal line load (Fig. 11.1b) (Harr, 1966),

$$\sigma_z = \frac{Qx}{\pi}\left[\frac{m+1}{2m}\left\{\frac{(z-d)^2}{r_1^4}-\frac{(d^2-z^2+6dz)}{r_2^4}+\frac{8dzx^2}{r_2^6}\right\}\right.$$

$$-\frac{m-1}{4m}\left\{\frac{1}{r_1^2}-\frac{1}{r_2^2}-\frac{4z(d+z)}{r_2^4}\right\}\Bigg] \tag{11.4}$$

$$\sigma_x = \frac{Qx}{\pi}\left[\frac{m+1}{2m}\left\{\frac{x^2}{r_1^4}+\frac{(x^2+8dz+6d^2)}{r_2^4}+\frac{8dz(d+z)^2}{r_2^6}\right\}\right.$$

$$\left.+\frac{m-1}{4m}\left\{\frac{1}{r_1^2}+\frac{3}{r_2^2}-\frac{4z(d+z)}{r_2^4}\right\}\right] \tag{11.5}$$

$$\tau_{xz} = \frac{Q}{\pi}\left[\frac{m+1}{2m}\left\{\frac{(z-d)x^2}{r_1^4}+\frac{(2dz+x^2)(d+z)}{r_2^4}-\frac{8dz(d+z)x^2}{r_2^6}\right\}\right.$$

$$\left.+\frac{m-1}{4m}\left\{\frac{(z-d)}{r_1^2}+\frac{(3z+d)}{r_2^2}-\frac{4z(z+d)^2}{r_2^4}\right\}\right] \tag{11.6}$$

In the above expressions:

P = Vertical line load

Q = Horizontal line load

x = Distance of the point under consideration for obtaining stresses from vertical axis passing through point 'O' (Fig. 11.1)

z = Depth of the point below ground surface

$$m = \frac{1-\mu}{\mu}$$

μ = Poisson's ratio

$$r_1^2 = x^2 + (z-d)^2 \tag{11.7a}$$

$$r_2^2 = x^2 + (z+d)^2 \tag{11.7b}$$

It may be noted that in the case $d = 0$, stress Eqs (11.1) to (11.6) will reduce to Eqs (7.1) to (7.6) as given in Chapter 7 (Art. 7.2).

11.3 Method of Super-position

Figure 11.2a shows a section of a strip footing subjected to a uniformly distributed vertical load of intensity q_v. For getting the solution of the footing located at a depth d, uniformly distributed load may be converted into concentrated loads, say in 'n' equal parts. Then magnitude of each concentrated load, say P (Fig. 11.2b), will be:

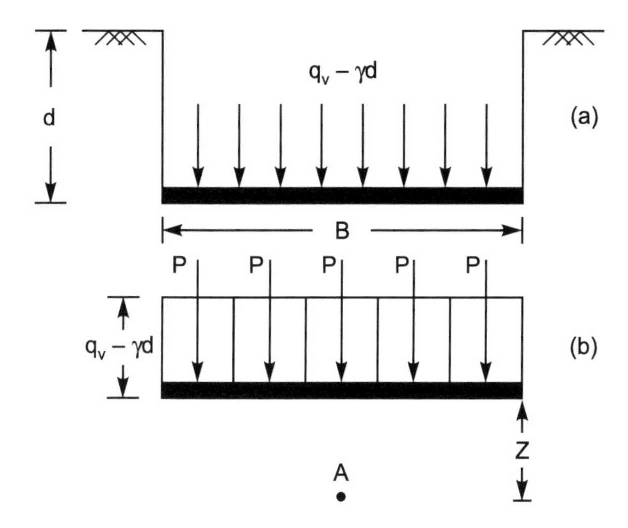

Fig. 11.2. Uniformly distributed vertical acting on a footing located at a certain depth below around surface.

$$P = \frac{(q_v - \gamma d)B}{n} \tag{11.8}$$

If it is desired to obtain stresses (σ_z, σ_x and τ_{xz}) at point '*A*' (Fig. 11.2b), the same can be done by using Eqs (11.1), (11.2) and (11.3) by putting relevant values of *x* and *z*. This exercise is done for each vertical concentrated load '*P*' (Fig. 11.2b); values of total stresses (i.e. of each σ_z, σ_x and τ_{xz}) are then obtained by numerical addition.

Similarly, Fig. 11.3a shows a section of strip footing subjected to a uniformly distributed horizontal load of intensity q_h. Proceeding as illustrated above:

$$Q = \frac{q_h.B}{n} \tag{11.9}$$

Stresses σ_z, σ_x and τ_{xz} at point '*A*' are obtained using Eqs (11.4), (11.5) and (11.6) for each horizontal concentrated load of magnitude '*Q*'. Total value of stresses are obtained by numerical addition.

If the variation of vertical and horizontal load intensities is other than uniform (say, triangular increasing or triangular decreasing or parabolic, etc.), the principle of addition can be applied by converting this distribution into equivalent vertical and horizontal loads (i.e. *P* and *Q*). Accuracy of the method depends on the value of '*n*'. It may be seen that if the distribution of loading intensities is as shown in Figs 11.4a and 11.4b, values of P_1, P_2 ... and Q_1, Q_2 ... are obtained by computing the area of corresponding trapeziums.

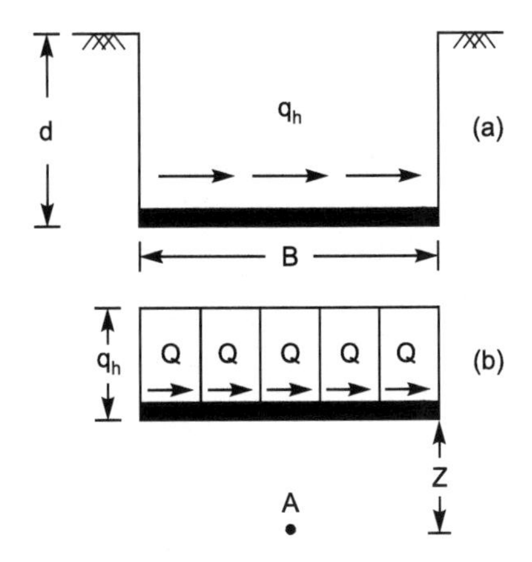

Fig. 11.3. Representation of horizontal distributed load into equivalent concentrated horizontal loads.

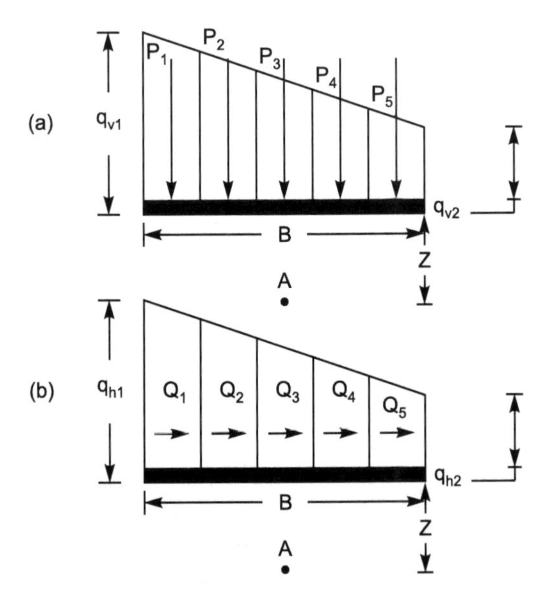

Fig. 11.4. Representation of trapezoidally varying vertical and horizontal loading into respective concentrated loads.

11.4 Shallow Strip Footing on Clay (Static Case)

As mentioned in the earlier chapters, the effect of base roughness on the pressure-settlement characteristic of a footing is marginal. Therefore, only the analysis of a smooth-based footing is presented herein. Further, because the average settlement of a flexible footing (A_S/B) is almost equal to the uniform settlement of a rigid footing of same width, the footing is considered flexible.

11.4.1 Assumptions

- Soil mass below depth '*d*' from ground surface is assumed as semi-infinite and isotropic
- The footing base is considered perfectly smooth and flexible. Therefore, contact pressure distribution will be uniform (Fig. 11.2)
- The soil mass below the base of the footing is divided into a large number of strips (Fig. 11.5) in which stresses and strains are assumed to be uniform along any vertical section

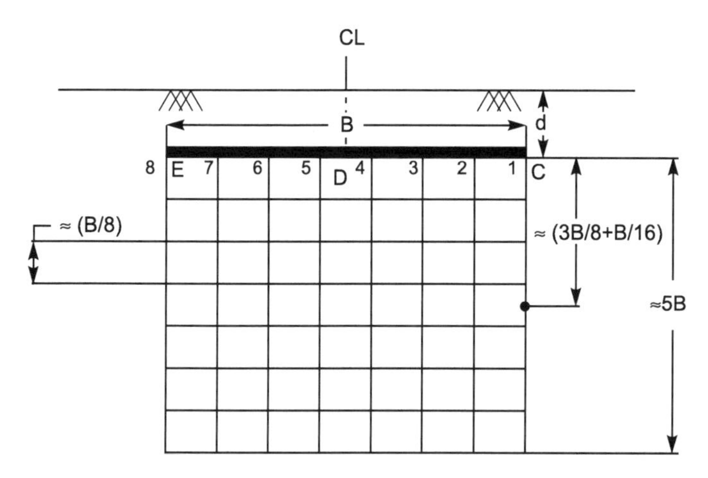

Fig. 11.5. The division of soil stratum into a number of layers.

- Stresses in each layer are computed using Eqs (11.1) to (11.3) for a pressure intensity equal to $q_v - \gamma d$; q_v is the pressure intensity acting on the footing for which settlement is desired; γ is the unit weight of the soil; and amount of q_v is greater than γd
- Strains have been computed using the constitutive law of soil, and considering no slippage at the interface of layers of the soil mass

11.4.2 Analysis

Analysis for obtaining the pressure-settlement characteristics of an actual shallow footing of width *B* located at a depth *d* below ground surface is required to carry out the following steps:

- Divide the soil strata into thin layers upto significant depth ($\approx$ 5.0 *B*) below the base of footing. Thickness of each layer may be taken as equal to $B/8$ with *B* being the width of footing (Fig. 11.5)

- Base contact pressure is considered uniform to adopt a value pressure intensity q_v. Divide this into n small parts, so that each part may be considered to have a concentrated vertical load of magnitude $\frac{(q_v - \gamma d).B}{n}$
- Select about eight points on the base of the footing, including points A, B and C. Further procedure is to determine the values of settlements at these points. For this, consider a vertical section passing through any of the selected points
- For determining the settlement of a selected point at the base of footing, determine the settlement of each layer at that point. For example, consider fourth layer and vertical section passing through point A depth of the centre of this layer below the base of footing $\left(\frac{3B}{8} + \frac{B}{16}\right)$. Determine the stresses (σ_z, σ_x and τ_{xz}) at this point, using for all the concentrated vertical loads. Using the principle of addition, the final values of σ_z, σ_x and τ_{xz} are obtained. Determine the principal stresses σ_1 and σ_3 and their directions with respect to vertical, i.e. θ_1 and θ_3 in the usual manner
- In the case of clay, the value of principal strain (ε_1) in the direction of major principal stress may be obtained as per Eq. (11.10)

$$\varepsilon_1 = \frac{a(\sigma_1 - \sigma_3)}{1 - b(\sigma_1 - \sigma_3)} \tag{11.10}$$

where a and b are Kondner's hyperbola constants

$$\varepsilon_3 = -\mu_2.\varepsilon_1 \tag{11.11}$$

$$\mu_2 = \frac{-\sigma_3 + \mu_1 \sigma_1}{\sigma_1 - \mu_1 \sigma_3} \tag{11.12}$$

$$\mu_1 = \frac{\mu}{1 - \mu} \tag{11.13}$$

μ is Poisson's ratio

- The vertical strain at the selected point will then be

$$\varepsilon_z = \varepsilon_1 \cos^2 \theta_1 + \varepsilon_3 \cos^2 \theta_3 \tag{11.14}$$

- On multiplying ε_z with the thickness of the strip, settlement of the strip at the point under consideration is obtained. Evaluation of the total settlement along any vertical section is done by numerically integrating the quantity, i.e.

$$S_t = \int_0^N \varepsilon_z dz \tag{11.15}$$

where S_t = total settlement along i^{th} vertical section

ε_z = vertical strain at depth z below the base of footing

dz = thickness of strips at depth z

N = number of strips in which the soil strata upto a significant depth is divided

In this way the values of total settlements along other vertical sections passing through the base of footing are obtained. The settlement pattern will be as shown in Fig. 11.6.

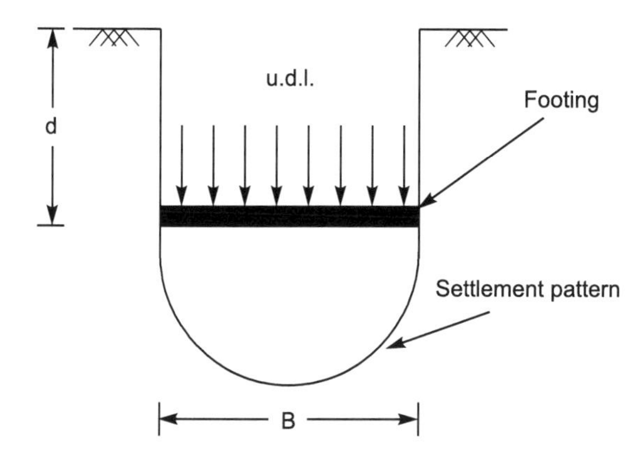

Fig. 11.6. Settlement diagram of a shallow footing subjected to vertical uniformly distributed loading.

The above procedure is applicable for a flexible footing resting on clay. For getting the settlement of an equivalent rigid footing, the area of settlement diagram (Fig. 11.6) is divided by the width of the footing. The above procedure is repeated for other values of pressure intensities q_v. In this way, complete pressure-settlement characteristics of an actual footing of width B resting on clay, as defined by Kondner's soil parameters a and b, is obtained. It, therefore, enables us to proportion the footing completely.

11.5 Strip Footing on Clay (Seismic Case)

When a structure is located in a seismic zone, it is subjected to additional forces due to vibrations generated by an earthquake. There are two methods of analysis; namely, pseudo-static analysis based on seismic coefficients and dynamic analysis. Herein pseudo-static analysis is presented. If A_h and A_v are respectively the horizontal and vertical seismic coefficients, the total vertical load acting on a strip footing will be $V(1 \pm A_v)$ and horizontal load be equal to $V.B.A_h$ (Fig. 11.7a) acting at a certain height h above the base of the footing. V is the total weight of the superstructure and the foundation system. Equivalent force is shown in Fig. 11.7b. Therefore, the footing may be considered to be subjected to pressure intensities as shown in Figs 11.7c and 11.7d i.e.,

For $\frac{e}{B} \leq \frac{1}{6}$

$$V' = V(1 \pm A_v) \tag{11.15a}$$

$$M = V.A_h.(h + d) \tag{11.15b}$$

$$e = \frac{M}{V'} \tag{11.16}$$

$$p_{max} = \frac{V'}{A}\left(1 + \frac{6e}{B}\right) \tag{11.17a}$$

$$p_{min} = \frac{V'}{A}\left(1 - \frac{6e}{B}\right) \tag{11.17b}$$

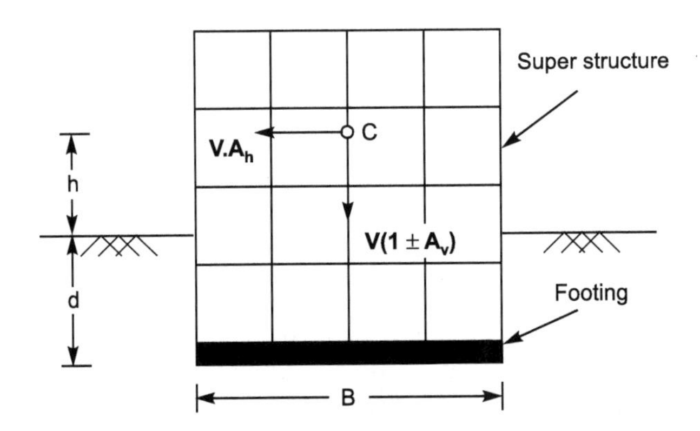

(a) Forces acting on the superstructure-footing system in seismic case

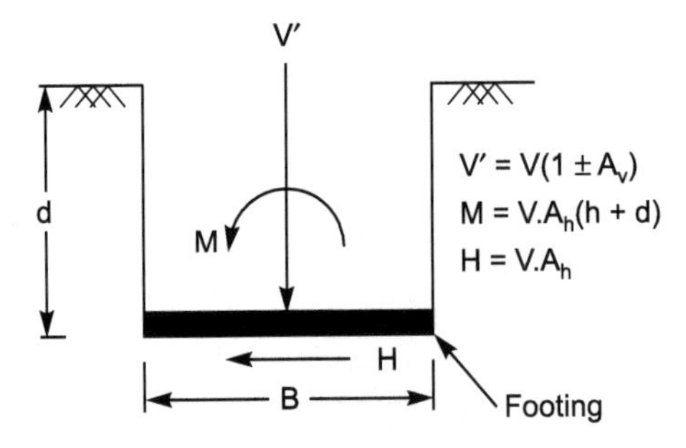

(b) Equivalent force diagram

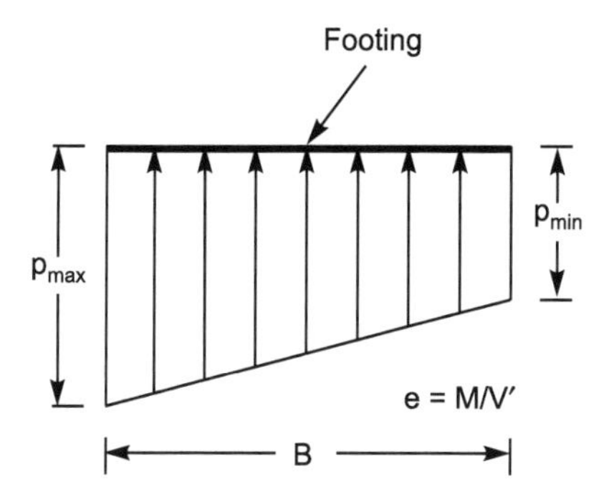

(c) Vertical pressure diagram when $e \leq \frac{B}{6}$

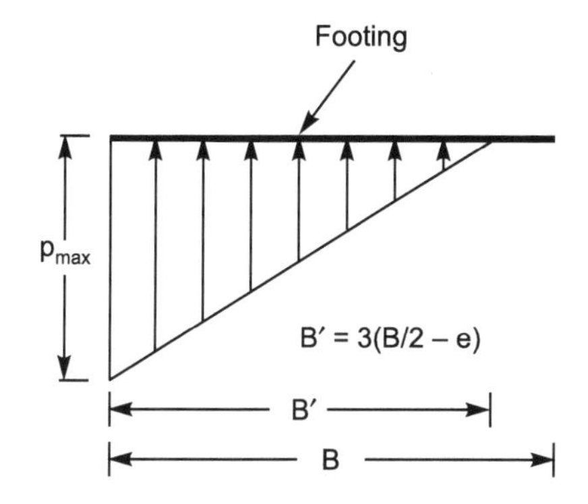

(d) Vertical pressure diagram when $e > \frac{B}{6}$

Fig. 11.7. Forces and moments acting on a shallow footing in seismic case.

For $\frac{e}{B} > \frac{1}{6}$ (Fig. 11.7d)

$$p_{max} = \frac{V'}{B}\left(\frac{4B}{3B - 6e}\right) \tag{11.18a}$$

$$p_{min} = 0 \tag{11.18b}$$

$$B' = 3\left(\frac{B}{2} - e\right) \tag{11.18c}$$

Procedure for the analysis is exactly the same as discussed above in Art. 11.4.2, keeping the following additional points in view:

Final values of stresses (σ_z, σ_x and τ_{xz}) as obtained by numerically adding the stress caused by:

- Vertical pressure diagram is shown in Fig. 11.7c when $e/B \le 1/6$ or vertical pressure diagram shown in Fig. 11.7d, for $e/B > 1/6$
- Horizontal pressure diagram is shown in Fig. 11.8.

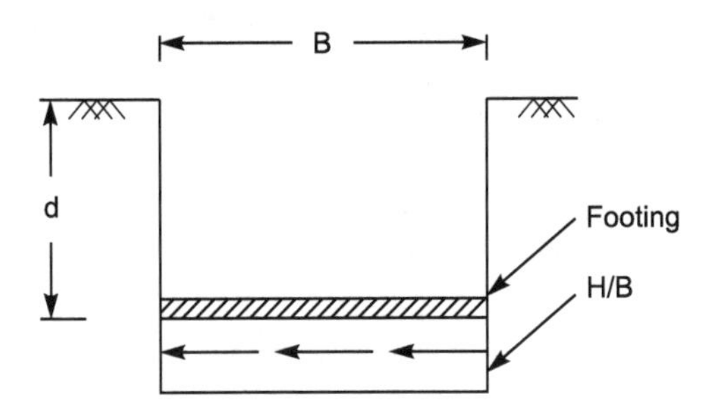

Fig. 11.8. Horizontal pressure diagram.

Readers may note that the only difference in the analysis presented in Art 10.2 of Chapter 10 is in the expression of M (i.e. Eq. (11.16b)). Change in this expression will be reflected in further analysis.

11.6 Results and Interpretation (Static Case)

As per the detailed discussion given in Art 4.4 of Chapter 4, pressure versus average settlement width ratio curves of a surface footing is improved with increase in $1/a$ and $1/b$ values (Fig. 4.14). Further these curves are independent of the width of the footing (Fig. 4.13). Similar curves are shown in Figs 11.9 and 11.10 for d/B ratios equal to 0.5 and 1.0.

Values of ultimate bearing capacity obtained from Fig. 4.14 and Fig. 11.9 by inter-section method are given in Table 11.1.

In Column No 5, the ratios of ultimate bearing capacity values $\frac{(q_u)_{d/B}}{(q_u)_{d/B=0}}$ are given for $d/B = 0$, 0.5 and 1. It is quite evident that these values are independent of $1/a$ and $1/b$ values, and are dependent only on d/B ratio. Regression analysis gives the following relationship:

$$\frac{(q_u)_{d/B}}{(q_u)_{d/B=0}} = 1 + 0.262\,(d/B) \tag{11.19}$$

Values of settlements S_u for pressure intensities q_u are listed in Col. 6 of Table 11.1. Settlement ratios $\frac{(S_u)_{d/B}}{(S_u)_{d/B=0}}$ are given in Col. 7. Here it may be noted that the settlement ratios are independent of $1/a$ and $1/b$ values and dependent on d/B ratio. On regression analysis, the following equation is obtained:

$$\frac{(S_u)_{d/B}}{(S_u)_{d/B=0}} = 1 - 0.096\,(d/B) \tag{11.20}$$

Table 11.1. Values of bearing capacity and settlement ratios for F.O.S = 1

Depth (m)	$1/a$ (kN/m^2)	$1/b$ (kN/m^2)	$(q_u)_{d/B=0}$ (kN/m^2)	$(q_u)_{d/B=0}/(q_u)_{d/B=0}$	$(S_u)_{d/B=0}$ (mm)	$(S_u)_{d/B=0}/(S_u)_{d/B=0}$
(1)	(2)	(3)	(4)	(5)	(6)	(7)
0	5000	35	45	1	33.6031	1
0	7000	50	64.57	1	35.0841	1
0	9000	60	79.97	1	39.7568	1
0	12000	90	108.14	1	44.4701	1

Depth (m)	$1/a$ (kN/m^2)	$1/b$ (kN/m^2)	$(q_u)_{d/B=0.5}$ (kN/m^2)	$(q_u)_{d/B=0.5}/(q_u)_{d/B=0}$	$(S_u)_{d/B=0.5}$ (mm)	$(S_u)_{d/B=0.5}/(S_u)_{d/B=0}$
(1)	(2)	(3)	(4)	(5)	(6)	(7)
0.5	5000	35	50.5	1.1222	31.9424	0.9506
0.5	7000	50	72.5	1.1228	33.3445	0.9504
0.5	9000	60	90	1.1254	37.7771	0.9502
0.5	12000	90	121.5	1.1235	42.2578	0.9503

Depth (m)	$1/a$ (kN/m^2)	$1/b$ (kN/m^2)	$(q_u)_{d/B=1}$ (kN/m^2)	$(q_u)_{d/B=1}/(q_u)_{d/B=0}$	$(S_u)_{d/B=1}$ (mm)	$(S_u)_{d/B=1}/(S_u)_{d/B=0}$
(1)	(2)	(3)	(4)	(5)	(6)	(7)
1	5000	35	56.62	1.2582	30.3678	0.9037
1	7000	50	81.33	1.2596	31.7032	0.9036
1	9000	60	101.29	1.2666	35.9098	0.9032
1	12000	90	136.827	1.2653	40.1817	0.9036

As in design, settlements are required for the factors of safety lying between 2 and 3, their values obtained by keeping in mind this point in view. In Table 11.2, Col. 4 indicates the values of pressure intensities corresponding to the factor of safety equalling 2.0. These are obtained by dividing the value of q_u (Table 11.2) by 2.0. In Col. 5, the values of corresponding settlements $(S)_{d/B}$ are given which are obtained from pressure-settlement curves given in Figs 4.14, 11.9 and 11.10 respectively. Values of settlement ratios $\frac{(S)_{d/B=0.5}}{(S)_{d/B=0}}$ and $\frac{(S)_{d/B=1}}{(S)_{d/B=0}}$ are given in Col. 6.

Similarly values of $\frac{(S)_{d/B=0.5}}{(S)_{d/B=0}}$ and $\frac{(S)_{d/B=1}}{(S)_{d/B=0}}$ and for factors of safety 2.5 and 3.0 are given in Tables 11.3 and 11.4 respectively. It is evident from

Tables 11.2, 11.3 and 11.4 that the settlement ratios are independent of the soil properties (i.e. $1/a$ and $1/b$ values), and depend on d/B ratios. Regression analysis gives the following equation to obtain the settlement ratio:

$$\frac{(S)_{d/B}}{(S)_{d/B=0}} = 1 + 0.064\,(d/B) - 0.033(d/B)^2 \qquad (11.21)$$

Table 11.2. Values of bearing capacity and settlement ratios for F.O.S = 2

Depth (*m*)	$1/a$ (kN/m^2)	$1/b$ (kN/m^2)	$q_{(d/B=0)}$ (kN/m^2)	$S_{(d/B=0m)}$ (*mm*)	$S_{(d/B=0)}/S_{(d/B=0)}$
0	5000	35	22.500	7.3902	1.0000
0	7000	50	32.285	7.5888	1.0000
0	9000	60	39.985	7.4132	1.0000
0	12000	90	54.070	7.5675	1.0000

Depth (*m*)	$1/a$ (kN/m^2)	$1/b$ (kN/m^2)	$q_{(d/B=0)}$ (kN/m^2)	$S_{(d/B=0m)}$ (*mm*)	$S_{(d/B=0.5)}/S_{(d/B=0)}$
0.5	5000	35	25.250	7.5453	1.0210
0.5	7000	50	36.250	7.7534	1.0217
0.5	9000	60	45.000	7.5966	1.0247
0.5	12000	90	60.750	7.7347	1.0221

Depth (*m*)	$1/a$ (kN/m^2)	$1/b$ (kN/m^2)	$q_{(d/B=0)}$ (kN/m^2)	$S_{(d/B=0m)}$ (*mm*)	$S_{(d/B=1.0)}/S_{(d/B=0)}$
1	5000	35	28.310	7.5619	1.0232
1	7000	50	40.665	7.7758	1.0246
1	9000	60	50.645	7.6505	1.0320
1	12000	90	68.414	7.7954	1.0301

Table 11.3. Values of bearing capacity and settlement ratios for F.O.S = 2.5

Depth (*m*)	$1/a$ (kN/m^2)	$1/b$ (kN/m^2)	$q_{(d/B=0)}$ (kN/m^2)	$S_{(d/B=0m)}$ (*mm*)	$S_{(d/B=0)}/S_{(d/B=0)}$
0	5000	35	18.000	5.4563	1.0000
0	7000	50	25.828	5.5997	1.0000
0	9000	60	31.988	5.4472	1.0000
0	12000	90	43.256	5.5496	1.0000

Depth (*m*)	$1/a$ (kN/m^2)	$1/b$ (kN/m^2)	$q_{(d/B=0)}$ (kN/m^2)	$S_{(d/B=0m)}$ (*mm*)	$S_{(d/B=0.5)}/S_{(d/B=0)}$
0.5	5000	35	20.200	5.5810	1.0229
0.5	7000	50	29.000	5.7315	1.0235
0.5	9000	60	36.000	5.5914	1.0265
0.5	12000	90	48.600	5.6835	1.0241

(Contd.)

Depth (m)	$1/a$ (kN/m^2)	$1/b$ (kN/m^2)	$q_{(d/B=0)}$ (kN/m^2)	$S_{(d/B=0)}$ (mm)	$S_{(d/B=1.0)}/S_{(d/B=0)}$
1	5000	35	22.648	5.6065	1.0275
1	7000	50	32.532	5.7614	1.0289
1	9000	60	40.516	5.6429	1.0359
1	12000	90	54.731	5.7400	1.0343

Table 11.4. Values of bearing capacity and settlement ratios for F.O.S = 3

Depth (m)	$1/a$ (kN/m^2)	$1/b$ (kN/m^2)	$q_{(d/B=0)}$ (kN/m^2)	$S_{(d/B=0)}$ (mm)	$S_{(d/B=0)}/S_{(d/B=0)}$
0	5000	35	15.000	4.3323	1.0000
0	7000	50	21.523	4.4449	1.0000
0	9000	60	26.657	4.3141	1.0000
0	12000	90	36.047	4.3906	1.0000

Depth (m)	$1/a$ (kN/m^2)	$1/b$ (kN/m^2)	$q_{(d/B=0)}$ (kN/m^2)	$S_{(d/B=0m)}$ (mm)	$S_{(d/B=0.5)}/S_{(d/B=0)}$
0.5	5000	35	16.833	4.4352	1.0237
0.5	7000	50	24.167	4.5532	1.0244
0.5	9000	60	30.000	4.4316	1.0272
0.5	12000	90	40.500	4.5005	1.0250

Depth (m)	$1/a$ (kN/m^2)	$1/b$ (kN/m^2)	$q_{(d/B=0)}$ (kN/m^2)	$S_{(d/B=0m)}$ (mm)	$S_{(d/B=1.0)}/S_{(d/B=0)}$
1	5000	35	18.873	4.4609	1.0297
1	7000	50	27.110	4.5826	1.0310
1	9000	60	33.763	4.4772	1.0378
1	12000	90	45.609	4.5502	1.0363

Notes:

(i) q_u is ultimate bearing capacity obtained by intersection tangent method

(ii) S_u is settlement of the footing at pressure intensity equal to q_u

(iii) In all computations, width of the footing is considered as 1.0 m

11.7 Results and Interpretation (Seismic Case)

In Chapter 10, the procedure is given to obtain the pressure versus maximum settlement, pressure versus minimum settlement and pressure versus tilt characteristics of a surface footing ($d/B = 0$) located in a seismic region. For footings located at a certain depth 'd' below ground surface, the procedure is given in this very chapter. For the soil having $1/a = 9000$ kN/m^2 and $1/b = 60$ kN/m^2, pressure versus maximum settlement, pressure versus minimum settlement and pressure versus tilt characteristics of a surface footing for different values of horizontal seismic coefficient (A_h = 0.05 , 0.10 and 0.15) are given in Figs 10.6, 10.7 and 10.8 respectively. These figures also give the variations in characteristics with respect to h/B ratio (Biswas et al., 2016; Biswas, 2017).

Adopting the procedure given in this chapter, pressure versus maximum settlement, pressure versus minimum settlement and pressure versus tilt characteristics were obtained for d/B = 0.5 and d/B = 1.0 as mentioned below:

(i) For A_h = 0.05 and d/B = 0.5 in Figs 11.11a, 11.11b and 11.11c
(ii) For A_h = 0.10 and d/B = 0.5 in Figs 11.12a, 11.12b and 11.12c
(iii) For A_h = 0.15 and d/B = 0.5 in Figs 11.13a, 11.13b and 11.13c
(iv) For A_h = 0.05 and d/B = 1 in Figs 11.14a, 11.14b and 11.14c
(v) For A_h = 0.10 and d/B = 1 in Figs 11.15a, 11.15b and 11.15c
(vi) For A_h = 0.15 and d/B = 1 in Figs 11.16a, 11.16b and 11.16c

Proceeding exactly in the same way as done in Chapter 10 for d/B = 0, independent correlations are obtained for d/B = 0.5 and d/B = 1.0 for predicting ultimate bearing capacity, settlement and tilt of the footing. These are given as below (Biswas, 2017):

For $d/B = 0.5$

$$[(q_u)_{Ah}/(q_u)_{Ah=0}] = 1 - [(-2.8A_h^2 + 0.78\,A_h + 0.047)*(h/B)] \qquad (11.22)$$

$$[(S_{max})_{Ah}/(S_{avg})_{Ah=0}] = [(-0.200A_h^2 + 0.0100\,A_h - 0.004)*(h/B)^2$$

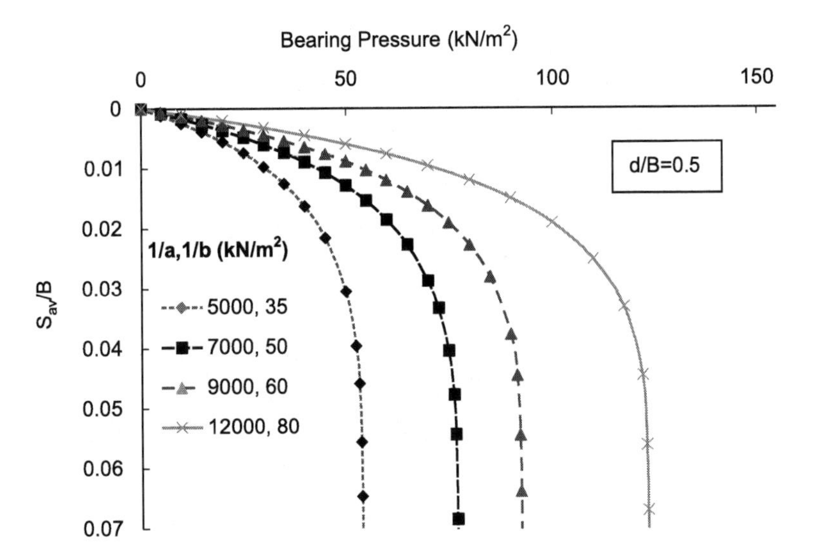

Fig. 11.9: Bearing pressure versus S_{av}/B curves for d/B = 0.5 (Static case).

$$+ (-1A_h^2 - 0.09\,A_h - 0.026)*(h/B) + 1] \qquad (11.23)$$

$$[(S_{min})_{Ah}/(S_{avg})_{Ah=0}] = [(-6.20A_h^2 - 2.11A_h + 0.093)*(h/B)^2$$

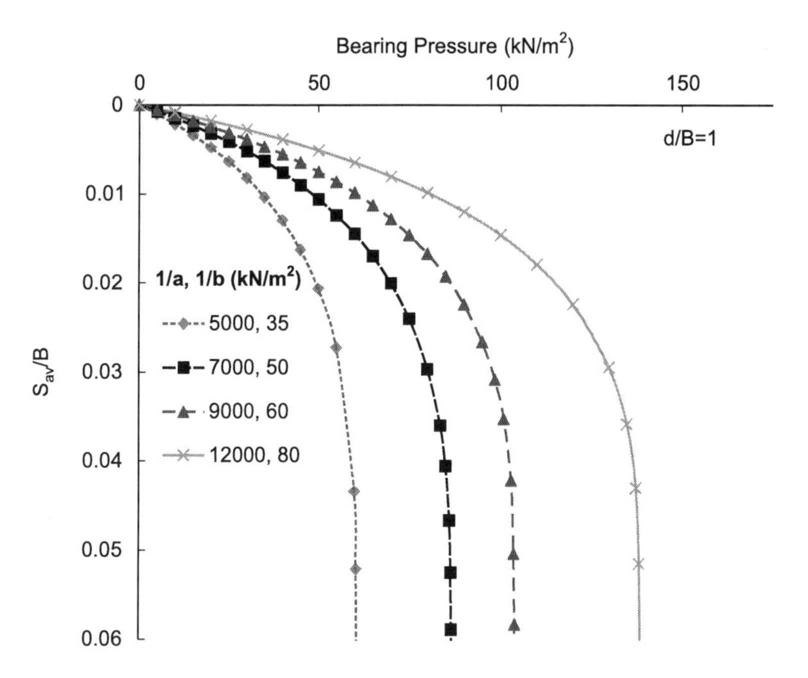

Fig. 11.10: Bearing pressure versus S_{av}/B curves for $d/B = 1$ (Static case).

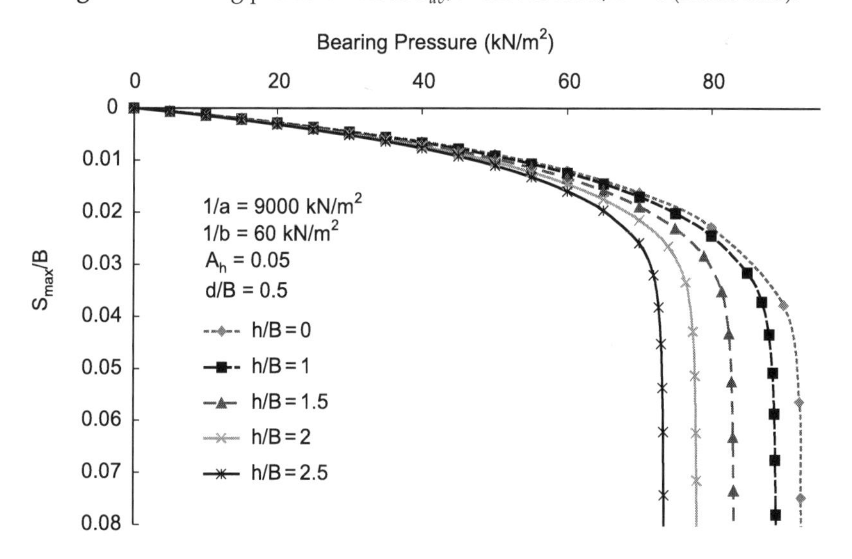

Fig. 11.11a. Bearing pressure versus S_{max}/B curves for $A_h = 0.05$ and $d/B = 0.5$.

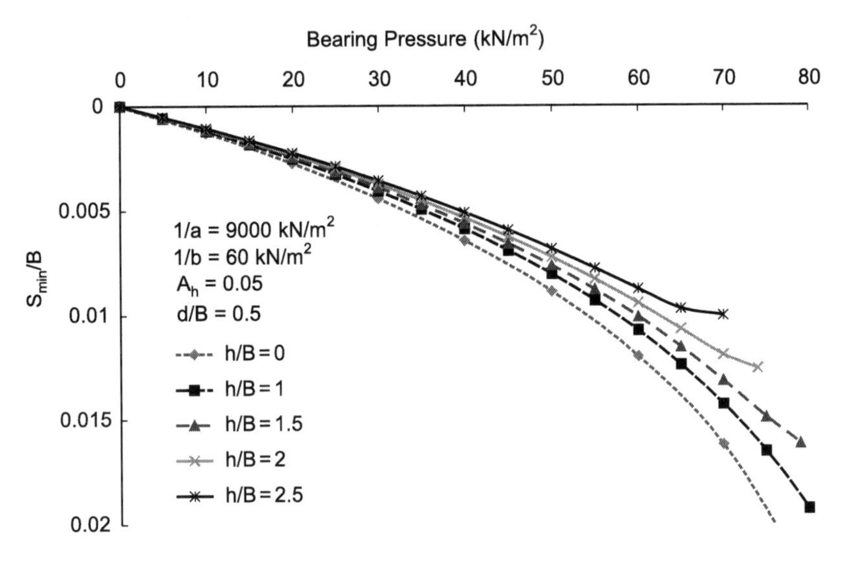

Fig. 11.11b. Bearing pressure versus S_{min}/B curves for $A_h = 0.05$ and $d/B = 0.5$.

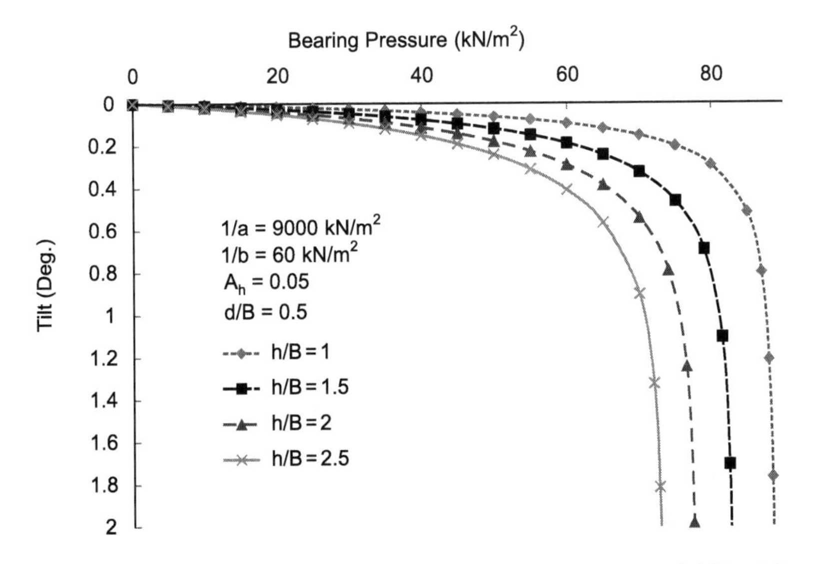

Fig. 11.11c. Bearing pressure versus tilt curves for $A_h = 0.05$ and $d/B = 0.5$.

$$+ (22.8A_h^2 - 7.16\, A_h + 0.151)^*(h/B) + 1] \quad (11.24)$$

For $d/B = 1$

$$[(q_u)_{Ah}/(q_u)_{Ah=0}] = 1 - [(-5.08A_h^2 + 1.278\, A_h + 0.036)^*(h/B)] \quad (11.25)$$

$$[(S_{max})_{Ah}/(S_{avg})_{Ah=0}] = [(0.094A_h^2 - 0.155A_h + 0.0029)^*(h/B)^2$$
$$+ (-0.9A_h^2 - 0.319A_h - 0.0331)^*(h/B) + 1] \quad (11.26)$$

$$[(S_{min})_{Ah}/(S_{avg})_{Ah=0}] = [(-5.38A_h^2 - 1.969A_h + 0.0826)*(h/B)^2 + (21.52A_h^2 - 6.958A_h + 0.1249)*(h/B) + 1] \quad (11.27)$$

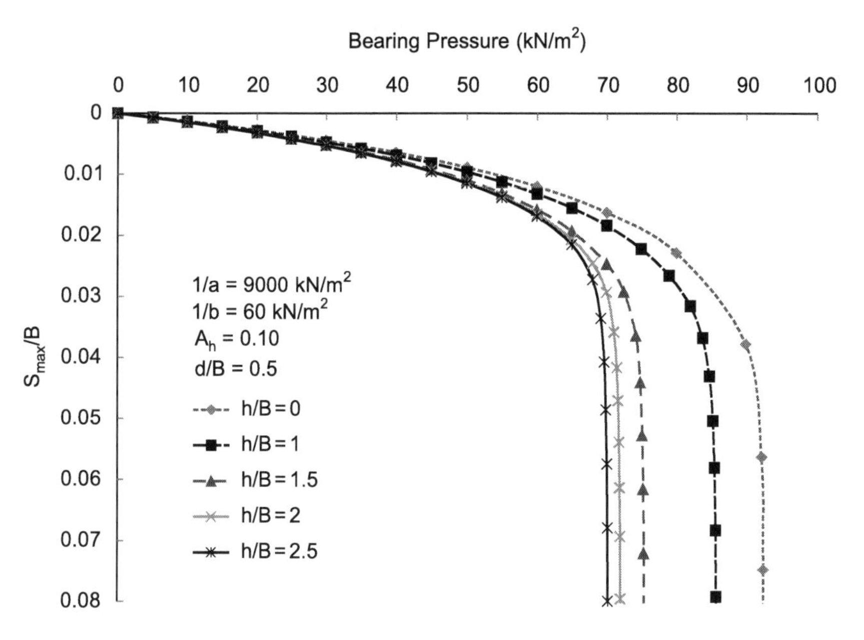

Fig. 11.12a. Bearing pressure versus S_{max}/B curves for $A_h = 0.10$ and $d/B = 0.5$.

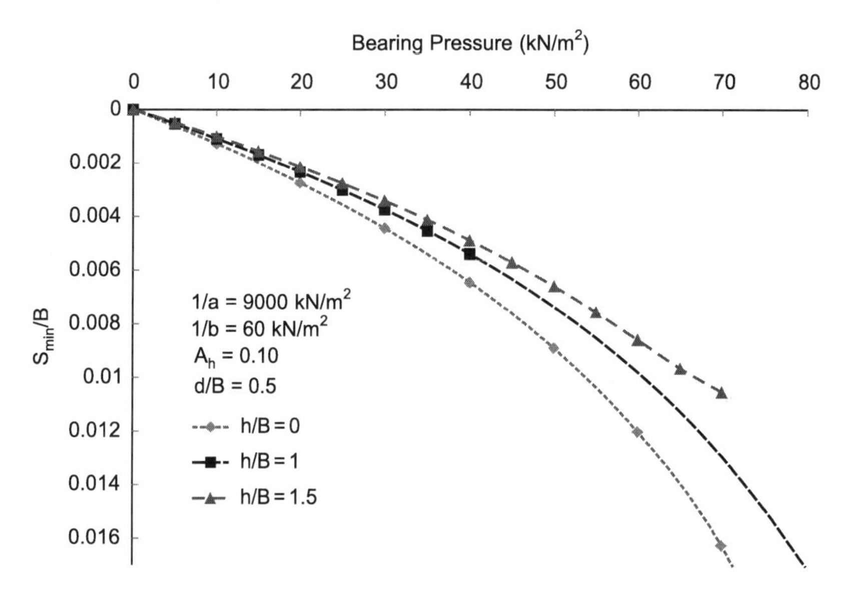

Fig. 11.12b. Bearing pressure versus S_{min}/B curves for $A_h = 0.10$ and $d/B = 0.5$.

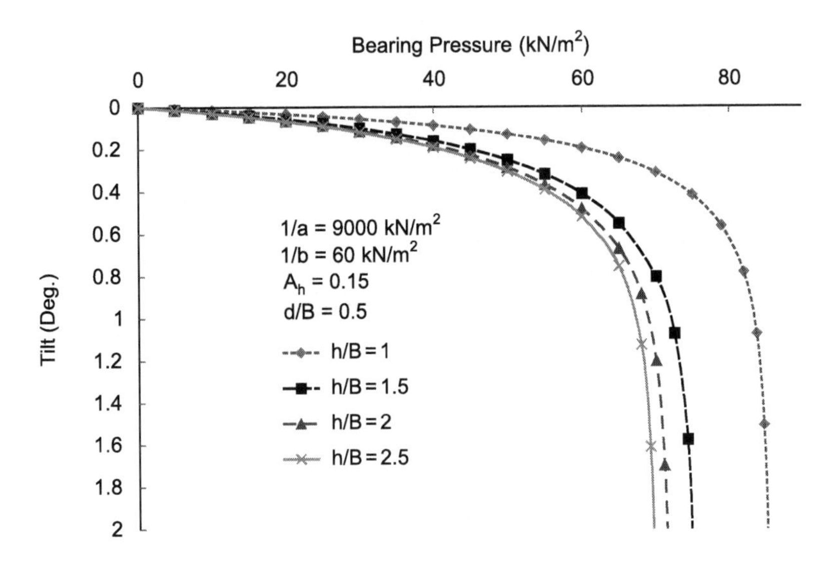

Fig. 11.12c. Bearing pressure versus tilt curves for $A_h = 0.10$ and $d/B = 0.5$.

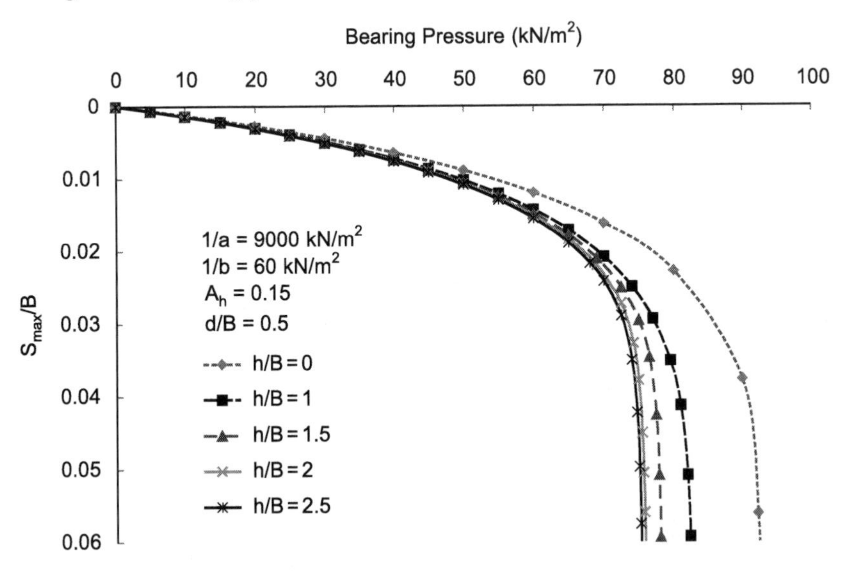

Fig. 11.13a. Bearing pressure versus S_{max}/B curves for $A_h = 0.15$ and $d/B = 0.5$.

Covering the complete range of d/B ratio in shallow foundation, the above equations can be summarized as below:

$$[(q_u)_{Ah}/(q_u)_{Ah=0}] = 1 - [(a_1 A_h^2 + a_2 A_h + a_3)^*(h/B)] \tag{11.28}$$

$$[(S_{max})_{Ah}/(S_{avg})_{Ah=0}] = [(b_1 A_h^2 + b_2 A_h + b_3)^*(h/B)^2 + (b_4 A_h^2 + b_5 A_h + b_6)^*(h/B) + 1] \tag{11.29}$$

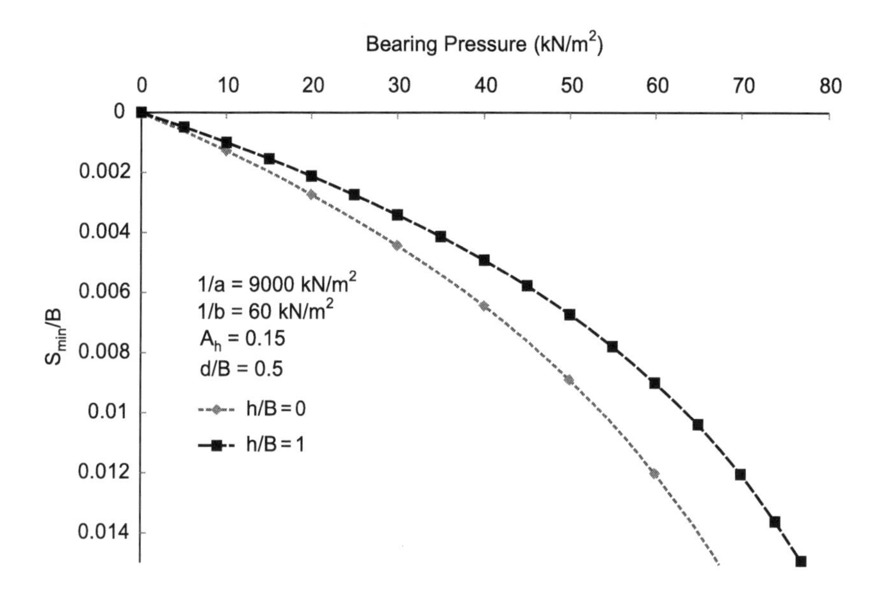

Fig. 11.13b. Bearing pressure versus S_{min}/B curves for A_h = 0.15 and d/B = 0.5.

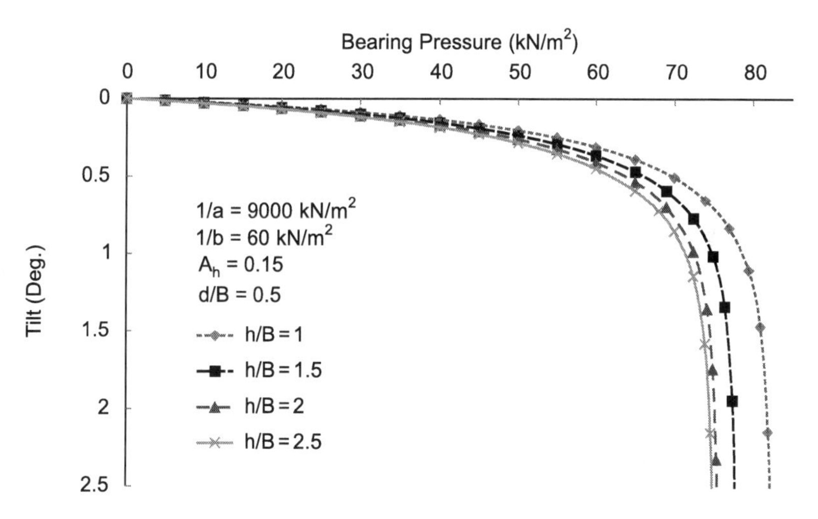

Fig. 11.13c. Bearing pressure versus tilt curves for A_h = 0.15 and d/B = 0.5.

$$[(S_{min})_{Ah}/(S_{avg})_{Ah=0}] = [(c_1A_h^2 + c_2A_h + c_3)*(h/B)^2 + (c_4A_h^2 + c_5A_h + c_6)*(h/B) + 1] \quad (11.30)$$

In Eqs (11.28), (11.29) and (11.30), a_1, a_2, a_3, b_1, b_2, b_3, b_4, b_5, b_6, c_1, c_2, c_3, c_4, c_5 and c_6 are parameters which depend on d/B ratio. On regression analysis, following relations were obtained:

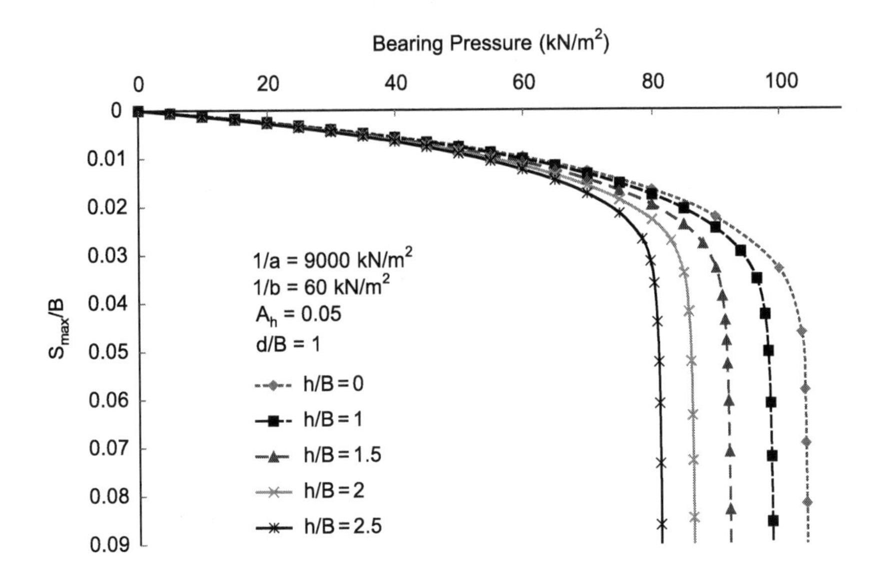

Fig. 11.14a. Bearing pressure versus S_{max}/B curves for $A_h = 0.05$ and $d/B = 1$.

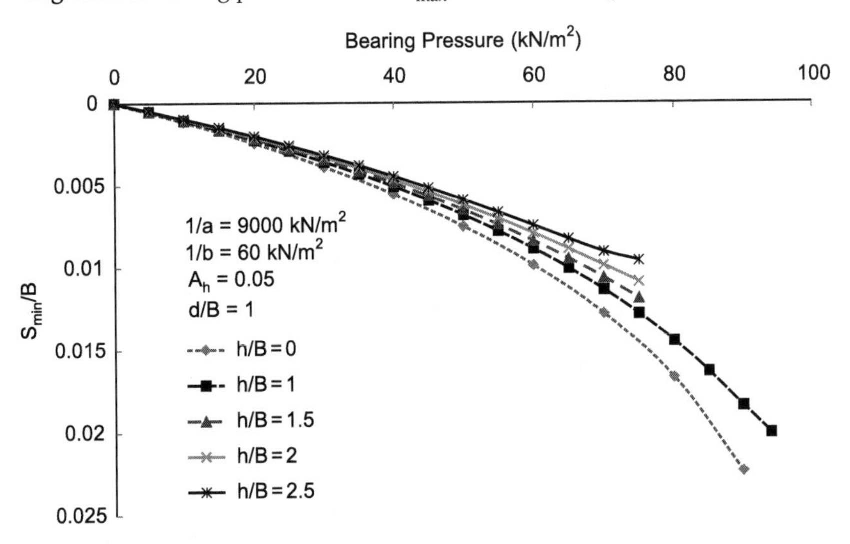

Fig. 11.14b. Bearing pressure versus S_{min}/B curves for $A_h = 0.05$ and $d/B = 1$.

$$a_1 = \{6.48(d/B)^2 - 14.28(d/B) + 2.72\} \quad (11.31)$$

$$a_2 = \{2.176(d/B)^2 - 2.276(d/B) + 1.378\} \quad (11.32)$$

$$a_3 = \{-0.0306(d/B)^2 + 0.0239(d/B) + 0.0427\} \quad (11.33)$$

$$b_1 = \{-4.732(d/B)^2 + 7.686(d/B) - 2.86\} \quad (11.34)$$

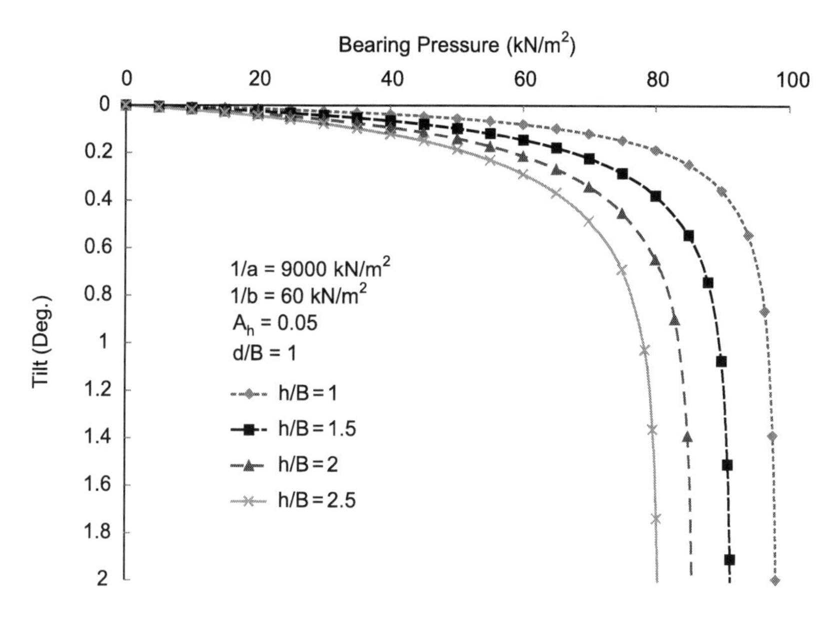

Fig. 11.14c. Bearing pressure versus tilt curves for $A_h = 0.05$ and $d/B = 1$.

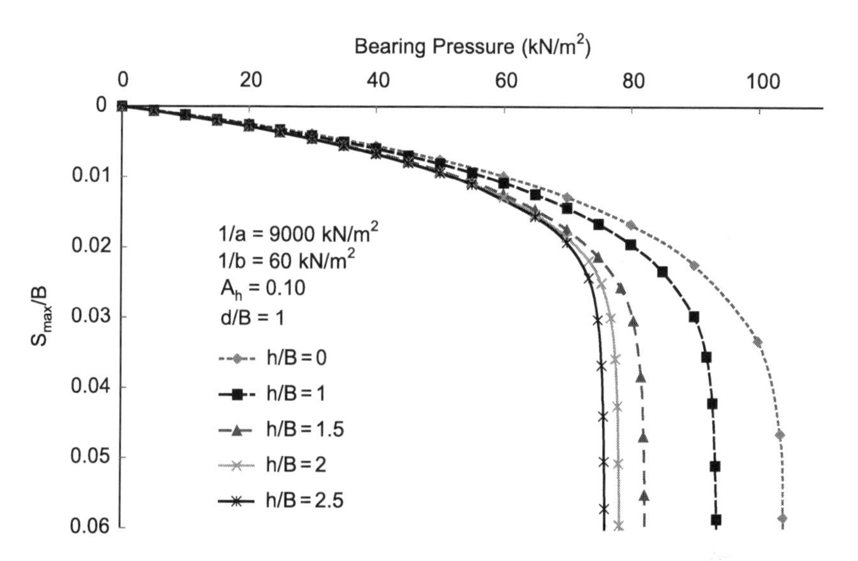

Fig. 11.15a. Bearing pressure versus S_{max}/B curves for $A_h = 0.10$ and $d/B = 1$.

$$b_2 = \{-0.3422(d/B)^2 + 0.1833(d/B) + 0.0039\} \tag{11.35}$$

$$b_3 = \{0.0296(d/B)^2 - 0.0306(d/B) + 0.0039\} \tag{11.36}$$

$$b_4 = \{-6.28(d/B)^2 + 9.62(d/B) - 4.24\} \tag{11.37}$$

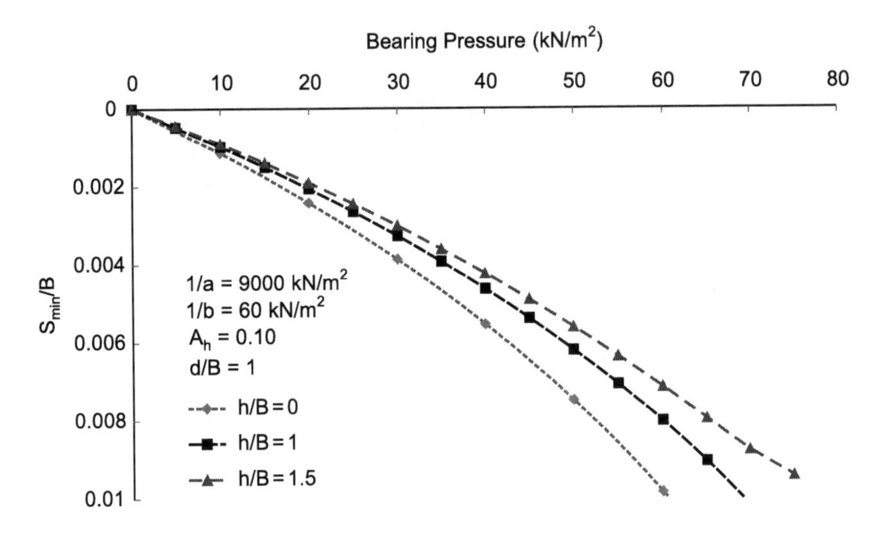

Fig. 11.15b. Bearing pressure versus S_{min}/B curves for $A_h = 0.10$ and $d/B = 1$.

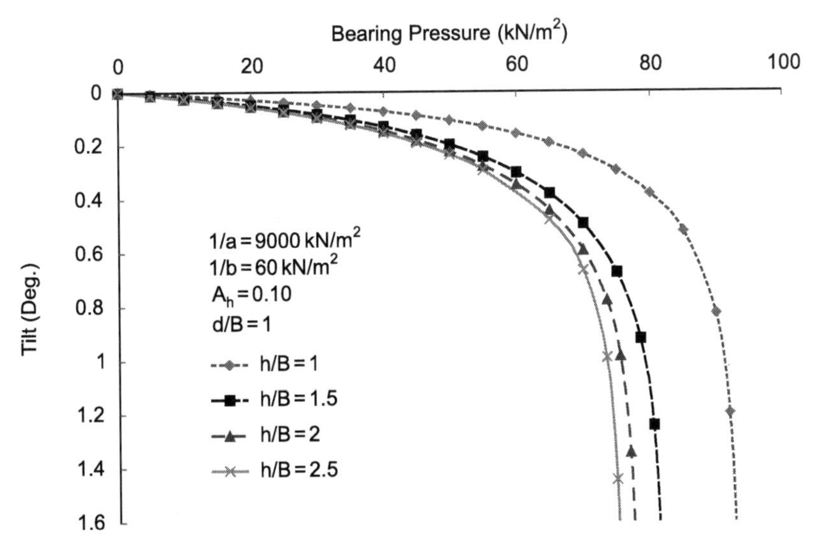

Fig. 11.15c. Bearing pressure versus tilt curves for $A_h = 0.10$ and $d/B = 1$.

$$b_5 = \{0.07(d/B)^2 - 0.559(d/B) + 0.172\} \quad (11.38)$$

$$b_6 = \{-0.0936(d/B)^2 + 0.1262(d/B) - 0.0657\} \quad (11.39)$$

$$c_1 = \{18.76(d/B)^2 - 26.5(d/B) + 2.36\} \quad (11.40)$$

$$c_2 = \{3.452(d/B)^2 - 4.896(d/B) - 0.525\} \quad (11.41)$$

$$c_3 = \{-0.2074(d/B)^2 + 0.2903(d/B) - 0.0003\} \quad (11.42)$$

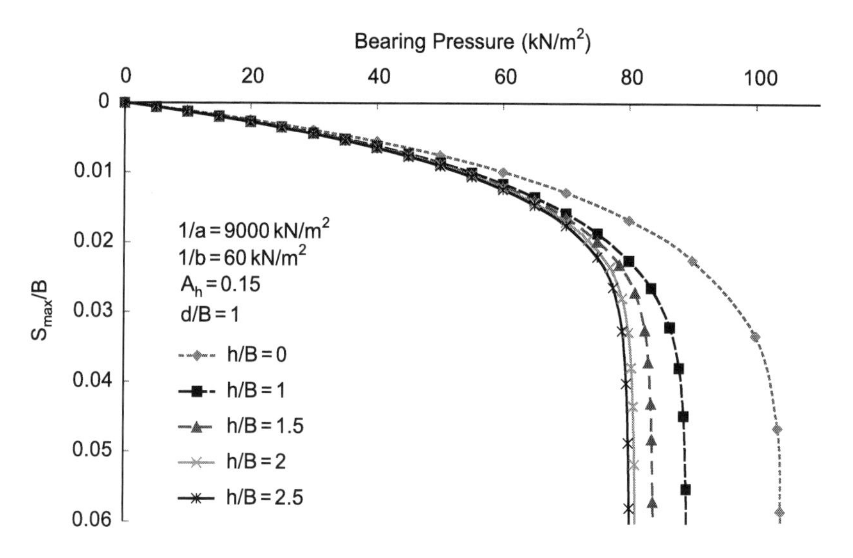

Fig. 11.16a. Bearing pressure versus S_{max}/B curves for $A_h = 0.15$ and $d/B = 1$.

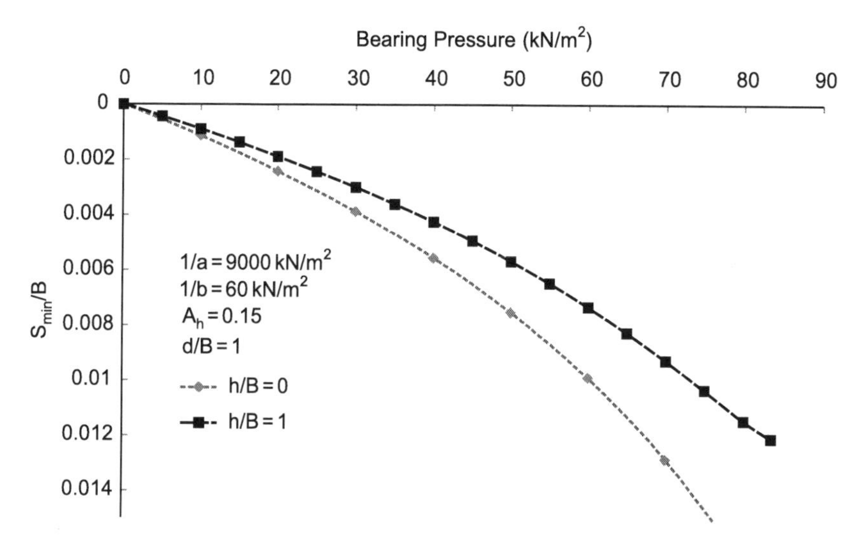

Fig. 11.16b. Bearing pressure versus S_{min}/B curves for $A_h = 0.15$ and $d/B = 1$.

$$c_4 = \{-30.84(d/B)^2 + 43.7(d/B) + 8.66\} \tag{11.43}$$

$$c_5 = \{2.7924(d/B)^2 - 3.7846(d/B) - 5.9658\} \tag{11.44}$$

$$c_6 = \{-0.3602(d/B)^2 + 0.4881(d/B) - 0.003\} \tag{11.45}$$

Hence using the Eqs (11.28), (11.29) and (11.30), one can obtain the values of seismic bearing capacity, maximum settlement and minimum

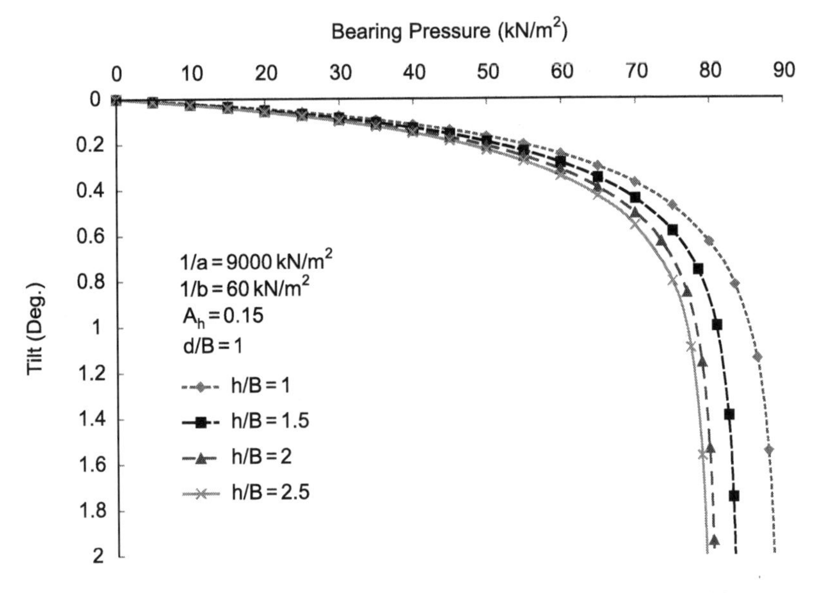

Fig. 11.16c. Bearing pressure versus tilt curves for $A_h = 0.15$ and $d/B = 1$.

settlement if the values of static bearing capacity and static settlement of the same footing are known, and these may be obtained using the procedure given in Chapter 4 or using conventional approaches given in most textbooks on soil mechanics and foundation engineering. Knowing the values of maximum and minimum settlements, the tilt of the footing can be obtained by dividing the difference between the two by the effective width of the footing.

Illustrative Examples

Example 11.1

Determine the amount of vertical load that a strip footing of width 1.0 m resting on saturated clay ($1/a = 12000$ kN/m^2, $1/b = 80$ kN/m^2) located at a depth of 0.5 m below ground surface. Consider a factor of safety against shear failure as 3.0 m and permissible settlement equal to 20 mm.

Solutions

1. Using the methodology described in Chapter 4, obtain the pressure-settlement characteristics of a strip footing of width 1.0 m resting on the ground surface (Fig. 11.17). By intersection tangent method, the ultimate bearing capacity works out as 108.14 kN/m^2. Therefore, safe bearing capacity of the surface footing in static case is $\dfrac{108.14}{3} = 36$ kN/m^2.

2. $d/B = 0.5/1.0 = 0.5$
   Using Eq. (11.19)

$$\frac{(q_u)_{d/B=0.5}}{(q_u)_{d/B=0}} = 1 + 0.262 \times 0.5 = 1.131$$

   Therefore, $(q_u)_{d/B=0.5} = 1.131 \times 108.14 = 122.30 \text{ kN/m}^2$

   Use a factor of safety equal to 3.0

$$\text{Safe bearing capacity} = \frac{122.30}{3.0} = 40.7 \text{ kN/m}^2$$

3. For a pressure intensity equal to 36 kN/m^2, pressure-settlement characteristics give a settlement equal to 4.39 mm.
   From Eq. (11.21)

$$\frac{(S)_{d/B}}{(S)_{d/B=0}} = 1 + 0.064\left(\frac{d}{B}\right) - 0.033\left(\frac{d}{B}\right)^2$$

$$(S)_{d/B=0} = [1 + 0.064(0.5) - 0.033(0.5)^2] \times (4.39) = 4.5 \text{ mm}$$

The above problem can also be solved by obtaining the pressure-settlement characteristics of a strip footing of width 1.0 m having d/B ratio equal to 0.5, keeping $1/a = 12000$ kN/m^2, $1/b = 80$ kN/m^2. The procedure for this is given in this very chapter. By doing this, pressure-settlement characteristics are obtained as shown in Fig. 11.17.

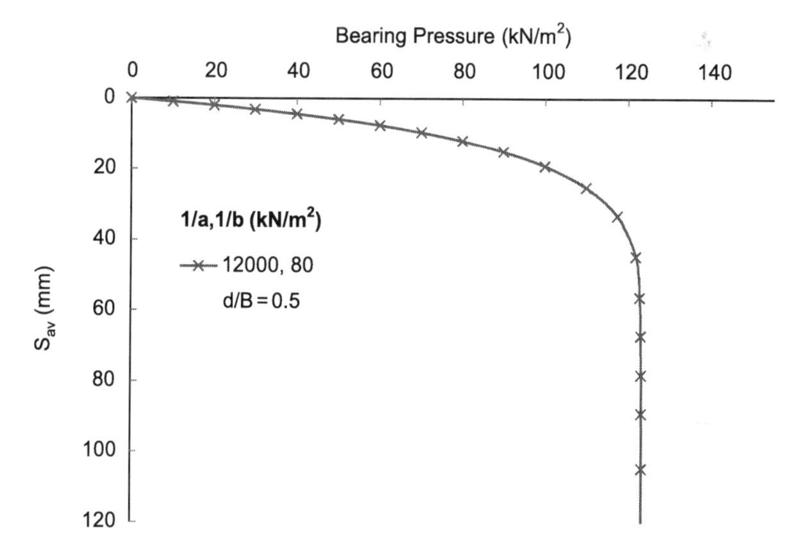

Fig. 11.17. Relation between bearing pressure and average settlement.

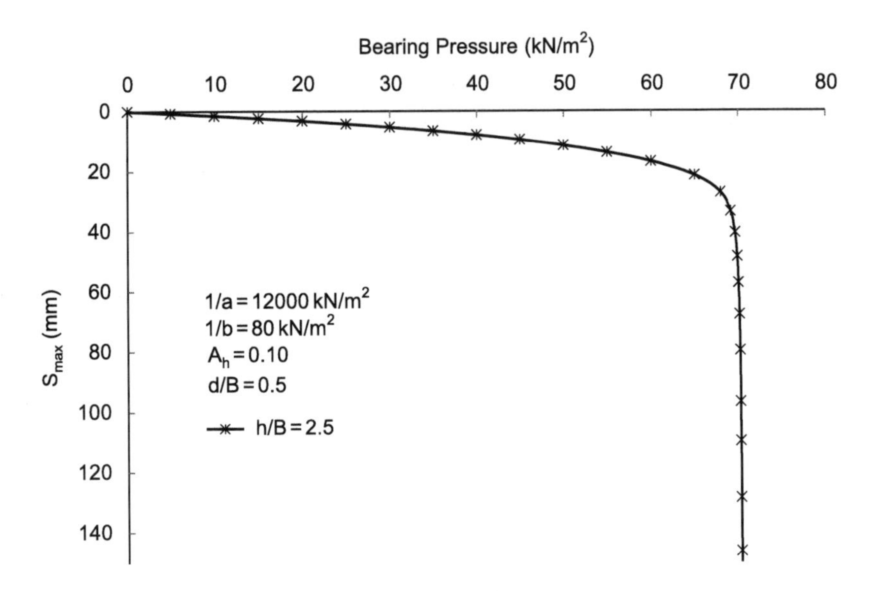

Fig. 11.18. Bearing pressure versus S_{max} curves for $A_h = 0.05$ and $h/B = 1$.

Intersection tangent method gives the value of the ultimate bearing capacity as 121.5 kN/m². Thus, safe bearing capacity will be (121.5/3) = 40.5 kN/m². Corresponding to this pressure intensity, Fig. 11.18 gives a settlement of 4.5 mm.

It may be noted that the settlement increases with increase in d/B ratio because pressure intensity acting on the footing is 40.7 kN/m² instead of 36 kN/m². For the footing having d/B ratio equal to 0.5 and pressure intensity equal to 36 kN/m², the settlement is 4.1 mm (Fig. 11.18), which is less than 4.5 mm. However, the effect is not significant.

Example 11.2

Solve Example 11.1 considering that the structure is located in a seismic zone having horizontal seismic coefficient as 0.1, considering that the horizontal seismic force acts at a height of 2.5 m from the base of the footing.

Solution

1. Width of footing, $B = 1.0$ m
   Depth of footing, $d = 0.5$ m
   Height of point application of horizontal seismic force = 2.5 m
   Therefore, $d/B = 0.5$, and $h/B = 2.5/1.0 = 2.5$
2. As in Step 1 of Example 11.1

$$(q_u)_{A_h = 0} = 108.14 \text{ kN/m}^2$$

This value corresponds to the surface footing, i.e. $d/B = 0$. From Eq. (11.28)

3. $$\frac{(q_u)_{A_h}}{(q_u)_{A_h=0}} = 1-\left[a_1 A_h^2 + a_2 A_h + a_3\right]\left(\frac{h}{B}\right)$$

where

$$a_1 = \{6.48(d/B)^2 - 14.28(d/B) + 2.72\}$$
$$= \{6.48(0.5)^2 - 14.28(0.5) + 2.72\}$$
$$= 1.62 - 7.14 + 2.72 = -2.800$$
$$a_2 = \{2.176(d/B)^2 - 2.276(d/B) + 1.378\}$$
$$= \{2.176(0.5)^2 - 2.276(0.5) + 1.378\}$$
$$= 0.544 - 1.138 + 1.378 = 0.784$$
$$a_3 = \{-0.0306(d/B)^2 + 0.0239(d/B) + 0.0427\}$$
$$= \{-0.0306(0.5)^2 + 0.0239(0.5) + 0.0427\}$$
$$= -(7.65 \times 10^{-3}) + 0.01195 + 0.0427 = 0.047$$

$$\therefore \quad \frac{(q_u)_{A_h}}{(q_u)_{A_h=0}} = 1 - [(-2.80) \times (0.1)^2 + 0.784 \times (0.1) + 0.047]\,(2.5)$$
$$= 1 - [-0.028 + 0.0784 + 0.047]\,(2.5) = 0.7565$$

$$\therefore \quad (q_u)_{A_h} = 0.7565 \times 108.14 = 81.8 \text{ kN/m}^2$$

$$\text{Safe bearing capacity} = \frac{81.8}{3} = 27.27 \text{ kN/m}^2$$

Therefore, it may be noted that the safe bearing capacity due to seismicity decreases from 40.7 kN/m^2 to 27.27 kN/m^2, considering $A_h = 0.1$ and $h/B = 2.5$.

4. For settlement and tilt computations using Eq. (11.28), maximum settlement is obtained as below:

$$\frac{(S_{\max})_{A_h}}{(S_{\text{avg}})_{A_h=0}} = \left[\left(b_1 A_h^2 + b_2 A_h + b_3\right)\times\left(\frac{h}{B}\right)^2 + \left(b_4 A_h^2 + b_5 A_h + b_6\right)\times\left(\frac{h}{B}\right) + 1\right]$$

where

$$b_1 = \{-4.732(d/B)^2 + 7.686(d/B) - 2.86\}$$
$$= \{-4.732(0.5)^2 + 7.686(0.5) - 2.86\}$$

$$= -1.183 + 3.843 - 2.86 = -0.2$$

$$b_2 = \{-0.3422(d/B)^2 + 0.1833(d/B) + 0.0039\}$$

$$= \{-0.3422(0.5)^2 + 0.1833(0.5) + 0.0039\}$$

$$= -0.0855 + 0.09165 + 0.0039 = 0.01$$

$$b_3 = \{0.0296(d/B)^2 - 0.0306(d/B) + 0.0039\}$$

$$= \{0.0296(0.5)^2 - 0.0306(0.5) + 0.0039\}$$

$$= 0.0074 - 0.0156 + 0.0039 = -0.004$$

$$b_4 = \{-6.28(d/B)^2 + 9.62(d/B) - 4.24\}$$

$$= \{-6.28(0.5)^2 + 9.62(0.5) - 4.24\}$$

$$= -1.57 + 4.81 - 4.24 = -1.0$$

$$b_5 = \{0.07(d/B)^2 - 0.559(d/B) + 0.172\}$$

$$= \{0.07(0.5)^2 - 0.559(0.5) + 0.172\}$$

$$= 0.0175 - 0.2795 + 0.172 = -0.09$$

$$b_6 = \{-0.0936(0.5)^2 + 0.1262(0.5) - 0.0657\}$$

$$= \{-0.0936(0.5)^2 + 0.1262(0.5) - 0.0657\}$$

$$= -0.0234 + 0.0631 - 0.0657 = -0.026$$

$$\frac{(S_{\max})_{A_h}}{(S_{\text{avg}})_{A_h=0}} = \left[\left\{-0.2\times(0.1)^2 + 0.01\times 0.1 - 0.004\right\}\times\left(\frac{2.5}{1.0}\right)^2\right.$$

$$\left. + \left\{-1.0\times(0.1)^2 - 0.09\times(0.1) - 0.026\right\}\times\left(\frac{2.5}{1}\right) + 1\right]$$

$$= [(-0.002 + 0.001\times 0.1 - 0.004)\times(2.5)^2$$

$$-(-0.01 + 0.009 - 0.026)\times(2.5) + 1]$$

$$= -0.03125 + 0.1125 + 1 = 1.08125$$

From Example 11.1, $(S_{\text{avg}})_{A_h=0} = 4.39$ mm

[For pressure intensity = 40.7 kN/m^2 (Fig. 11.17)]

Therefore,

$$(S_{\max})_{A_h} = 1.08125 \times 4.39 = 4.75 \text{ mm}$$

Readers should note that $(S_{max})_{A_h}$ = 4.75 mm corresponds to a pressure intensity of 21.8 kN/m^2. For this, pressure intensity $(S_{avg})_{A_h=0}$ is equal to 4.39 mm (Fig. 11.18).

Hence, the value of maximum settlement is significantly higher.

$$\frac{(S_{min})_{A_h}}{(S_{avg})_{A_h=0}} = \left[\left(c_1 A_h^2 + c_2 A_h + c_3\right) \times \left(\frac{h}{B}\right)^2 + \left(c_4 A_h^2 + c_5 A_h + c_6\right) \times \left(\frac{h}{B}\right) + 1\right]$$

$$c_1 = \{18.76(d/B)^2 - 26.5(d/B) + 2.36\}$$
$$= \{18.76(0.5)^2 - 26.5(0.5) + 2.36\}$$
$$= 4.69 - 13.25 + 2.36 = -6.17$$
$$c_2 = \{3.452(d/B)^2 - 4.896(d/B) - 0.525\}$$
$$= \{3.452(0.5)^2 - 4.896(0.5) - 0.525\}$$
$$= 0.863 - 2.448 - 0.525 = -2.11$$
$$c_3 = \{-0.2074(d/B)^2 + 0.2903(d/B) - 0.0003\}$$
$$= \{-0.2074(0.5)^2 + 0.2903(0.5) - 0.0003\}$$
$$= -0.0515 + 0.145 - 0.0003 = 0.0933$$
$$c_4 = \{-30.84(d/B)^2 + 43.7(d/B) + 8.66\}$$
$$= \{-30.84(0.5)^2 + 43.7(0.5) + 8.66\}$$
$$= -7.71 + 21.85 + 8.66 = 22.8$$
$$c_5 = \{2.7924(d/B)^2 - 3.7846(d/B) - 5.9658\}$$
$$= \{2.7924(0.5)^2 - 3.7846(0.5) - 5.9658\}$$
$$= 0.6981 - 1.8923 - 5.9658 = -7.16$$
$$c_6 = \{-0.3602(d/B)^2 + 0.4881(d/B) - 0.003\}$$
$$= \{-0.3602(0.5)^2 + 0.4881(0.5) - 0.003\}$$
$$= -0.09005 + 0.244 - 0.003 = 0.151$$

$$\frac{(S_{min})_{A_h}}{(S_{avg})_{A_h=0}} = \left[\{-6.17 \times (0.1)^2 - 2.11 \times 0.1 + 0.0933\} \times \left(\frac{2.5}{1.0}\right)^2 + \{22.8 \times (0.1)^2 - 7.16 \times 0.1 + 0.151\} \times \left(\frac{2.5}{1}\right) + 1\right]$$

$$= [(-0.062 - 0.211 + 0.093) \times (2.5)^2$$

$$+ (0.228 - 0.716 + 0.151) \times (2.5) + 1]$$

$$= -1.125 - 0.8425 + 1 = -0.9675$$

$$(S_{min})_{A_h} = -0.9675 \times 4.39 = -4.247 \text{ mm}$$

$$\therefore \quad B' = \left(\frac{S_{max}}{S_{max} + S_{min}} \right) B = \left(\frac{4.75}{4.75 + 4.247} \right) \times (1000) = 527.9 \text{ mm}$$

$$\tan t = \left(\frac{S_{max}}{B'} \right) = \left(\frac{4.75}{527.9} \right) = 8.997 \times 10^{-3}$$

$$t = 0.5154 \text{ degree}$$

Practice Problems

1. Explain with reasons the effect of depth of footing on bearing capacity and settlement of a strip footing resting on saturated clay. Consider that the structure is in a non-seismic region.
2. Considering the pseudo-static approach, show the forces and moments acting on the base of footing. Draw the typical settlement pattern of the footing, explaining all the terms clearly.
3. Enumerating clearly all the assumptions involved, describe the procedure for obtaining settlement of a rigid footing resting on saturated clay and located in non-seismic region.
4. Explain stepwise the procedure for obtaining maximum-settlement and minimum-settlement in the case of a rigid strip footing resting on saturated clay and located in a seismic region. Estimate the tilt of footing considering the effect of eccentricity caused by the seismic force.
5. Determine the amount of vertical load on a strip footing of width 1.25 m resting on saturated clay ($1/a$ = 12000 kN/m^2, $1/b$ = 90 kN/m^2) located at a depth of 0.5 m below ground surface. Consider a factor of safety against shear failure as 3.0 m and permissible settlement equal to 20 mm.
6. Solve the problem 5, if the footing is located in seismic region having $A_h = 0.1$.

REFERENCES

Biswas, T., Saran, S. and Shanker, D. (2016a). Analysis of a Strip Footing Using Constitutive Law. *Geosciences*, **6(2):** 41-44.

Biswas, T., Saran, S. and Shanker, D. (2016b). Behavior of a Strip Footing Resting on Clay Using Non-linear Constitutive Law. *Geosciences*, **6(3):** 75-77.

Biswas, T. (2017). Pseudo-static Analysis of a Strip Footing Using Constitutive Laws of Soils. Ph.D. Thesis, IIT Roorkee, Roorkee.

Harr, M.E. (1966). Fundamentals of Theoretical Soil, Mechanics. McGraw-Hill. New York, USA.

Melan, E. (1919). Die Druckverteilung durch eine elastische Schicht. *Beton und Eisen*, **8(9):** 83-85.

Melan, E. (1932). Der Spannungszustand der durch eine Einzelkraft im Innern beanspruchten Halbscheibe. *ZAMM – Journal of Applied Mathematics and Mechanics/Zeitschrift für Angewandte Mathematik und Mechanik*, **12(6):** 343-346.

Saran, S. (2006). Analysis and Design of Substructures Limit State Design. Oxford and IBH Publishing Co., New Delhi.

Terzaghi, K. (1943). Theoretical Soil Mechanics. John Wiley and Sons, Inc., New York.

Terzaghi, K. and Peck, R.B. (1967). Soil Mechanics in Engineering Practice. John Wiley and Sons, Inc., New York.

Subject Index